Transparency in Biology

Kohei Soga · Masakazu Umezawa · Kyohei Okubo
Editors

Transparency in Biology

Making the Invisible Visible

Springer

Editors
Kohei Soga
Materials Science and Technology
Tokyo University of Science
Tokyo, Japan

Masakazu Umezawa
Materials Science and Technology
Tokyo University of Science
Tokyo, Japan

Kyohei Okubo
Materials Science and Technology
Tokyo University of Science
Tokyo, Japan

ISBN 978-981-15-9626-1 ISBN 978-981-15-9627-8 (eBook)
https://doi.org/10.1007/978-981-15-9627-8

This Springer imprint is published by the registered company Springer Nature Singapore Pte Ltd.
The registered company address is: 152 Beach Road, #21-01/04 Gateway East, Singapore 189721, Singapore

Preface

Thirty years ago, when I was a graduate student, my classmate told me that you can see the skin through with several centimeters depth! I did not pay much attention on it. By 15 years ago, on my research works, I had nothing to do with biology or bioimaging. I had been mainly working on fluorescence materials for optical communication. One day, a biomaterial's scientist complained me that he could not make fluorescence bioimaging probe luminescent. I was interested in it and started to collaborate with him and a biologist for upconversion fluorescent material for bioimaging. The upconversion material is to emit visible light with near-infrared (NIR) excitation. Eventually, I'd noticed that we do not need "visible" light for bioimaging. Anyway, people do not observe cells or tissues with naked eyes. They are using cameras. Thanks to the optical communication research. Optical communication is using NIR light for communication because of the transparency of glass fibers with the NIR light. I was familiar with the materials and devices for the NIR optics. I started to utilize the NIR for bioimaging in 2005.

Ten years ago, a biologist visited me for discussing the NIR bioimaging. He is the inventor of a tissue clearing chemical. Five years ago, I joined his project on bioimaging. Three years ago, I was attending a conference, where he gave a talk with its title "How do the cells feel on a dish?" I was impressed with the title. What is the matter that I understand its feeling and mind? I am a materials scientist as well as a spectroscopist. Yes, I understand the minds of light and matters.

In these 15 years, by joining the community of researches on bioimaging, I have experienced the importance of the communication among the different specialty fields speaking different terms. Of course, we have to talk to medical doctors and biological scientists for understanding what they want, the users' mind. For designing and producing fluorescence probes, as well delivering them, chemistry is important. The light mind can be understood by physics. For the management of devices, we need electronics and computer science. It is a recent trend to use mathematical algorithms involving machine learning to analyze the captured data. Truly, for understanding "bioimaging," we need to involve people from various scientific fields.

During 15 years, I have concluded that one of a big issues for bioimaging is the "transparency." Also, there are so many misunderstandings of the "light mind." We need basic physics for understanding the light mind. As well, the basic one is pretty enough. Do not mix up the things. Do not misunderstand. I could not find a book which describes the "transparency in biology" properly.

This book consists of four parts and 11 chapters. The authors are from physics, chemistry and biology background. The first part is describing the essence of "transparency." All of the physics for transparency for biology is involved in this part. It lightly introduces the tissue clearing chemical but concludes that for live imaging with transparency, we need to use the NIR light. The second part reviews the NIR fluorescence imaging and its probes. The third part deals with special bioimaging by the NIR fluorescence imaging involving thermometry and computed tomography. The last part introduces the applications of the NIR photonics, especially for clinics involving photodynamic therapy (PDT), hyperspectral imaging (HSI) and multimodal imaging with currently medically used imaging modalities. The HSI is not a fluorescence imaging. It is a probe-free imaging method by using very weak absorption by molecules in the NIR range.

The light mind is unexpectedly simple. I hope this book will make people to understand the simple mind of light for transparency to use and develop more advanced bioimaging.

Tokyo, Japan Kohei Soga
June 2020

Contents

Part I
Introduction

Chapter 1
Introduction to Transparency

Kohei Soga

Abstract Seeing is believing. For recognizing biological phenomena, everybody wants to visualize some information from a biological system consists of molecules, organelles, cells, organs and bodies. The techniques are called "imaging" to give contrast to the information and show on two or more dimensional "images." Why things are transparent? Why things are "not" transparent? For answering this question, the "mind of light" for propagation is explained. The phenomena are simple. They are only absorption and bending. They are controlled by the charges of molecules and ions in substances. Light is electromagnetic wave. Things can simply be described by physics. In this chapter, we deal with the light as an electromagnetic wave meets substances as distributed charges. All of the approaches are described with classical physics. The transparency is, in other word, elimination of optical loss. If light can propagate straight ahead without any disturbance, things are transparent. As mentioned above, the loss reasons are only two for biological tissues, absorption and bending. The absorption problem is solved by breaching or avoiding the wavelength of absorption peaks. How we can solve the problem of the bending? The chapter will describe all physics background for understanding the mind of light.

Keywords Light · Transparency · Bioimaging · Scattering · Refraction · Reflection

1.1 Transparency

Biological tissues can be transparent as shown in Fig. 1.1. For optical imaging, everybody seeks for transparency. Especially for in vivo imaging, the tissues between the objective and the eyes (or observing devices as cameras) disturb the observation. The optical loss with the tissues is originated from only two phenomena of the light that comes from the objective, *bending* and *absorption*. By improving the illumination or the camera sensitivity, the darkening due to the *absorption* can be solved. On the

---

K. Soga (✉)
Department of Materials Science and Technology, Faculty of Industrial Science and Technology, Tokyo University of Science, 6-3-1 Niijuku, Katsushika, Tokyo 125-8585, Japan
e-mail: mail@ksoga.com

K. Soga et al. (eds.), *Transparency in Biology*,
https://doi.org/10.1007/978-981-15-9627-8_1

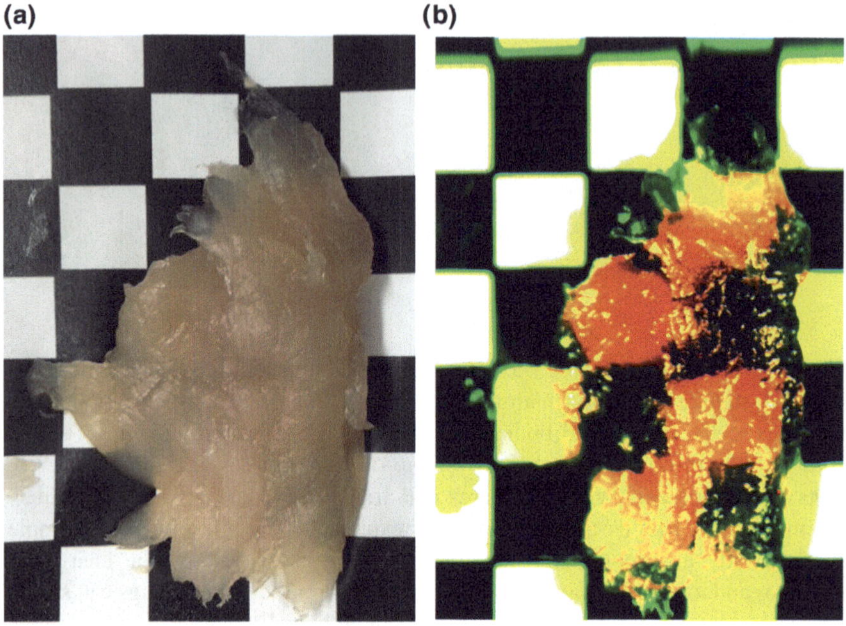

Fig. 1.1 Visible (**a**) and near-infrared (**b**) image of chicken breast. Color channels are assigned to express 1066-nm image with red, and 1281-nm one with green and 1981-nm one with blue in (**b**)

other hand, the *bending*, which involves reflection, refraction or scattering, enhances the loss of contrast. That is why the "bending phenomenon" of light is troublesome for optical imaging.

As everybody knows, light propagates straight ahead. As long as the straight propagation is achieved, we call the situation to be "transparent." This is the starting point for understanding the transparency. For solving the problems of transparency, we need to know *why matters are not transparent*. The darkness reason is the *absorption*. The opacity reason is the *bending*. Once we know *how the light feels*, we can understand how light wants to behave. We need to know the "light mind." Light is an electromagnetic wave. In physics, the "light mind" can be completely explained by Maxwell's equations, which are disliked even by the students studying physics. The solution of the equations will be done in the latter part. Then, we can understand the *bending* phenomena. Even more, we can understand *why light propagates straight ahead*.

1.2 Bending Phenomena: Reflection, Refraction and Scattering

The bending phenomena are categorized as reflection, refraction and scattering. Essentially, there is no difference among these three. If you take care of metals or semiconductors, you need to think of the behaviors of the electric carriers. The biological tissues do not have them. Then, the reflection only can be caused on the interface of matters with different refractive indices. As on the surface of lenses (or even you can imagine the images of a matter in a water filled in a glass) light refracts. Again, it is caused on the interface of matters with different refractive indices. Now, as for the scattering, when the direction of the bending is random, it is called "scattering." University professors like to talk about the famous Rayleigh scattering because it is well formalized with a correspondence to the minus fourth powers of the wavelength. Furthermore, it is easy to experience it by watching the blue sky. Forget about the Rayleigh scattering. Think about the thickness of the air in the sky. It is 500 km. The Rayleigh scattering is caused by molecules. The molecular density of gasses is 1000 times less than condensed matter. Okay. Let us divide the thickness by 1000. Even though we need 500 m thickness for observing the color change. The Rayleigh scattering is a very weak phenomenon and only observable with quite a distance of the propagation of light. It is nothing to do with the biological tissues. So, forget the Rayleigh scattering. The scattering by biological tissues is only caused by the change of refractive indices along with the light propagation. If you want to name the phenomenon, it is called Mie scattering. If you google the "Mie scattering," you can meet with many complicated equations and beautiful view graphs. Unfortunately, all of those understanding are for simply-shaped matters as spheres. It does not explain the scattering by the biological tissues with complicated shapes. Now, please understand here only that "light bends with the change of refractive index." In the following part, the reasons of straight propagation of light and the bending phenomena will be explained starting with Maxwell's equations. If you are bored to making toy with the equation, you can skip to the part of Sect. 1.2.2.

1.2.1 Maxwell's Equations

The following equations are called Maxwell's equations and classically describe all features of light as an electromagnetic wave.

$$\boldsymbol{\nabla} \cdot \boldsymbol{D} = \rho \tag{1.1}$$

$$\boldsymbol{\nabla} \cdot \boldsymbol{B} = 0 \tag{1.2}$$

$$\boldsymbol{\nabla} \times \boldsymbol{E} = -\frac{\partial \boldsymbol{B}}{\partial t} \tag{1.3}$$

$$\nabla \times \boldsymbol{H} = \frac{\partial \boldsymbol{D}}{\partial t} + \boldsymbol{J} \tag{1.4}$$

$$\boldsymbol{D} = \varepsilon \boldsymbol{E} \tag{1.5}$$

$$\boldsymbol{B} = \mu \boldsymbol{H} \tag{1.6}$$

where $\boldsymbol{D}$ and $\boldsymbol{B}$ are the electric and magnetic flux densities, ρ is the charge density, $\boldsymbol{J}$ is the current density, $\boldsymbol{E}$ and $\boldsymbol{H}$ are the electric and magnetizing fields and ε and μ are the permittivity and permeability, respectively. One may be more familiar with the equation for vacuum (free space), where ε and μ are replaced to be ε_0 and μ_0 and the ρ and $\boldsymbol{J}$ are set to be zero. For biological tissues, we can apply the almost same modification as the case of vacuum except the permittivity, ε. Now the equations can be modified as follows.

$$\nabla \cdot \boldsymbol{D} = 0 \tag{1.7}$$

$$\nabla \times \boldsymbol{H} = \frac{\partial \boldsymbol{D}}{\partial t} \tag{1.8}$$

$$\boldsymbol{B} = \mu_0 \boldsymbol{H} \tag{1.9}$$

Let us derive basic differential equations for light wave. By operating one more $\nabla \times$ from the left to Eq. (1.3), also by using Eq. (1.6), we obtain

$$\nabla \times (\nabla \times \boldsymbol{E}) = \nabla \times \left(-\frac{\partial \boldsymbol{B}}{\partial t}\right) = -\frac{\partial}{\partial t}(\nabla \times \boldsymbol{B}) = -\frac{\partial}{\partial t}(\nabla \times \mu \boldsymbol{H}) \tag{1.10}$$

By applying a mathematical relation

$$\nabla \times (\nabla \times \boldsymbol{A}) = \nabla \cdot (\nabla \cdot \boldsymbol{A}) - \nabla^2 \boldsymbol{A}, \tag{1.11}$$

we modify the left side of Eq. (1.10) to be

$$\nabla \times (\nabla \times \boldsymbol{E}) = \nabla \cdot (\nabla \cdot \boldsymbol{E}) - \nabla^2 \boldsymbol{E} = -\nabla^2 \boldsymbol{E} \tag{1.12}$$

where $\nabla \cdot \boldsymbol{E} = 0$ because of Eqs. (1.5) and (1.7).

$$\nabla \cdot \boldsymbol{E} = \frac{1}{\varepsilon} \nabla \cdot \boldsymbol{D} = 0 \tag{1.13}$$

By using Eqs. (1.5), (1.6) and (1.8), the right side of Eq. (1.10) can be rewritten to be

$$-\frac{\partial}{\partial t}(\boldsymbol{\nabla} \times \mu \boldsymbol{H}) = -\mu \frac{\partial}{\partial t}\left(\frac{\partial \boldsymbol{D}}{\partial t}\right) = -\varepsilon\mu \frac{\partial^2}{\partial t^2}\boldsymbol{E}. \quad (1.14)$$

By inserting Eqs. (1.12) and (1.14) into Eq. (1.10), we obtain the following equation.

$$\boldsymbol{\nabla}^2 \boldsymbol{E} - \varepsilon\mu \frac{\partial^2}{\partial t^2}\boldsymbol{E} = 0 \quad (1.15)$$

In the same way,

$$\boldsymbol{\nabla}^2 \boldsymbol{B} - \varepsilon\mu \frac{\partial^2}{\partial t^2}\boldsymbol{B} = 0 \quad (1.16)$$

Let us think about the meaning of Eqs. (1.15) and (1.16). Both of the $\frac{\partial}{\partial t}$ and $\boldsymbol{\nabla} = \begin{pmatrix} \frac{\partial}{\partial x} \\ \frac{\partial}{\partial y} \\ \frac{\partial}{\partial z} \end{pmatrix}$ are partial differentiations. That means, for one differentiation (e.g., $\frac{\partial}{\partial t}$), the term of the other ($\boldsymbol{\nabla}$) is constant. Can you read a periodical symmetry for an operation "second derivative" on the equations? By two times, differentiations make the result to be in the original functional form for both $\boldsymbol{E}$ and $\boldsymbol{B}$. This essential property of the equations implies that the solutions for them have periodicity for both space and time. One can propose the following solutions for them.

$$\boldsymbol{E}(\boldsymbol{r}, t) = \boldsymbol{E}_0 \cos(\boldsymbol{k} \cdot \boldsymbol{r} - \omega t + \phi) \quad (1.17)$$

$$\boldsymbol{B}(\boldsymbol{r}, t) = \boldsymbol{B}_0 \cos(\boldsymbol{k} \cdot \boldsymbol{r} - \omega t + \phi) \quad (1.18)$$

where $\boldsymbol{r}$ is the position vector, $\boldsymbol{r} = \begin{pmatrix} x \\ y \\ z \end{pmatrix}$ and $\boldsymbol{k} = \begin{pmatrix} k_x \\ k_y \\ k_z \end{pmatrix}$, which is called the wave vector with the size of the wavenumber. Now, if we assume the light propagates in the direction of z-axis, we can set the $\boldsymbol{k}$ vector to be

$$\boldsymbol{k} = \begin{pmatrix} 0 \\ 0 \\ k \end{pmatrix} \quad (1.19)$$

where $\boldsymbol{k} = |k| = {}^{2\pi}/_{\lambda}$ with the wavenumber k and the wavelength λ.

Now, let us characterize the light. The results can show us the following important characters.

$$\boldsymbol{E} \perp \boldsymbol{B} \tag{1.20}$$

$$B_z = E_z = 0 \tag{1.21}$$

Formula (1.20) says always the electric and magnetic fields are perpendicular to each other. Equation (1.21) says there is no z component; i.e., always the fields are perpendicular to the propagating direction as a transverse wave.

As a normal property of propagating wave, the light speed, c, the frequency, ν, and the wavelength, λ, are keeping the following relationship.

$$c = \nu\lambda = \frac{\omega\lambda}{2\pi} = \frac{\omega}{k} \tag{1.22}$$

where the $\omega = 2\pi\nu$ is the angular frequency. By inserting (1.17) and (1.18) to the original differential Eqs. (1.15) and (1.16), we notice that the parameters must fulfill the following relation.

$$k^2 = \varepsilon\mu\omega^2 \tag{1.23}$$

The combination of Eqs. (1.20) and (1.21) yields into a very important result;

$$c = \frac{1}{\sqrt{\varepsilon\mu}}. \tag{1.24}$$

In fact, for the light propagation in vacuum (a free space), the light speed can be expressed as

$$c_0 = \frac{1}{\sqrt{\varepsilon_0\mu_0}} \tag{1.25}$$

For biological tissues, we can assume that they are non-magnetic substances, where $\mu = \mu_0$. Then, the light speed in a biological tissue can be rewritten as

$$c = \frac{1}{\sqrt{\varepsilon\mu_0}} \tag{1.26}$$

1.2.2 How Refractive Index is Decided?

We all learn that the light refraction (bending) obeys to Snell's law. Essentially, the refraction occurs due to the change of the light speed, c, on the interface of matters with a different permittivity. In Fig. 1.2, two media A and B with different

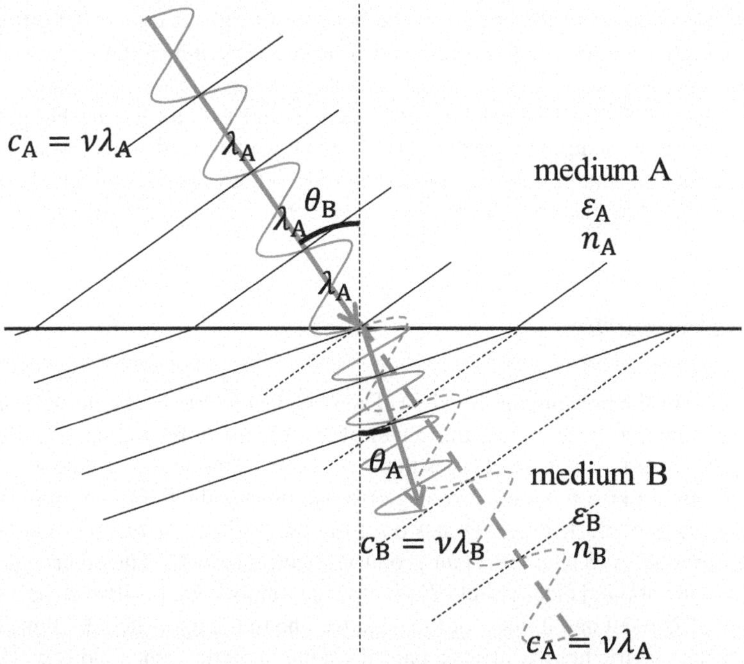

Fig. 1.2 Light propagation from medium A into medium B with a different permittivity, ε_A and ε_B. Refraction should maintain Snell's law for the continuity of wave on the interface

permittivity, ε_A and ε_B, are facing. According to Eq. (1.26), light speeds are different between the media. They are $c_A = \frac{1}{\sqrt{\varepsilon_A \mu_0}}$ and $c_B = \frac{1}{\sqrt{\varepsilon_B \mu_0}}$, respectively. The figure shows the propagation of light during Δt in each media. Essentially, the frequency, ν, does not change in both media. The number of the modulation of both media is $\nu\Delta t$. The propagation length is $c_A\Delta t$ and $c_B\Delta t$, respectively. Now because of the difference of the permittivity, the propagation length must change due to the change of the light speed. For the consistency of the wave face, the *light must bend*. If it does not bend as expressed with dashed lines, the propagation length does not change. The unchanged length conflicts with the change of light speed. Accordingly, the bending, as defined as *refraction*, occurs with the angle relation as expressed below. The relation between n and θ is called Snell's law.

$$n_{AB} = \frac{\sin\theta_A}{\sin\theta_B} = \frac{n_B}{n_A} = \frac{\lambda_A}{\lambda_B} = \frac{c_A}{c_B} = \frac{\sqrt{\varepsilon_B \mu_0}}{\sqrt{\varepsilon_A \mu_0}} \tag{1.27}$$

The n_{AB} is called relative refractive index. n_A and n_B are called absolute refractive indices and defined as follows referring the permittivity and permiability of vacuum.

$$n_A = \frac{\sqrt{\varepsilon_A \mu_0}}{\sqrt{\varepsilon_0 \mu_0}} = \frac{\sqrt{\varepsilon_A}}{\sqrt{\varepsilon_0}}, \quad n_B = \frac{\sqrt{\varepsilon_B \mu_0}}{\sqrt{\varepsilon_0 \mu_0}} = \frac{\sqrt{\varepsilon_B}}{\sqrt{\varepsilon_0}} \tag{1.28}$$

Now, we understand the reason of the bending of light is due to the difference of the refractive indices that is originated from the variation of the permittivity of matters.

The (absolute) refractive index is 1 for vacuum and 1.00 for the air. For gaseous matters, the index changes its 4th digit (10^{-4}). On the other hand, that of water is 1.33 for water and more than 1.4 for vegetable oils. What is the value for biological tissues? It is around 1.43 as discussed later. *Then, what is the reason of the permittivity?*

1.2.3 Permittivity

Let us think of the meaning of Eq. (1.5). The $\boldsymbol{E}$ is the electric field coming from the outside of a matter. Here, we assume a pair of charges as shown in Fig. 1.3, which is called polarization. What are charges then? A positive charge is a producing source of electric field like a fountain. The negative one inhales the field as a drain. When you have the pair, electric field comes out from the positive charge and it goes into the negative one. As a result, the pair produces electric field $\boldsymbol{E}'$. The other important fact is that the applied electric field $\boldsymbol{E}$ gives force to move the positive charge to the direction of $\boldsymbol{E}$ itself and it does for the negative one to the opposite direction. There for, once the electric field $\boldsymbol{E}$ meets a pair of charge in matters, the field is weakened to be $\boldsymbol{E} - \boldsymbol{E}'$. The effect is reflected to the permittivity, ε. The effective density of the electric field is reduced to $\boldsymbol{D}$, and the degree of the reduction is expressed by the ε. So-called relative permittivity is defined as

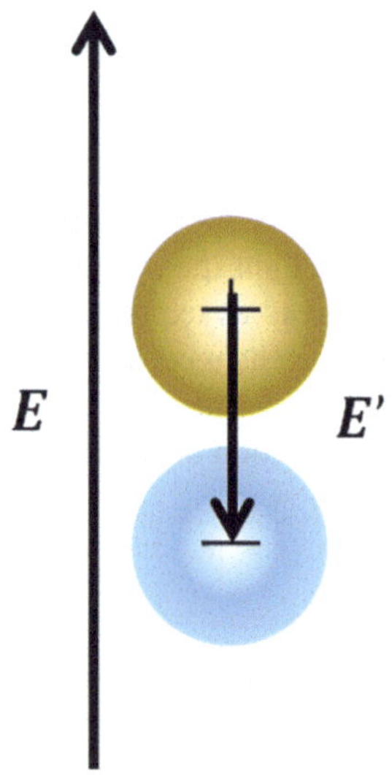

Fig. 1.3 Pair of charges

$$\varepsilon_\mathrm{r} = \frac{\varepsilon}{\varepsilon_0}. \tag{1.29}$$

Now, by referring Eqs. (1.28) and (1.29), the absolute refractive index can be expressed for a matter by the following equation.

$$n = \sqrt{\frac{\varepsilon}{\varepsilon_0}} = \sqrt{\varepsilon_\mathrm{r}}. \tag{1.30}$$

Refractive indices, *n*, are listed in Table 1.1. For more information on optical constants, you can refer a reliable database on Web [1]. The value for the vacuum and air is 1. For the most of the gasses, the density of molecules is thousand times smaller compared to condensed matters. It means that the source of the change of the light speed, the polarization, is too dilute to be effective. The difference of the *n* for gasses changes from that of vacuum only with the minus fourth digits. Water has the refractive index of 1.33.

As popularly known, the major component of the biological tissues is the water. What bends the light? Generally, the most major component of the light-bending agent is the membranes. The membranes are made of phospholipid, typically, phosphatidylcholines. As shown in Fig. 1.4a, it has almost nonpolar tails with 15–17 CH_2 and strongly polar phosphate head. Water molecules have strong polarity with hydrogen side $H^{\delta+}$ and $O^{\delta-}$. They like to be attaching together by facing the positive and negative charges. The polar part of the phospholipid likes to be together with the water molecule. On the other hand, the nonpolar tails do not have any likeliness to charges. The tail parts are excluded from the hydrophilic system with water and the head parts. As a result, the tails face together to form a layer structure as shown in Fig. 1.4b. This is the major structure of the membrane. How thick is the membrane? Typically, it is in single nanometer order [8]. Light has the wavelength of hundreds of nanometers or more. It is too thin!! Light do not mind the membrane itself. Assume a space with 500 hundred nm size. The normal cell size is about 10 μm. While, the virus size is several tens nanometers. With Mitochondria, Golgi and so on as well as virus give membrane structures with various orientation and various density to the space, which causes the variation of the refractive index from space to space. Membrane increases the index of space around 1.43, which is much higher than that of saline, 1.35.

Figure 1.5 shows the phase-contrasted image of cells [9]. The phase-contrasted image shows the part of higher refractive index brighter. As seen from the left figure, the bright shining line can be found along with the cell membrane. The right figure explains the light propagation. When parallel beams of light comes from the top, each beam propagates along with the membrane and finally comes out with random orientations. The random orientation will totally lose a chance for forming an image. This is the exact mechanism of the loss of contrast due to scattering with biological tissues. Can we avoid this loss?

Table 1.1 Refractive indices of typical substances

Category	Substance	Refractive index	Wavelength (nm)	References
Gas	Hydrogen	1.000132	589.29	[1]
Gas	Air	1.000293	589.29	[1]
Gas	Carbon dioxide	1.001	589.29	[1]
Inorg. mater	MgF_2	1.3777	589.29	[2]
Inorg. mater	CaF_2	1.4338	589.29	[2]
Inorg. mater	Fused silica	1.458	589.29	[1]
Inorg. mater	BK7[a]	1.5167	589.29	[2]
Inorg. mater	Soda lime silica glass[b]	1.5233	589.29	[2]
Inorg. mater	NaCl	1.5441	589.29	[2]
Inorg. mater	Al_2O_3	1.7681	589.29	[2]
Inorg. mater	Diamond	2.4166	589.29	[2]
Inorg. mater	TiO_2	2.6129	589.29	[2]
Polymer	PDMS[a]	1.4302	589.29	[2]
Polymer	PLA[a]	1.4585	589.29	[2]
Polymer	PMMA[a]	1.4905	589.29	[2]
Polymer	Polystyrene, PS	1.5915	589.29	[2]
Pure liquid	Methanol	1.3312	589.29	[2]
Pure liquid	Water	1.333	589.29	[1]
Pure liquid	Acetonitrile	1.3421	589.29	[2]
Pure liquid	Acetone	1.3591	589.29	[2]
Pure liquid	Ethanol	1.361	589.29	[1]
Pure liquid	Propanol	1.3768	589.29	[2]
Pure liquid	Butanol	1.399	589.29	[2]
Pure liquid	Cyclohexane	1.4264	589.29	[2]
Pure liquid	Hexane	1.4264	589.29	[2]
Pure liquid	Carbon tetrachloride	1.461	589.29	[1]
Pure liquid	Oleic acid	1.463	589.29	[3]
Pure liquid	Glycerol	1.4722	589.29	[2]
Pure liquid	DMSO[a]	1.4771	589.29	[2]
Pure liquid	Benzene	1.501	589.29	[1]
Pure liquid	Carbon disulfide	1.628	589.29	[1]
Solution	10% glucose aq	1.3477	589.29	[1]
Solution	10% NaCl aq	1.3514	589.29	[4]
Solution	20% glucose aq	1.3635	589.29	[1]
Solution	20% NaCl aq	1.3692	589.29	[4]
Solution	DPPC[b] sat. ethanol sol	1.39	589.29	[5]

(continued)

Table 1.1 (continued)

Category	Substance	Refractive index	Wavelength (nm)	References
Solution	60% glucose aq	1.4394	589.29	[1]
Tissues	Colon	1.3430	589.29	[2]
Tissues	Dermis	1.41	~1300	[6]
Tissues	Stratum corneum	1.51	~1300	[6]
Tissues	Epidermis	1.34–1.37	~1300	[6]
Tissues	Liver tumor	1.35–1.37	450–1551	[7]
Tissues	Liver (healthy)	1.36–1.38	450–1551	[7]
Tissues	Mesenteric adipose	1.467	~1300	

[a]BK7: optical glass; soda lime silica glass: normal glass; PDMS: polydimethylsiloxane; PLA: polylactic acid; PMMA: poly(methyl methacrylate); DMSO: dimethyl sulfoxide; DPPC[b]: 1,2-dipalmitoyl-sn-glycero-3-phosphocholine

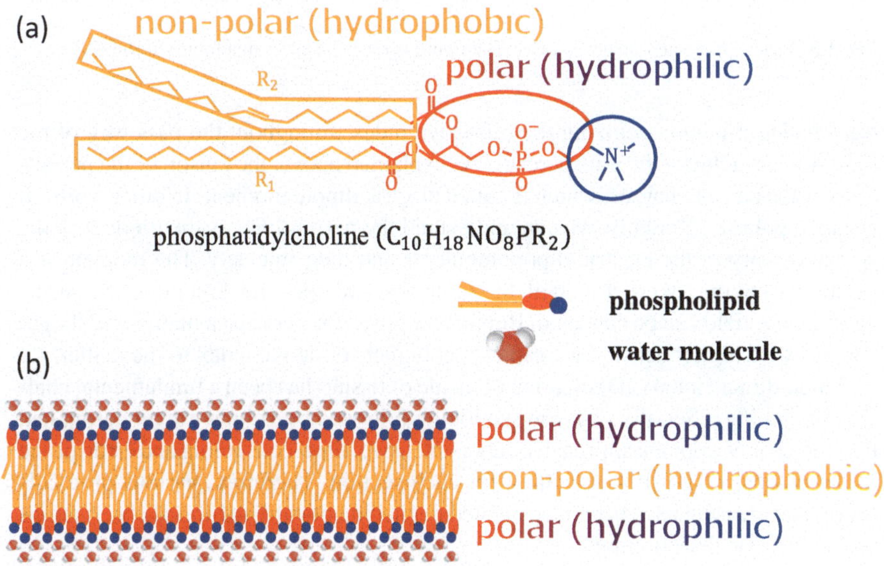

Fig. 1.4 Molecular structure of phosphatidylcholines (**a**) and layered structure of a membrane made of phospholipids (**b**). Nonpolar parts are facing together. The polar heads are all facing out to the outside of membrane to interact with water molecules

1.2.4 Chemical Tissue Clearing

How can we have transparency of a tissue? As for the absorption, mostly the molecules to absorb the incoming light can be decomposed. Otherwise, simply we should increase the sensitivity or light intensity. Can we avoid the *bending*? Yes. Now we know the bending reason. Since the change of the refractive index causes

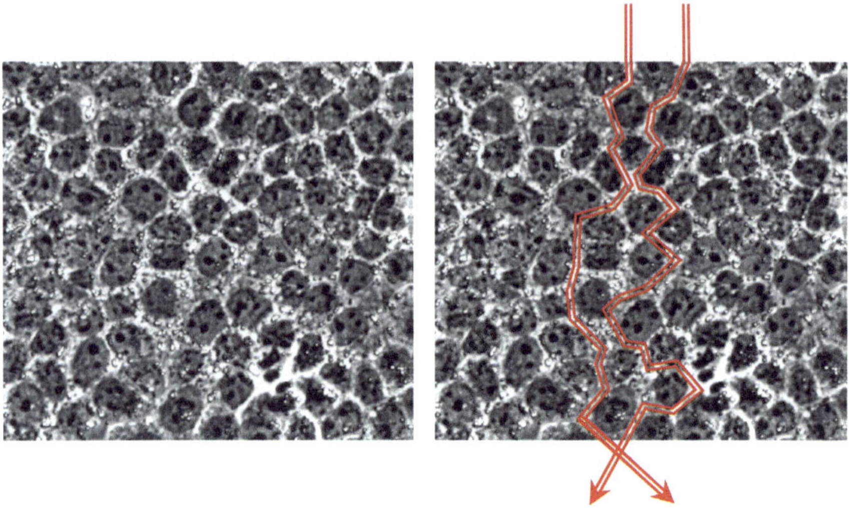

Fig. 1.5 Phase-contrasted image of cells. The figure is reproduced from a figure in Ref. [9]

the *bending*, by homogenizing the refractive index throughout the pass way of the light, we can achieve the transparency. As explained above, the reason of the permittivity is the pair of charges, which is called electric dipole moment. In other words, it is called polarity. Roughly, we can understand the permittivity as the product of the spatial density of the electric dipole moments and their intensity. The moment is a vector. By having two vectors with the same size and opposite directions, the generated electric field can be canceled. But just roughly, if we consist a matter with larger charges with higher density, we can expect higher refractive index to the matter.

Three-dimensional (3D) imaging of inside of tissues has been a fundamental challenge in biology. Serially physical sectioning followed by volumetric rendering had been done; however, the approach is labor-intensive, time-consuming and error-prone in the process of position matching for each section. Therefore, various protocols for optical clearing have been investigated to enable optically sectioned observations without physical sectioning.

In early stage of the research, hydrophobic organic agents were applied to the clearing. In this method, samples are incubated in a mixture of benzyl alcohol and benzyl benzoate after dehydration of the sample with ethanol and hexane [10]. The use of tetrahydrofuran or dibenzyl ether instead of alcohol for dehydration improved the clearing of whole tissue samples such as the brains of adult mice [11]. This approach has large limitations by damaging tissue structure and quenching fluorescence proteins visualizing molecular phenomena in tissue samples.

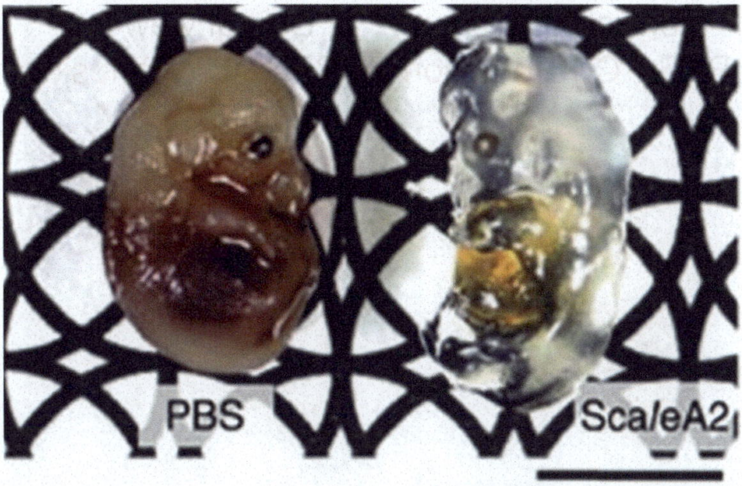

Fig. 1.6 Mouse embryos taken with a black and white pattern as background with a scale bar presenting 5 mm. Left, embryo placed in phosphate-buffered saline (PBS) after fixation with 4% paraformaldehyde (PFA). Right, embryo incubated in Sca*l*eA2 solution for 2 weeks after fixation with 4% PFA. Reprinted from Ref. [12] with permission from Springer Nature, 2020

To overcome these problems, clearing methods based on aqueous agents were developed. In 2011, a Sca*l*e protocol based on hyperhydrated compounds was introduced as an approach for fluorescence imaging of mouse brains [12]. As demonstrated with whole fetus samples of mice, the scattering of light by biological tissues was dramatically suppressed by 2-week incubation in Sca*l*eA2 cocktail as shown in Fig. 1.6. The distinctive aspect of this method likely relies on the capability of urea to enter into tightly folded regions of proteins with high refractive index by forming an osmotic gradient to pull out water [13]. The CUBIC method also utilizes a hyperhydration urea-based mechanism with a high refractive index sucrose solution in the clearing process. In addition, a 2-week CUBIC protocol achieved optical clearing of whole body of mice by combination with tissue a decoloring process (Fig. 1.7) [14]. When the tissues are incubated in high concentration aqueous solution of phosphoric acid, which consists of small molecules (or ions) to cause high refractive index, the scattering by tissues are reduced rapidly by 60-min incubation (Fig. 1.8) [5]. Further researches are under conduction to find easier and simpler methods to obtain clearer 3D images of isolated tissues and whole bodies of small animals.

Unfortunately, the clearing processing kills the cells. High clarity image of a tissue with live cells is difficult to obtain by this method. However, by using near-infrared light, you can possibly achieve the transparency for live biological tissues or bodies as explained next.

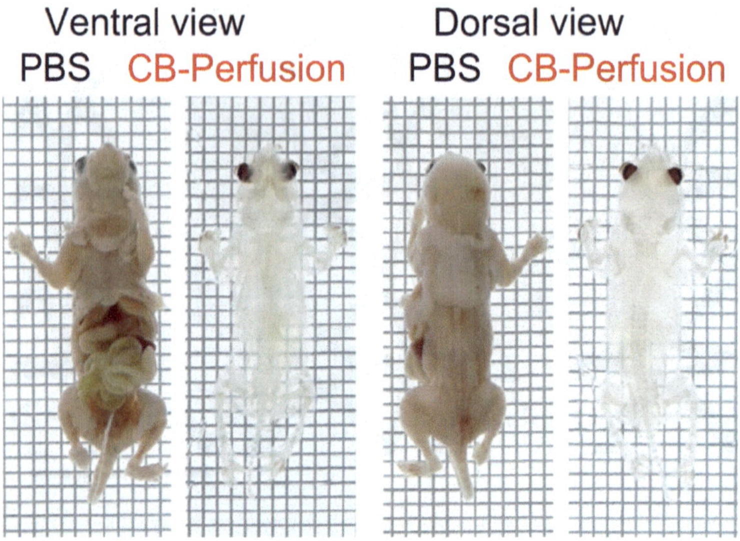

Fig. 1.7 Whole-body clearing of a small (6-day-old) mouse. Bright-field images of whole body incubated in PBS after fixation with PFA or subjected to CUBIC-perfusion protocol are shown. The apertures of the grid are 1 mm. Reprinted from Ref. [14] with permission from Elsevier, 2020

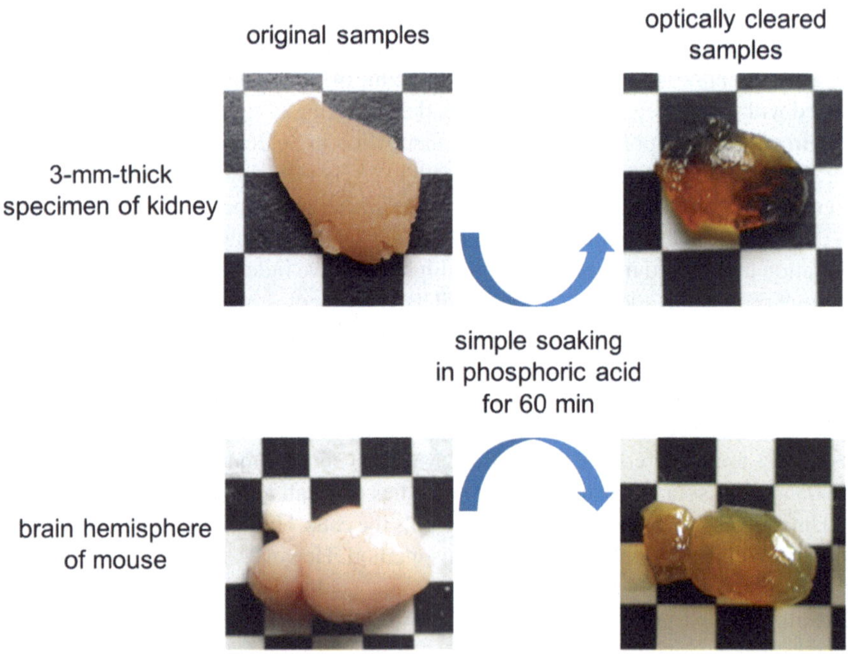

Fig. 1.8 Mouse tissues (kidney specimen and brain hemisphere) fixed with 4% PFA for 48 h followed by incubation in 14.2 M phosphoric acid for 60 min. The apertures of the background lattice are 4 mm. Reprinted from Ref. [5] with Creative Commons Attribution License

1.3 NIR Light as a Transparent Light

1.3.1 Lorenz Model

Now, we know that the light propagates straight ahead if there is no change of the permittivity, ε, with a speed $c = 1/\sqrt{\varepsilon\mu_0}$. For understanding more of the "light mind," we should understand how "matters" look like for the light. Light itself is nothing but electric field. Once the light takes a look of NaCl crystal, it does not even see Na or Cl as atoms or ions. It is only interested in charges, + and −, as fountains of electric field. For simplifying the "matters," the Lorenz model describes them as a pair of charges, $+q$ and $-q$, displaced each other with a distance X as shown in Fig. 1.9. Here, we use classical physics again, the famous Newton's equation. If you are not interested in the way to solve and derive the dependency of light speed onto wavelength or frequency, you can jump to Sect. 1.3.2.

As a model of matters for the light mind, we assume two charges, $+q$ and $-q$, with a displacement, X, connected with a spring constant of ω_0^2 as illustrated in Fig. 1.9. The spring corresponds to the atomic bonding in matters. By assuming a vector X from $-q$ to $+q$, this model is called an electric dipole moment

$$\boldsymbol{M} = q\boldsymbol{X}. \tag{1.31}$$

We use a classical physics approach. Let us go with Newton's equation. As the forces for the equation, we assume three forces, spring restoring force, $\omega_0^2 X$, resistance proportional to speed, $\Gamma_0 \frac{\mathrm{d}X}{\mathrm{d}t}$, and Coulomb force, qE.

$$m\frac{\mathrm{d}^2 X}{\mathrm{d}t^2} = \omega_0^2 X + \Gamma_0 \frac{\mathrm{d}X}{\mathrm{d}t} + qE \tag{1.32}$$

The light as an electromagnetic wave is introduced to this equation by replacing the electric field E into a dynamic wave of electric field, $E_0 \mathrm{e}^{-i\omega t}$

$$m\left(\frac{\mathrm{d}^2 X}{\mathrm{d}t^2} + \Gamma_0 \frac{\mathrm{d}X}{\mathrm{d}t} + \omega_0^2 X\right) = qE = qE_0 \mathrm{e}^{-i\omega t} \tag{1.33}$$

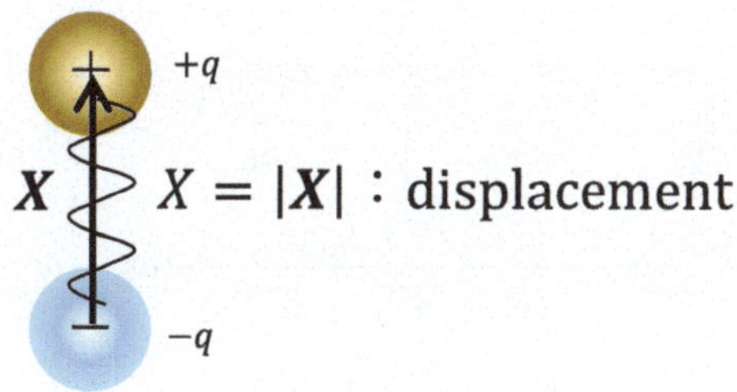

Fig. 1.9 A pair of charges, $+q$ and $-q$, displaced each other with a distance X. The vector X is oriented from the charge with $-q$ to that with $+q$

The general solution of the differential Eq. (3.1.3) can be written

$$X = X_0 \exp(-i\omega t). \tag{1.34}$$

By substituting Eq. (3.1.4) into Eq. (3.1.3), the condition to be satisfied by the amplitude X_0 is obtained as follows.

$$X_0 = \frac{qE_0/m}{\omega_0^2 - \omega^2 - i\omega\Gamma_0} = \frac{qE_0/m}{\left[\left(\omega_0^2 - \omega^2\right)^2 + \omega^2\Gamma_0^2\right]^{\frac{1}{2}}} e^{-i\phi} \tag{1.35}$$

with ϕ that meets the following condition.

$$\tan\phi = \frac{\omega\Gamma_0}{\omega_0^2 - \omega^2} \tag{1.36}$$

To connect the scheme of the electric dipole moment into "matters," let us assume that the moment M with a number N_0 exists a volume V. Here, we can write the polarization, P, as

$$P = \frac{N_0}{V} M = \frac{qXN_0}{V}. \tag{1.37}$$

Now, we can connect the electric dipole moment M and the permittivity ε with the following equations;

$$\boldsymbol{P} = \varepsilon_0 \chi \boldsymbol{E}, \tag{1.38}$$

$$\boldsymbol{D} = \varepsilon_0 \boldsymbol{E} + \boldsymbol{P} = \varepsilon_0 \boldsymbol{E} + \varepsilon_0 \chi \boldsymbol{E}, \text{ and} \tag{1.39}$$

$$\boldsymbol{D} = \varepsilon_0 (1 + \chi) \boldsymbol{E} = \varepsilon_0 \varepsilon_r \boldsymbol{E} = \varepsilon \boldsymbol{E}, \tag{1.40}$$

where the new parameter, χ, is called electric susceptibility.

By combining Eqs. (1.35) and (1.37)–(1.40),

$$\begin{aligned}
\varepsilon &= \varepsilon_0 + \frac{P}{E_0} = \varepsilon_0 + \frac{qX_0N_0/V}{E_0} = \varepsilon_0 + \frac{N_0q^2/mV}{\omega_0^2 - \omega^2 - i\omega\Gamma_0} \\
&= \varepsilon_0 + \frac{N_0q^2/mV}{\left(\omega_0^2 - \omega^2\right)^2 + (\omega\Gamma_0)^2}\left(\omega_0^2 - \omega^2 + i\omega\Gamma_0\right) \\
&= \varepsilon_0 + \frac{N_0q^2/mV}{\left(\omega_0^2 - \omega^2\right)^2 + (\omega\Gamma_0)^2}\left(\omega_0^2 - \omega^2\right) + i\frac{N_0q^2/mV}{\left(\omega_0^2 - \omega^2\right)^2 + (\omega\Gamma_0)^2}\omega\Gamma_0.
\end{aligned} \tag{1.41}$$

Here, we obtain the permittivity as a complex value.

$$\tilde{\varepsilon} = \varepsilon' + \varepsilon'' i \tag{1.42}$$

with

$$\varepsilon' = \varepsilon_0 + \frac{N_0 q^2/mV}{\left(\omega_0^2 - \omega^2\right)^2 + (\omega\Gamma_O)^2}\left(\omega_0^2 - \omega^2\right) \tag{1.43}$$

$$\varepsilon'' = \frac{N_0 q^2/mV}{\left(\omega_0^2 - \omega^2\right)^2 + (\omega\Gamma_O)^2}\omega\Gamma_O \tag{1.44}$$

Figure 1.10 shows the behavior of the complex permittivity under the variation of frequency. As we discussed, the real part, ε', corresponds to the frequency dependency of light speed and refractive index as shown in Eqs. (1.24) and (1.26). On the other hand, the imaginary part, ε'', does to the intensity of resonant absorption due to

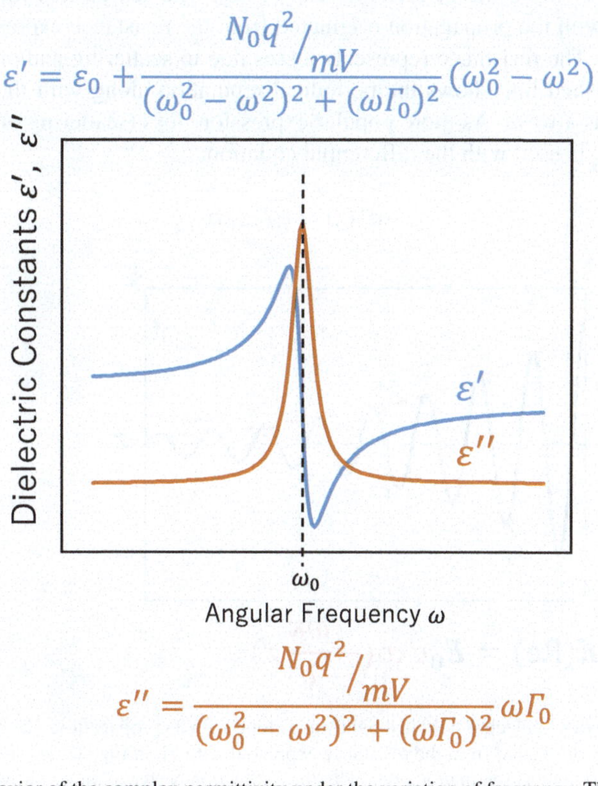

Fig. 1.10 Behavior of the complex permittivity under the variation of frequency. The real part, ε', is plotted with the blue line, and the imaginary part, ε'', is with the orange one

the resonance between light wave and the electric dipole moments. Technically, complex permittivity can be connected to more popular parameters, refractive index n and extinction coefficient κ for expressing complex refractive index, $\tilde{n}$, as

$$\tilde{n} = n + i\kappa \tag{1.45}$$

$$n^2 = \frac{1}{2\varepsilon_0}\left(\sqrt{\varepsilon'^2 + \varepsilon''^2} + \varepsilon'\right) \tag{1.46}$$

$$\kappa^2 = \frac{1}{2\varepsilon_0}\left(\sqrt{\varepsilon'^2 + \varepsilon''^2} - \varepsilon'\right). \tag{1.47}$$

By using the parameters, the light propagation to z direction can be expressed as

$$E = E_0 \exp\left(-\frac{\omega\kappa}{c}z\right) \exp\left\{-i\omega\left(t - \frac{n}{c}z\right)\right\}, \tag{1.48}$$

which can be obtained as the solution of the wave equation expressed as Eq. (1.15).

Figure 1.11 shows the behavior of Eq. (1.48). The red dashed line shows the decay along with the propagation originated from the resistance expressed with Γ_0 in Eq. (1.33). The resistance represents losses due to scattering and/or absorption. The green dashed line shows the periodical reputation along with the z direction propagation as a wave. As more popular expression, loss (scattering or absorption) coefficient, α, is used with the differential equation,

$$-\mathrm{d}I(z) = \alpha I(z)\mathrm{d}x \tag{1.49}$$

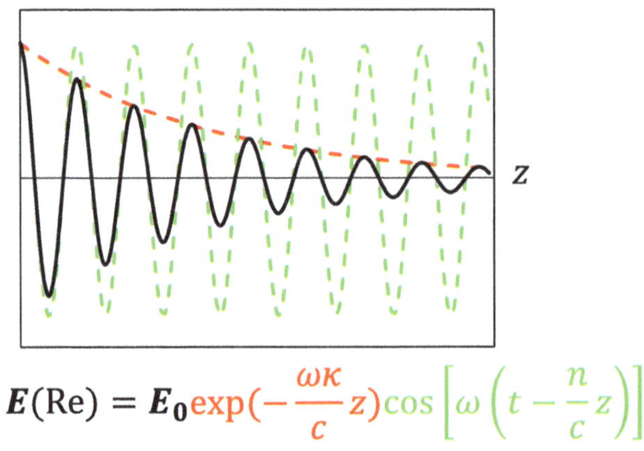

Fig. 1.11 Behavior of electric field in a light wave along with its propagation. The red dashed line shows the decay originated from the resistance expressed with Γ_0 in Eq. (1.33). The green dashed line shows the periodical reputation along with the z direction propagation as a wave

to express the intensity change of light for dx propagation. The solution of Eq. (1.49) is

$$I(z) = I_0 \exp(-\alpha z). \tag{1.50}$$

By comparing Eqs. (1.48) and (1.49), we obtain a relation,

$$\alpha = \frac{2\omega\kappa}{c} \tag{1.51}$$

to correlate the α and κ.

Let us take a look of some more concrete case of water. The light frequency dependency of the n and α is shown in Fig. 1.12. In case of the infrared absorption of water, major reason of the absorptions is three modes of molecular vibrations, ν_1, ν_2 and ν_3. Each reason can give a peak in both spectra. In the spectrum of α, you can observe the absorption peaks in a familiar form that corresponds to the orange line in Fig. 1.10. On the other hand, the refractive index, n, shows a tilt at each resonant frequency, ω_0, that corresponds to the blue line in Fig. 1.10. In real matters, we can observe the peak of α and the tilt of n in each wavelength region. Figure 1.13

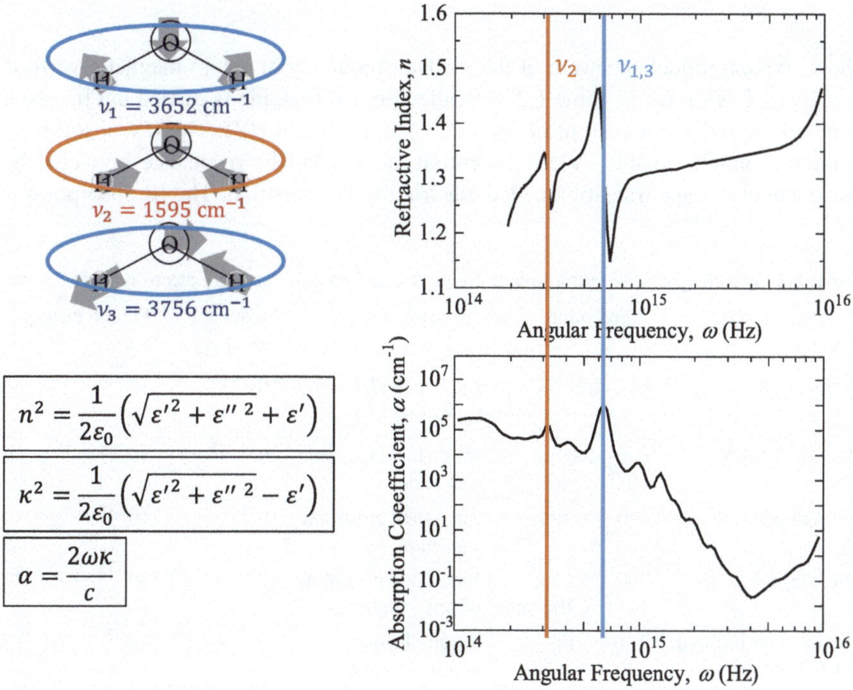

Fig. 1.12 Light frequency dependency of the n and α. The data is collected from Refs. [15–19] for reproduction of the plots

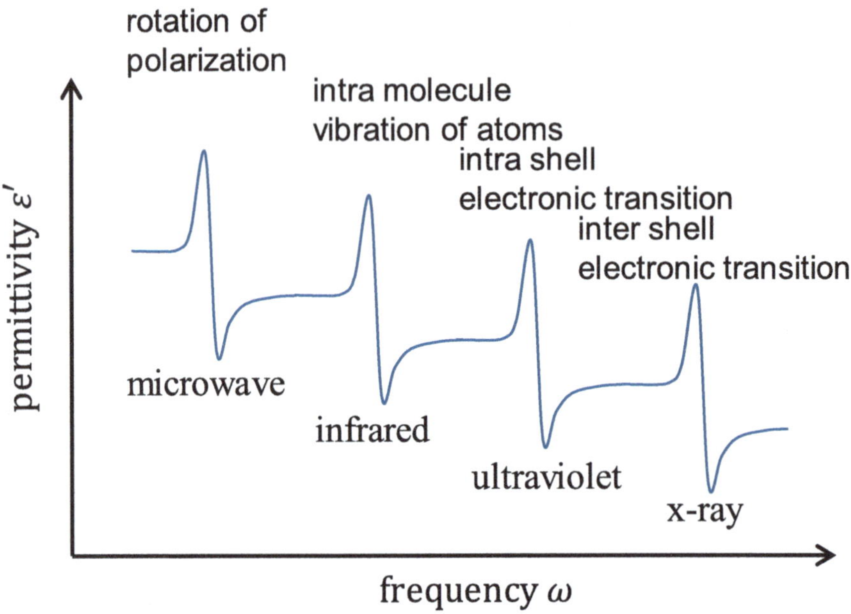

Fig. 1.13 Typical model spectrum of the ε'

shows typical model spectrum of the ε'. The frequency of electromagnetic wave is categorized as shown in Table 1.2. So-called optical imaging is carried out by using wave in the range between ultraviolet (UV) and infrared (IR). The UV absorption, sometime maybe visible (VIS) absorption, occurs by the resonance between the wave and electronic transition called electric dipole transition. The IR absorption is

Table 1.2 Wavelength, absorption source, frequency and photon energy of electromagnetic wave

Electromagnetic wave	Wavelength	Absorption source	Frequency s^{-1} = Hz	Photon energy J
X-ray, γ-ray	pm ~ nm $10^{-12\sim-9}$ m	Inter-shell electronic transition	$10^{17\sim20}$	$10^{-16\sim-13}$
Ultraviolet, UV	nm ~ 400 nm $10^{-9\sim-7}$ m	Intra-shell electronic transition	$10^{15\sim17}$	$10^{-18\sim-16}$
Visible, VIS	400–700 nm 10^{-7} m	Intra-shell electronic transition	10^{15}	10^{-18}
Near infrared, NIR	700–2000 nm $10^{-7} \sim 10^{-6}$ m	Inter-band transition of semiconductor	$10^{14} \sim 10^{15}$	$10^{-19} \sim 10^{-18}$
Mid and far infrared, MIR/FIR	2–100 μm	Intramolecular atomic vibration	$10^{12} \sim 10^{14}$	$10^{-21} \sim 10^{-19}$
Microwave	100 μm ~ 30 cm	Molecular rotation, electron spin	$10^{9} \sim 10^{12}$	$10^{-18} \sim 10^{-21}$

famous for characterizing atomic vibration in a molecule by using FT-IR equipment. Essentially, dynamic motion of electric dipole moments caused by the vibration of atoms resonantly absorbs the IR wave. At the resonant frequency, you can find the tilt of the refractive index, where the index changes steeply.

The resonant frequencies are different among substances. Assuming that two substances, 1 and 2, have the resonant frequency at the frequencies corresponding to 3000 and 3500 cm^{-1}, respectively, their refractive indices and absorption coefficients are simulated and plotted in Fig. 1.14. The difference of the refractive indices, Δn, is also shown. The difference gradually increases when approaching to the wavenumber that corresponds to the resonant frequency.

As more realistic case, Fig. 1.15 shows the refractive indices of oleic acid and water. Certainly, when approaching to the IR absorption range, the difference of the refractive indices (*the cause of bending!!*) increases, when increasing the wavelength. It is well known that the scattering intensity of matters increases when approaches to UV range by shortening the wavelength with the same reason. As well, it increases when approaching to the IR range.

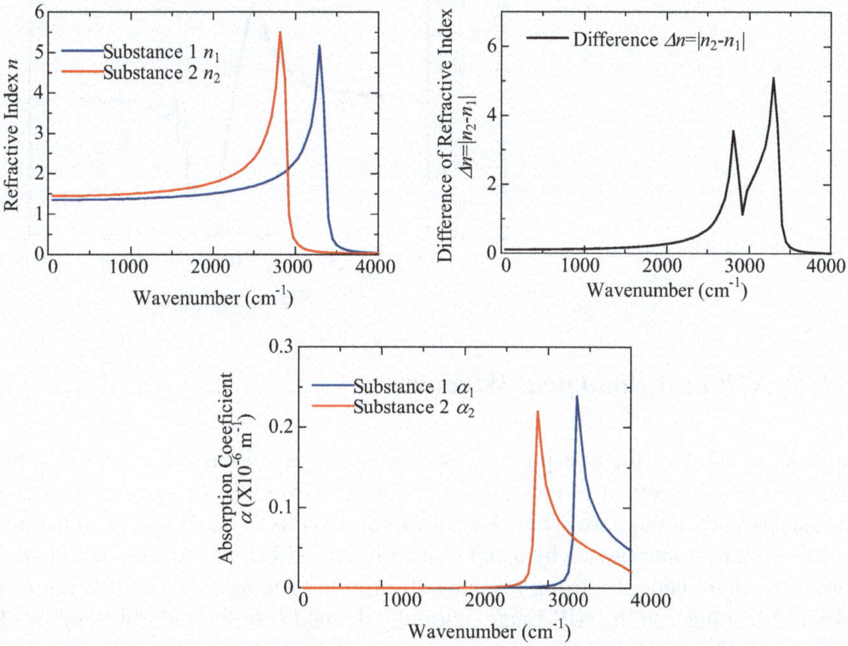

Fig. 1.14 Simulated refractive indices and absorption coefficients of two substances, 1 and 2. The resonant frequency at the frequencies corresponding to 3000 and 3500 cm^{-1}, respectively

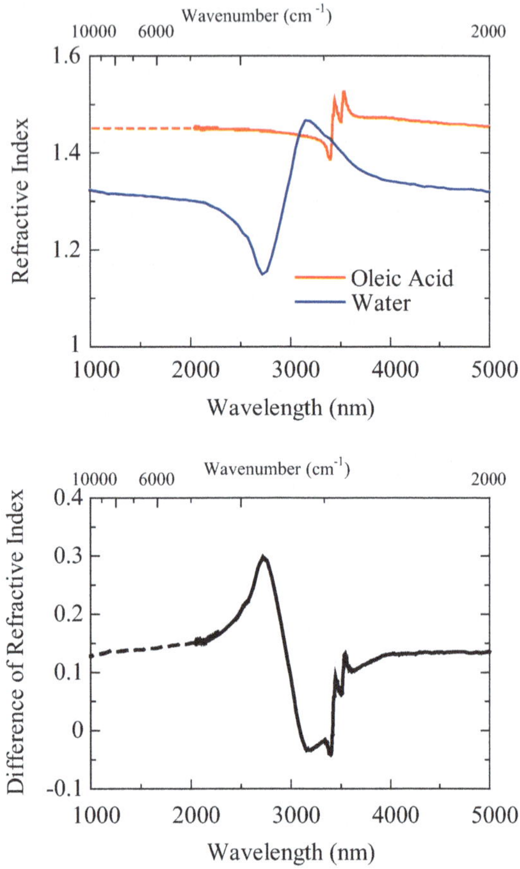

Fig. 1.15 Refractive indices of oleic acid and water reproduced from the data in [19]

1.3.2 NIR as a Biological Window

Now, let us think of the transparency of live tissues. As discussed in Sect. 2.4, by homogenizing the refractive index in tissues, we can avoid the *bending* phenomena for transparency, though we cannot keep the tissue or cells arrive. For "live" imaging, we can seek for transparency by using near-infrared (NIR) light. As discussed in the former section, there are strong resonant absorption of light in UV and IR regions around VIS range. In the NIR range, normally defined between the wavelengths of 800 and 2500 nm, the bending phenomena are relatively weak. Figure 1.16 shows the optical loss of human skin. The figure is reproduced from the literature reported in 1981 [17].

The loss consists of structureless gradual change expressed with a red dashed-dotted line and the peaks plotted with a blue dashed line. The former gradual change is due to the *bending* phenomena caused by the variation of refractive indices from parts to parts. The change increases in both regions approaching to UV and IR to form

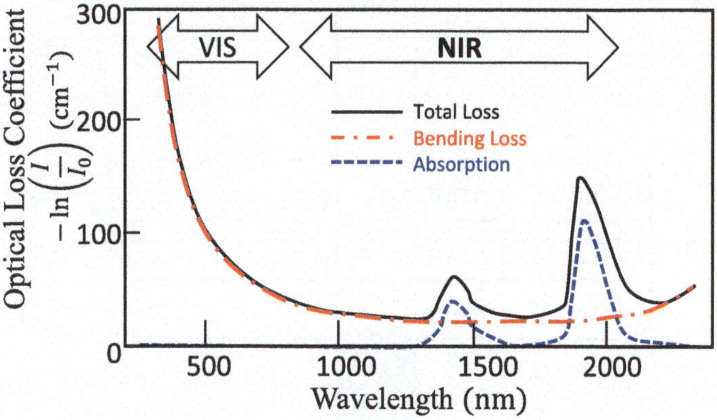

Fig. 1.16 Optical loss of human skin. The figure is reproduced from the literature reported in 1981 [20]

a valley in the NIR range. The structured peaks are due to the absorption originated from the overtone or combination tone of atomic vibrations in water molecules as discussed in Chap. 10. The fundamental vibration appears in the IR range. In Fig. 1.12, three vibrational modes, ν_1, ν_2 and ν_3, in a water molecule are shown. The correspondence to the wavelength of the overtones is 1331, 1369, 1869 and 1906 nm for the tones, $2\nu_3$, $2\nu_1$, $\nu_2 + \nu_3$ and $\nu_1 + \nu_2$, respectively. The values are for water molecules with gaseous state. In case of liquid water, the data for 28 °C is reported to be 3395 cm^{-1}, 1646 cm^{-1} and 3615 cm^{-1} for ν_1, ν_2 and ν_3, respectively [21]. According to it, the values are calculated to be 1383 nm, 1472 nm, 1900 nm and 1983 nm for $2\nu_3$, $2\nu_1$, $\nu_2 + \nu_3$ and $\nu_1 + \nu_2$, respectively, to have better matching with the results in Fig. 1.16.

1.3.3 Transmittance or Absorbance

For expressing the loss, which is better transmittance (or loss rate) or absorbance? Most of the people other than spectroscopists like to use the transmittance as *y*-axis because it is intuitive and easy to understand with a unit of %. However, if we are considering a phenomenon which is responsible for volume of light propagation, the transmittance, $\%T$, is not linear to the volume and we cannot deal the value as intensive variable for general expression with various situations. Assume that we have a plate with a thickness of 1 mm with 50 $\%T$ transmittance. Here, for simplicity, we neglect the reflection on the surface on the interface for simplicity. We only focus onto the light propagation in volume of the sample. For considering the optical loss with 3-mm thickness, we pile up the three plates with 1-mm thickness. As shown in Fig. 1.17, when light with an intensity I_0 comes in, the optical loss on the first plate is 50 % of the incident light to be $0.5I_0$ and the rest of $0.5I_0$ intensity light will come

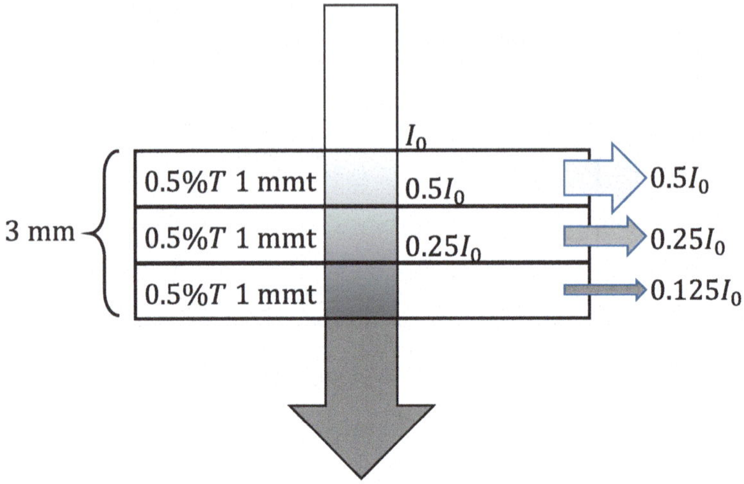

Fig. 1.17 Optical loss of three piled-up 1-mm thickness plates with 0.5%T each as an estimation model for the light propagation of 3 mm

into the second plate. In the second plate, incident light intensity is already $0.5I_0$ so that the loss is calculated to be $0.25I_0$. For the third one, the loss is $0.125I_0$. The amounts of the loss are different from plate to plate!

For solving this problem, we use the idea of differential calculus as shown in Fig. 1.18. Now, we think of sufficiently thin propagation length dx. The loss intensity

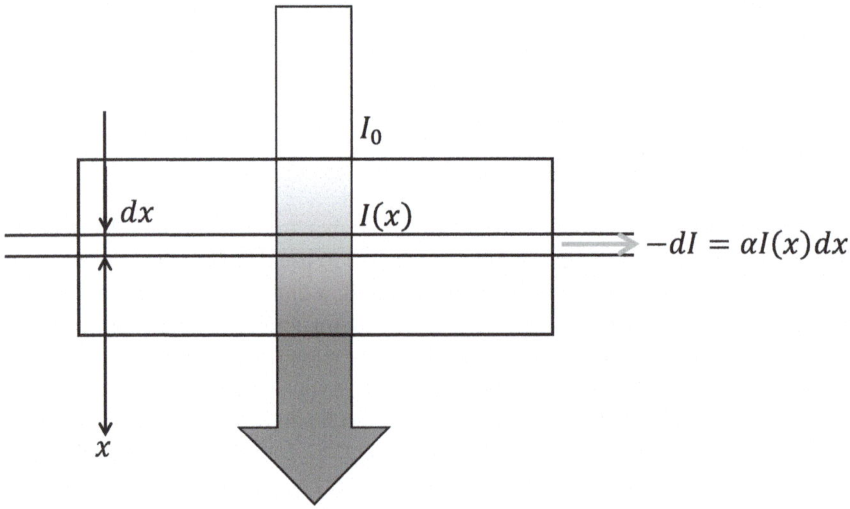

Fig. 1.18 Differential calculus model for calculating optical loss. Sufficiently thin propagation length dx is set with a loss rate of α at x with incident light intensity $I(x)$

is always proportional to the incident light intensity. We assume that the dx is thin enough to calculate the loss $-\mathrm{d}I$ to be

$$-\mathrm{d}I = \alpha I(x)\mathrm{d}x \tag{1.52}$$

$$\frac{\mathrm{d}I}{I(x)} = -\alpha \mathrm{d}x \tag{1.53}$$

By solving the differential Eq. (1.53), we obtain

$$I(x) = I_0 \exp(-\alpha x), \tag{1.54}$$

and

$$\alpha x = -\ln\left(\frac{I(x)}{I_0}\right). \tag{1.55}$$

The result corresponds to the solution in Sect. 1.3.1, Eq. (1.50). Important matter is the value of the left side of Eq. (1.55) is proportional to x, which is the propagation length or sample thickness. Assuming that the light beam is a parallel beam with a cross section of *S*, the volume that the light beam passed is proportional to x. Here, we can obtain α as an intensive variable individual from the volume or thickness. The α is normally expressed with a unit of cm^{-1} and called absorption coefficient. Chemists like to use extinction coefficient which comes from absorbance. Rather than taking the exponential base of *e*, the expression based on 10 can lead the results to express the absorbance to be

$$ax = -\log_{10}\left(\frac{I(x)}{I_0}\right) \tag{1.56}$$

with

$$I(x) = I_0 10^{-ax}, \tag{1.57}$$

where a is called extinction coefficient. The conversion between a and α can be written as

$$\alpha = \ln(10)a. \tag{1.58}$$

As well, the $\%T$ and the absorbance ax have the relation as follows.

$$ax = -\log_{10}\left(\frac{I}{I_0}\right) = -\log_{10}(100 \times \%T) \tag{1.59}$$

or

$$\%T = \frac{1}{100} \times \frac{I}{I_0} = \frac{1}{100} \times 10^{-ax} \quad (1.60)$$

1.3.4 Optical Loss in Biological Tissues

Here, for some more concrete ideas, optical loss spectra of typical mouse tissues are shown. Essentially, they consist of the structureless gradual change of the loss due to the bending phenomena and the absorption peaks due the overtone and the combination tones of water molecules. For avoiding to involve the reflection on the interfaces, absorbance spectra of samples of two different thicknesses were measured by a spectrometer. For the accuracy of the sample thickness, the sample holder illustrated in Fig. 1.19a was used. Then for eliminating the reflection loss on the interface of the sample and the air, the difference of the absorbance of the two spectra was calculated and divided by the thickness for converting into coefficient spectra as illustrated in Fig. 1.19b.

By using Eq. (1.58), finally, optical loss coefficient was calculated for adipose (abdominal white adipose tissue), brain (cerebrum), liver, muscle (peroneus muscle) and skin (abdominal skin) of mice. For separating the optical loss spectra into that with bending phenomena (mostly scattering) and absorption spectrum, at first the former spectrum was calculated to be subtracted from the raw data to generate the latter (absorption) spectrum. The results are shown in Fig. 1.20. As shown in Fig. 1.20b, the scattering is strongest with adipose. As you can read from Table 1.1, the refractive indices of lipids are relatively high. Adipose only has 10–30 % of water

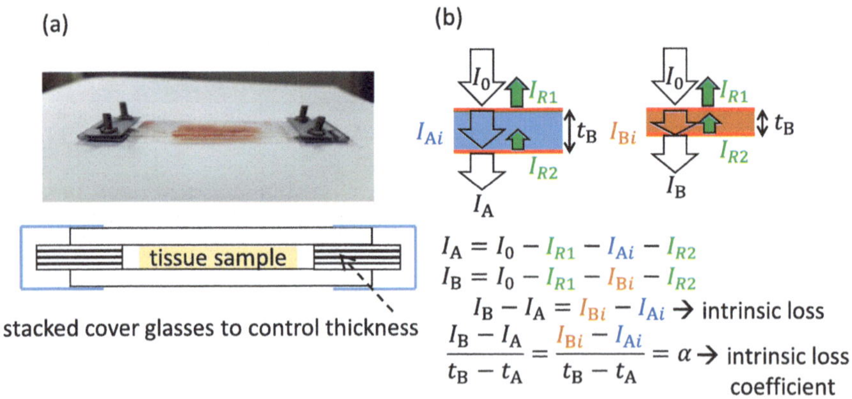

Fig. 1.19 Schematic for measuring the intrinsic optical loss spectra of tissues. (**a**) Sample setup. (**b**) Calculation method for eliminating reflection loss

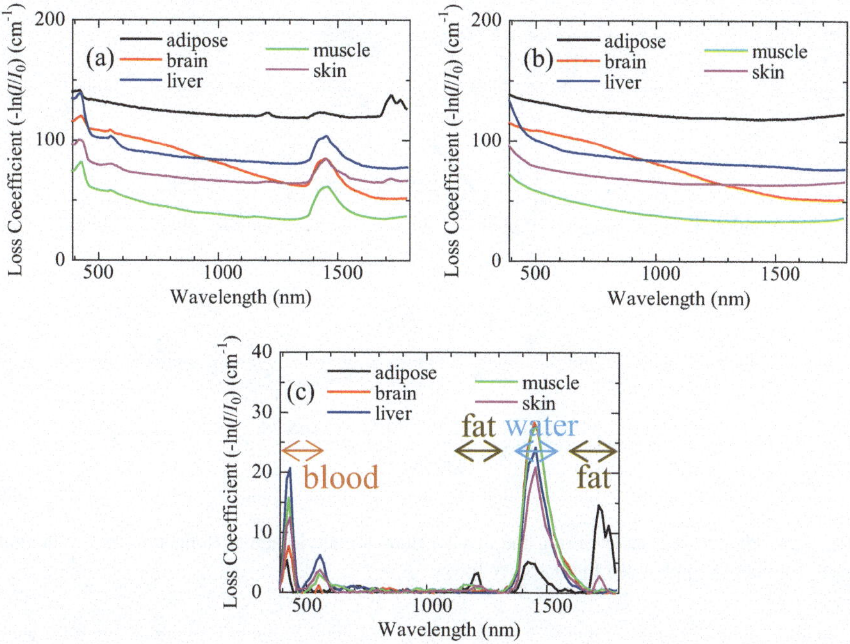

Fig. 1.20 Optical loss coefficient spectra of mouse tissues, (**a**) raw data, (**b**) loss due to bending and (**c**) absorption loss

content. The repeated structure with water and lipids generates strong scattering. All other organs contain 65–80% of water. Brain is on the next line in terms of scattering. It may be due to densely arranged nerve fiber (axon). Figure 1.20c shows that mainly the absorption peaks are assigned to water, blood and fat. Characteristic overtone absorption due to O–H stretching mode of water is observed around 1480 nm. Blood peaks appear in visible range. Fat shows its spectra at 1150 and 1750 nm. Multiplying by the thickness of tissue, we can calculate the optical loss of the optical pass length from the spectra. More precise discussion on the overtone and combination tone absorption in the NIR range will be made in Chap. 10.

1.3.5 Bending Phenomena in Small Animal Imaging

Figure 1.21 illustrates the phenomena that occurs during the in vivo imaging. It shows the case of fluorescence imaging. As seen from the figure, we must deal with the reflection and refraction when the excitation light comes in and the fluorescence light comes out. The *bending* phenomena on the interface of media 1 and 2 with refractive indices n_1 and n_2, respectively, are shown in Fig. 1.22. For easiness, simplified results for the case without any absorption or loss will be shown. If you need to involve them, you can convert the permittivity, ε, into the complex permittivity, $\tilde{\varepsilon}$. The reflection

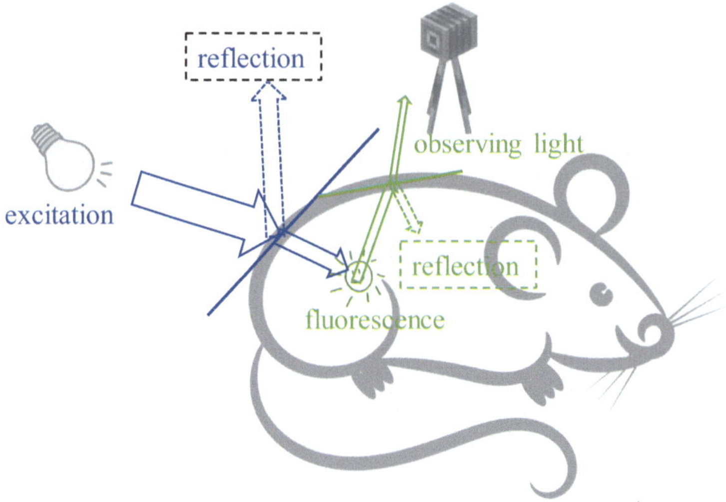

Fig. 1.21 Phenomena occur during the in vivo fluorescence imaging. Reflection and refraction occur for both excitation and fluorescence lights

Fig. 1.22 Bending phenomena on the interface of media 1 and 2 with refractive indices n_1 and n_2, respectively

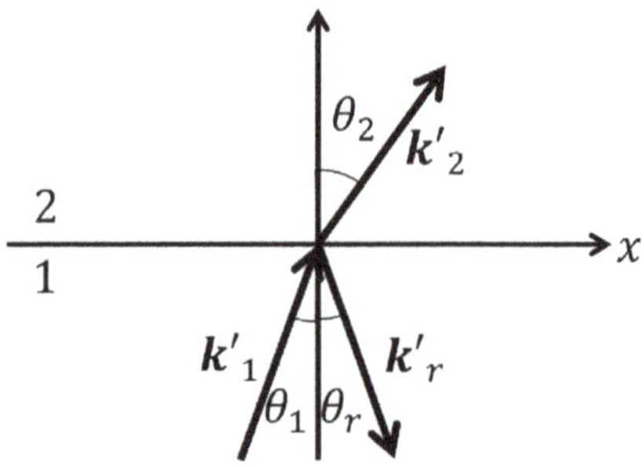

and refraction depend on the polarizations, s and p. For random polarization, you can average them for calculation. The reflectance, r, means the ratio of the reflected light intensity and the rest can be taken as transmittance t, the intensity ration of refracted light.

$$r_s = t_s - 1 = \frac{\cos\theta_1 - \sqrt{\left(\frac{\varepsilon_2}{\varepsilon_1}\right) - \sin^2\theta_1}}{\cos\theta_1 + \sqrt{\left(\frac{\varepsilon_2}{\varepsilon_1}\right) - \sin^2\theta_1}} \tag{1.61}$$

$$r_p = t_p - 1 = \frac{\left(\frac{\varepsilon_2}{\varepsilon_1}\right)\cos\theta_1 - \sqrt{\left(\frac{\varepsilon_2}{\varepsilon_1}\right) - \sin^2\theta_1}}{\left(\frac{\varepsilon_2}{\varepsilon_1}\right)\cos\theta_1 + \sqrt{\left(\frac{\varepsilon_2}{\varepsilon_1}\right) - \sin^2\theta_1}} \tag{1.62}$$

1.4 Summary

In this chapter, we tried to understand the physical basics of "transparency" by understanding the mind of light. Optical loss for losing the transparency can be simply understood by *absorption* and *bending*. The *bending* phenomena, reflection, refraction and scattering cause more serious problems to imaging. By approaching to the reason of *bending*, it is concluded that refractive index as well as permittivity is important to be studied. Transparency can possibly be achieved by chemical agents. However, for live imaging, we need to change the wavelength to the NIR.

Now, we reached to conclude that we can achieve deep tissue imaging alive by using the NIR light. This chapter will be followed by PART II. NIR Fluorescence Imaging and Phosphors and PART III. Application of NIR Light for Biomedical Photonics with Transparency. In PART II, characteristics and synthetic route of the NIR fluorescence imaging probes will be illustrated. In PART III, more demonstrative work by utilizing the transparency of the NIR light will be reviewed with more points of view for biological and medical applications. Some of the works proved that information of centimeter order depth can be visualized by using the NIR light.

References

1. Wikipedia, List of refractive indices. Available https://en.wikipedia.org/wiki/List_of_refractive_indices
2. Refractiveindex.Info. Available https://refractiveindex.info/
3. Pub chem. Available https://pubchem.ncbi.nlm.nih.gov/
4. C.-Y. Tan, Y.-X. Huang, Dependence of refractive index on concentration and temperature in electrolyte solution, polar solution, nonpolar solution, and protein solution. J. Chem. Eng. Data **60**, 2827 (2015). https://doi.org/10.1021/acs.jced.5b00018
5. M. Umezawa, S. Haruguchi, R. Fukushima et al., Rapid increase in transparency of biological organs by matching refractive index of medium to cell membrane using phosphoric acid. RSC Adv. **9**, 15269 (2019). https://doi.org/10.1039/C9RA01445D
6. G.J. Tearney, M.E. Brezinski, J.F. Southern et al., Determination of the refractive index of highly scattering human tissue by optical coherence tomography. Opt. Lett. **20**, 2258 (1995). https://doi.org/10.1364/OL.20.002258
7. P. Giannios, K.G. Toutouzas, M. Matiatou et al., Visible to near-infrared refractive properties of freshly-excised human-liver tissues: marking hepatic malignancies. Sci. Rep. **6**, 27910 (2016). https://doi.org/10.1038/srep27910
8. A. Briegel, D.R. Ortega, E.I. Tocheva et al., Universal architecture of bacterial chemoreceptor arrays. Proc. Natl. Acad. Sci. **106**, 17181 (2009). https://doi.org/10.1073/pnas.0905181106

9. B.R. Brückner, A. Janshoff, Importance of integrity of cell-cell junctions for the mechanics of confluent MDCK II cells. Sci. Rep. **8**, 14117 (2018). https://doi.org/10.1038/s41598-018-32421-2
10. H.-U. Dodt, U. Leischner, A. Schierloh et al., Ultramicroscopy: three-dimensional visualization of neuronal networks in the whole mouse brain. Nat. Methods **4**, 331 (2007). https://doi.org/10.1038/nmeth1036
11. A. Ertürk, K. Becker, N. Jährling et al., Three-dimensional imaging of solvent-cleared organs using 3DISCO. Nat. Protoc. **7**, 1983 (2012). https://doi.org/10.1038/nprot.2012.119
12. H. Hama, H. Kurokawa, H. Kawano et al., Sca*l*e: a chemical approach for fluorescence imaging and reconstruction of transparent mouse brain. Nat. Neurosci. **14**, 1481 (2011). https://doi.org/10.1038/nn.2928
13. I. Costantini, R. Cicchi, L. Silvestri et al., In-vivo and ex-vivo optical clearing methods for biological tissues: review. Biomed. Opt. Express **10**, 5251 (2019). https://doi.org/10.1364/BOE.10.005251
14. K. Tainaka, S.I. Kubota, T.Q. Suyama et al., Whole-body imaging with single-cell resolution by tissue decolorization. Cell **159**, 911 (2014). https://doi.org/10.1016/j.cell.2014.10.034
15. M.R. Querry, B. Curnutte, D. Williams, Refractive index of water in the infrared. J. Opt. Soc. Am. **59**, 1299 (1969). https://doi.org/10.1364/JOSA.59.001299
16. G.M. Hale, M.R. Querry, Optical constants of water in the 200-nm to 200-μm wavelength region. Appl. Opt. **12**, 555 (1973). https://doi.org/10.1364/AO.12.000555
17. M. Daimon, A. Masumura, Measurement of the refractive index of distilled water from the near-infrared region to the ultraviolet region. Appl. Opt. **46**, 3811 (2007). https://doi.org/10.1364/AO.46.003811
18. H. Hayashi, N. Watanabe, Y. Udagawa et al., The complete optical spectrum of liquid water measured by inelastic X-ray scattering. Proc. Natl. Acad. Sci. **97**, 6264 (2000). https://doi.org/10.1073/pnas.110572097
19. S.M. McGinty, M.K. Kapala, R.F. Niedziela, Mid-infrared complex refractive indices for oleic acid and optical properties of model oleic acid/water aerosols. Phys. Chem. Chem. Phys. **11**, 7998 (2009). https://doi.org/10.1039/B905371A
20. R.R. Anderson, J.A. Parrish, The optics of human skin. J. Invest. Dermatol. **77**, 13 (1981). https://doi.org/10.1111/1523-1747.ep12479191
21. P.A. Giguère, K.B. Harvey, On the infrared absorption of water and heavy water in condensed states. Can. J. Chem. **34**, 798 (1956). https://doi.org/10.1139/v56-103

Part II
NIR-Fluorescence Imaging and Phosphors

Chapter 2
Requirements of Nanostructures for Bioimaging Probes

Kohei Soga

Abstract Compared to other modalities of bioimaging, fluorescence bioimaging has advantages in sensitivity, resolution, dynamicity and polychromatism. One can easily find matters with very small concentration buried in complex system by their fluorescence. Resolution can go down below the wavelength nowadays by using a technique called super-resolution microscopy. The images can be color movies. For visible light fluorescence imaging, expression of chromophore by gene manipulation innovatively expanded the potential of the imaging. However, for near-infrared fluorescence for transparency to biological tissues, such method is not applicable to date. For achieving the NIR fluorescence bioimaging, it is required to prepare fluorescent probes to administrate from external bodies. For designing and preparing the probes, understanding of the behavior of nanostructured matters in biological system. This chapter starts with introduction to fluorescence bioimaging followed by bio-logistics into, in and from animal bodies. Also, some tips for giving targeting ability to the probes are discussed. The last section describes the facts of nano world in terms of size, surface charge and dispersion stability together with their characterization methods. Especially, the parts for dispersion stability and size control are important.

Keywords Near-infrared · Fluorescence · Nanostructure · Probe · Biologistics

2.1 Fluorescence Bioimaging

2.1.1 Merits of Fluorescence Imaging

The purpose of imaging is contrasting some information in biological system. There are various modalities as popular medical method like an X-ray or X-ray CT or MRI. Compared to those, what are the merits of fluorescence imaging? We can summarize

---

K. Soga (✉)
Department of Materials Science and Technology, Faculty of Industrial Science and Technology, Tokyo University of Science, 6-3-1 Niijuku, Katsushika, Tokyo 125-8585, Japan
e-mail: mail@ksoga.com

K. Soga et al. (eds.), *Transparency in Biology*,
https://doi.org/10.1007/978-981-15-9627-8_2

the merits to be sensitivity, resolution and dynamicity. Additionally, colors can be given to different objects.

Then think about the size of the stars. Is it visible? Aren't they too small? They are so far away. Thinking of the size of the star, they are invisible if you take diffraction limit into account. Why are they visible? Because it is luminescent. As long as things emit light, we can recognize the objects. Even if you are lost in desert at night, you can find a lodge far away as long as they light a lamp as in Fig. 2.1.

Sensitivity may contain various meaning. Here, assuming that you are told that a gold coin is in a desert (again *desert*). The weather is so hot, so dry. It's unbelievably huge. You're tired. Do you try to get the gold coin? Of course not. However, if you find a shine of the gold coin, you can easily find and earn it. This is the sense of sensitivity of fluorescence. Some scientists are challenging to find and trace one molecule in biological system. Even if you are having the object with a very small concentration in various other matters, you can find it out with the fluorescence.

People believed that smaller matter than visible light wavelength is invisible. The Nobel Prize in chemistry of 2017 was awarded to J. Dubochet, J. Frank and R. Henderson for the "high-resolution structure determination of biomolecules in solution [1]." By the super-resolution microscopy, the super-resolution can in principal achieve the resolution of nanometer level. As illustrated in Fig. 2.2, it uses blinking of fluorescence. Once the spot is blinking, it appears and disappears. If they all at once shine, they fill an area as a continuous field. That's the reason why we cannot

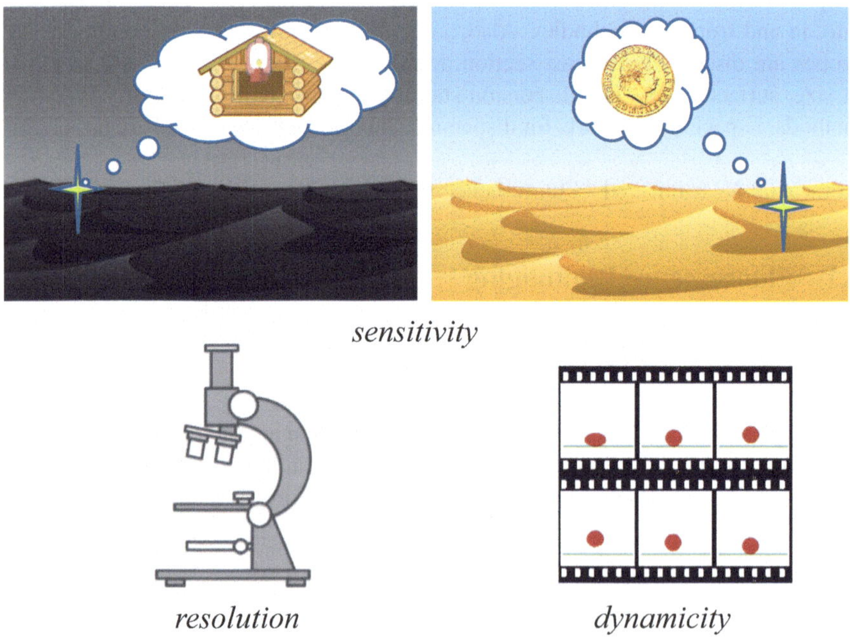

Fig. 2.1 Merits of fluorescence imaging; sensitivity, resolution and dynamicity

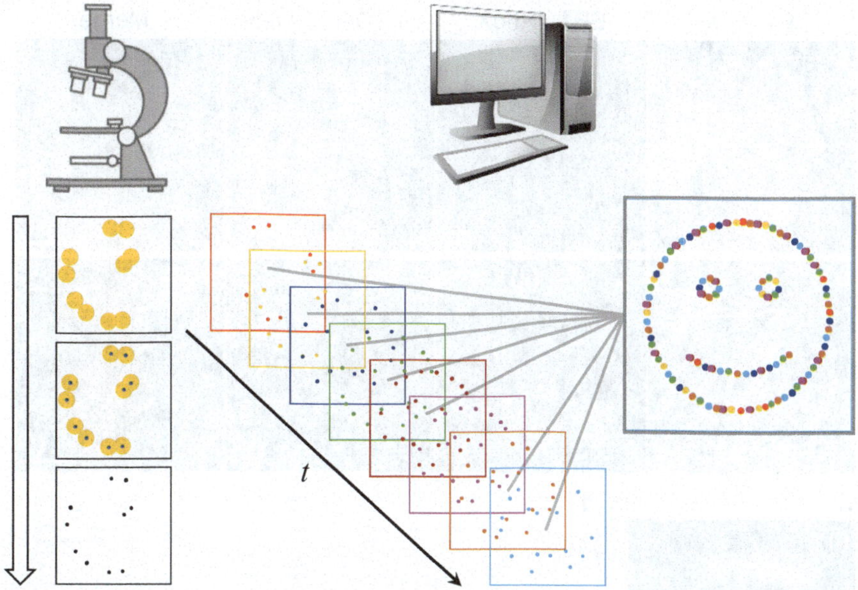

Fig. 2.2 How super high resolution can be achieved for that below wavelength?

recognize each shining point. However, with blinking, much less and scattered and separated dots appear. Each dot is a circle. By plotting the central point of the circle, the density of the points will depict the shape of the size less than the wavelength.

Figure 2.3 shows an example of such super-resolution imaging [2]. Much smaller structures than observation wavelength are clearly depicted.

As for dynamism, we are now using video rate of 30 fps (frames per second) with 2.1 MP (megapixel) resolution with a smartphone. For some purposes, we can choose 60 fps. Every 17 ~ 34 ms, the camera captures the brightness of more than 2 million dots with 16,777,216 color depth. The response time of human eyes is 50 ~ 100 ms. We can easily capture color images as a dynamic movie.

2.1.2 What is Fluorescence?

Let's start from a phenomenon, luminescence. Once things are luminescent, as explained above, we can recognize things even if they are too small. For the luminescence as a phenomenon to emit light, some energy source is required to emit light. According to the energy source, the luminescence phenomena are categorized into photo-luminescence, chemo-luminescence, thermo-luminescence and electro-luminescence. When electrons are excited by light, which has normally shorter wavelength than the luminescent light, the phenomenon is called fluorescence.

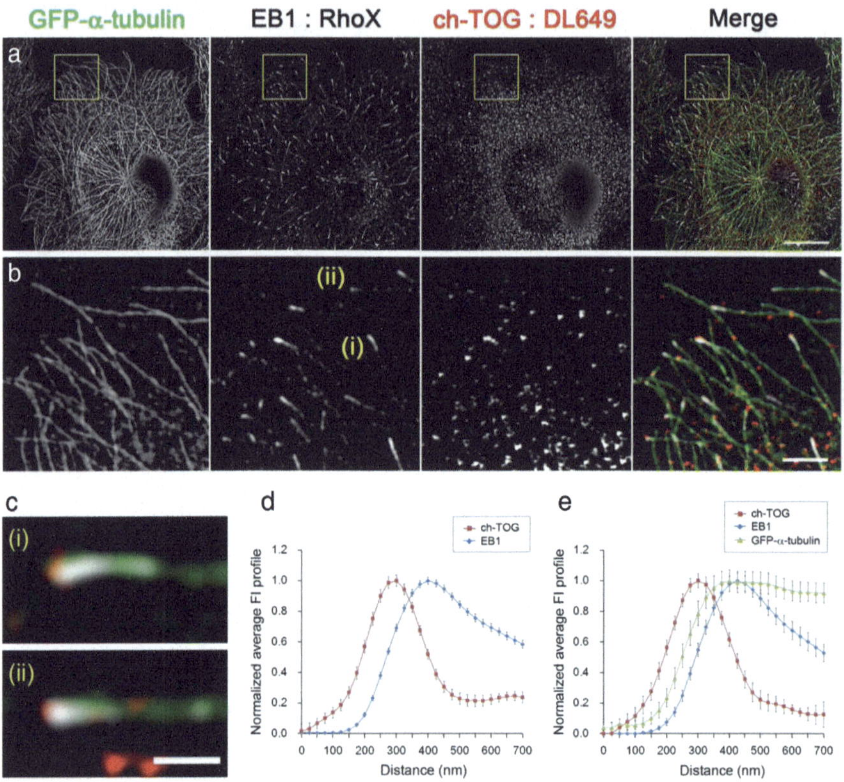

Fig. 2.3 An example of structured illumination microscopy with resolution below the fluorescence light wavelength. The microtubule structures in a HeLa cell are shown. The fluorescence wavelengths are between 500 and 700 nm, visible range. The scale bars for **a**, **b** and **c** are 10 μm, 2 μm and 500 nm, respectively. The nanostructure was analyzed in the range below 500 nm in **e** and **f**. Reprinted from Reference [2] with Creative Commons Attribution License

The chemo-luminescence is famous to be used by fireflies. Chemical energy created from ATP cycle is finally transferred to chromophore called luciferin. After numbers of energy transfer among the molecular groups, it finally emits green light as seen in the forest.

Fluorite, a mineral, is known to emit light by heating. When the heating is a cue to emit light, the luminescence is called thermos-luminescence. Nowadays, long afterglow materials are popular to be used for daily life. After photoexcitation, the excited electrons are trapped in shallow traps of of the excited state. The heat of room temperature will slowly pull the electrons from the traps to be luminescent. In this sense, the materials are also thermos-luminescent.

After the invention of blue light-emitting diode (LED), white light from LED replaces more and more illuminators around our daily life. Electrons in semiconductor are electrically moved from the ground state (valence band) of a part of one device to the excited state (conduction band) of the other part of the device for emitting light. This phenomenon is called electro-luminescence.

2.1.3 *How to Make the Objects Luminescent*

O. Shimomura, M. Chalfie and Roger Y. Tsien were awarded with the Nobel Prize in chemistry in 2008 for the "discovery and development of the green fluorescent protein, GFP." Some of the proteins are fluorescent naturally. Their work made it possible to artificially make a part of cells or tissues fluorescent by genomic manipulation. The work contributed to the progress of biological and medical sciences. However, they are not for near-infrared (NIR). The focus of this book is on "transparency." As introduced in Chap. 1, some chemicals can clear tissues. In fact, by using the clearing chemicals, some researchers have achieved millimeter order imaging of brain nerves as shown in Fig. 2.4 [3]. For the imaging of live tissue or live body, we need to use the NIR fluorescence for centimeter depth. As reviewed in the following chapters, organic dyes (Chap. 3), quantum dots (Chap. 4), carbon nanotubes (Chap. 5) and rare-earth-doped ceramic nanophosphors (Chap. 6) are luminescent. How can we make the target object to be luminescent? We have to introduce the phosphors, deliver and target the object. Especially for medical use, elimination (excretion) is important. In the next section, let's discuss bio-logistics.

2.2 Bio-Logistics: into, in and out from Animal Bodies

2.2.1 *Probe Introduction to Animal Body*

Now, let's discuss how we can introduce the fluorescent probes into a mouse body. Normally, the probes are introduced to the body through intravenous (IV) injection. Once the injected probes are coming into blood, they meet various kinds of disturbing matters. Blood consists of very ionic water with a 0.15 mol/L sodium chloride. And it contains a lot of proteins. As you may know, you can also find red, white and other blood cells as well as many kinds of proteins.

Then the probes start to circulate. After injected from the tail vein, once they are collected to the heart. Then from the heart, they are pumped out to whole part of the body. If their size is small enough, they can circulate without any hindrance. However, if their size is too large, they meet macrophages to be captured by a system called reticuloendothelial system (RES).

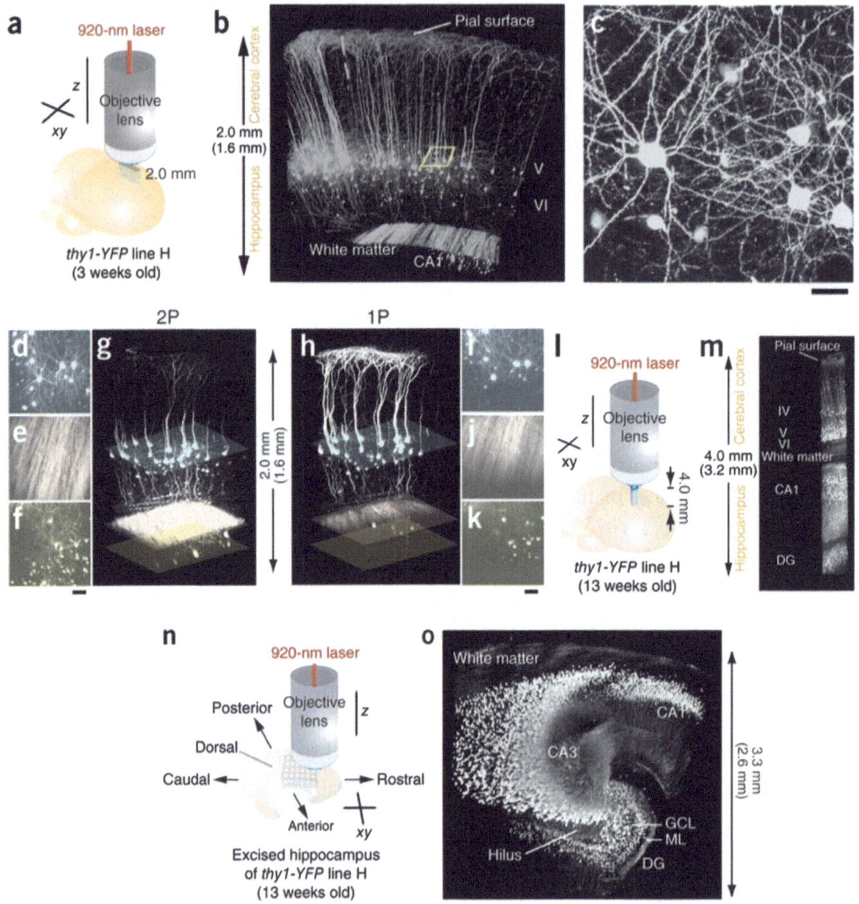

Fig. 2.4 YFP-expressing neurons in unsectioned brains and an excised hippocampus. Reprinted from Reference [3] with permission from Springer Nature, 2020

As for the interaction with the cells, if their size is larger than 100 nm, normal cells neglect them and nothing occurs except the case of phagosomes like macrophage. If the sizes are smaller than 100 nm, there are chances for the cells to take them into the cells by a mechanism called endocytosis as shown in Fig. 2.5. For phagosomes as macrophage, the situation is inversed. If the probe is smaller than 100 nm, macrophages do not recognize, while if it is larger, they recognize the probe as an extraneous matter. They englobe the probe into the inside and expose it to strong acid condition for decomposing it. Here, we can find a border of 100 nm. The size is important.

Also, the surface charge is important. Mostly, the surfaces of the cells are charged negatively. If the surface of the probe has neutral or negative charge, so far nothing occurs. On the other hand, if the charge is positive, it interacts with the cell membrane

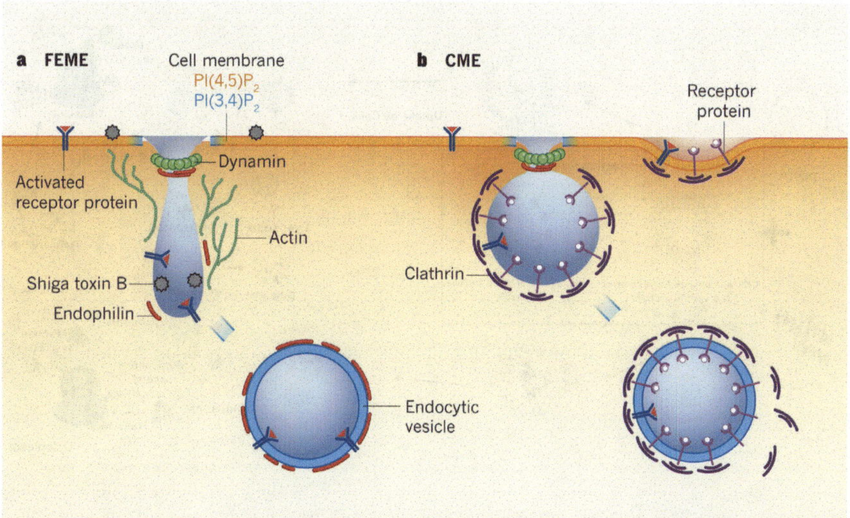

Fig. 2.5 Mechanisms of endocytosis. **a** Boucrot et al. 3 describe a method of cell entry dubbed fast endophilin-mediated endocytosis (FEME). **b** By contrast, in clathrin-mediated endocytosis (CME). Reprinted from Reference [4] with permission from Springer Nature, 2020

to merge. Sometimes, it kills the cell by destroying the membrane. So the positive charge matters. It is interactive and also toxic. Here, the size does not matter.

We can learn this phenomenon from the behavior of liposomes. Liposomes are capsules made of artificial membrane which consists of phospholipid. The size can be varied from several hundred nanometers to micron level. If the surface of the liposome is controlled to be negatively charged, almost nothing occurs even if it meets to cells. On the other hand, if the surface charge is positive, the merge of the membrane of the liposome and cell occurs. Then, the content of the liposome may be introduced into the cells even if the liposome sizes are much larger than 100 nm. The method is called liposome transfection (or lipofection) as illustrated in Fig. 2.6 [5].

2.2.2 Excretion

Especially in the case of medical imaging, we may have to remove the probe from the body. The process is called excretion. The most popular excretion route is called renal excretion through kidney. The kidney filter easily passes through matters with 10 nm size (some say 6 and some say 8 nm). As shown in Fig. 2.7, renal clearance of nano-substances is partly clarified as a hot topic [6]. With the size, the excretion occurs with a time of hour order. Another typical excretion is biliary excretion. In liver, there is a mechanism to capture hydrophobic matter by using bile acid. Bile acid

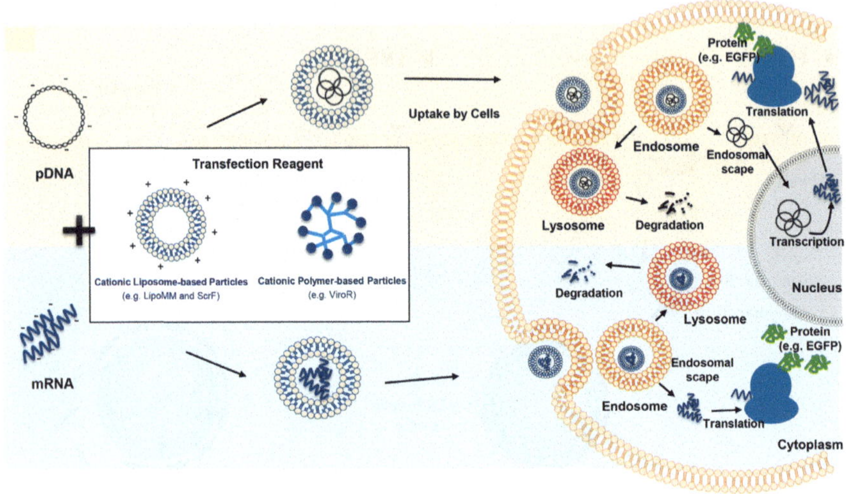

Fig. 2.6 Different paths to protein expression upon cell transfection with mRNA or pDNA. Reprinted from Reference [5] with Creative Commons Attribution License

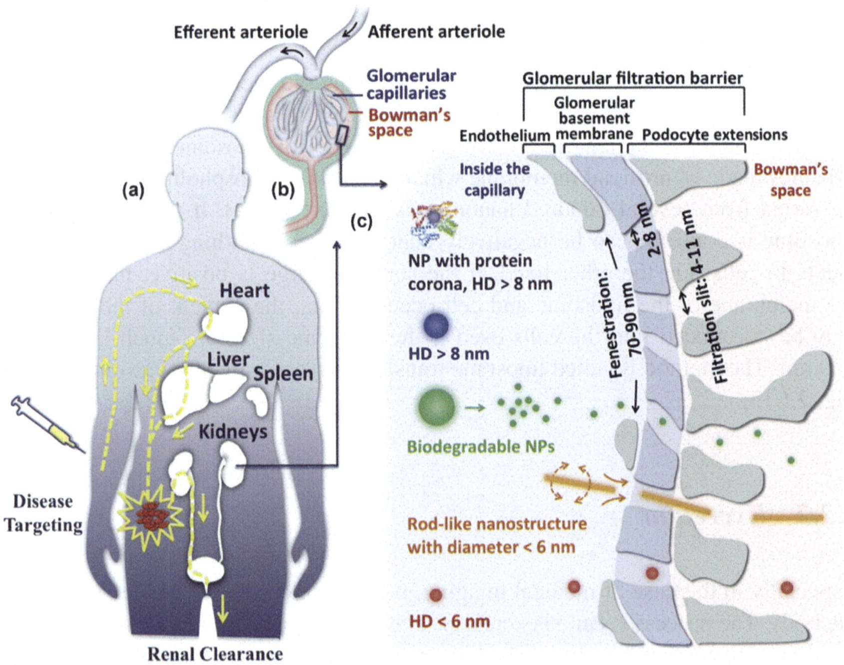

Fig. 2.7 Ideal disease targeting of renal clearable nanoparticles (NPs) in clinical practices: The NPs specifically target the diseases and untargeted ones are rapidly cleared out of the body through the urinary system. Reprinted from Reference [6] with permission from Elsevier, 2020

will surround the hydrophobic matters and send them into small intestine as shown in Fig. 2.8. Partly, they are reabsorbed from in the intestine and partly involved in feces to be excreted.

The circulation time is decided by the excretion. In case of the particles with size between 10 and 100 nm and with neutral or negative surface charge as well as hydrophilic surface, if the size does not change, it may circulate almost forever.

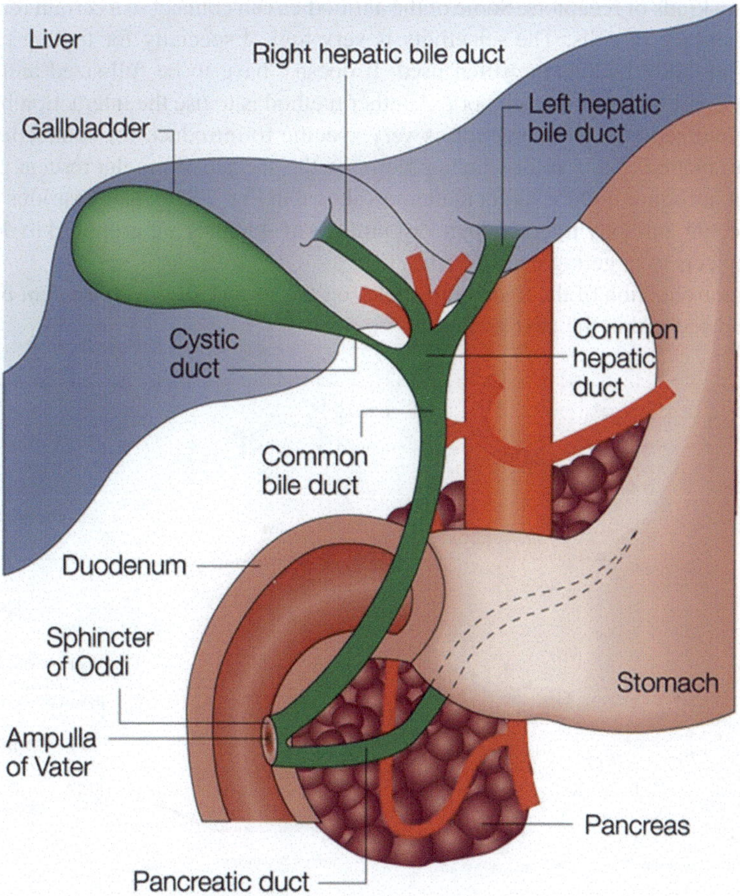

Fig. 2.8 Liver, bile duct and connection to intestine. Reprinted from Reference [7] with permission from Springer Nature, 2020

2.2.3 *Targeting*

We have to design the fluorescent probe by knowing the logistics in animal body. Also, for making a target object to be luminescence, we need to introduce ligand molecules. The role of ligand molecule is to stick onto a certain cell or matter in biological systems.

Normally, people use antigen–antibody reaction. On the surface of the cells, there are many kinds of receptors. Some of the antibodies can connect to a certain receptor on the surface of cells. The selectivity is very high. Especially for finding cancer organ, antibody ligands are often used. It doesn't have to be full-sized antibody. Maybe a part of antibody can work. Another method is to use the interaction biotin-avidin interaction. The interaction is very specific to introduce for connecting the probe to matters. So if you're having biotin on the surface of the fluorescent probe, you can introduce it to the target matter. As shown in Fig. 2.9, by using various kinds of molecule-antibody interaction, a vast number of schemes are proposed to deliver nano-objects to targeting part of bodies [8].

The introduction of the ligand molecules on the surface of the fluorescent probes will be discussed in the next section.

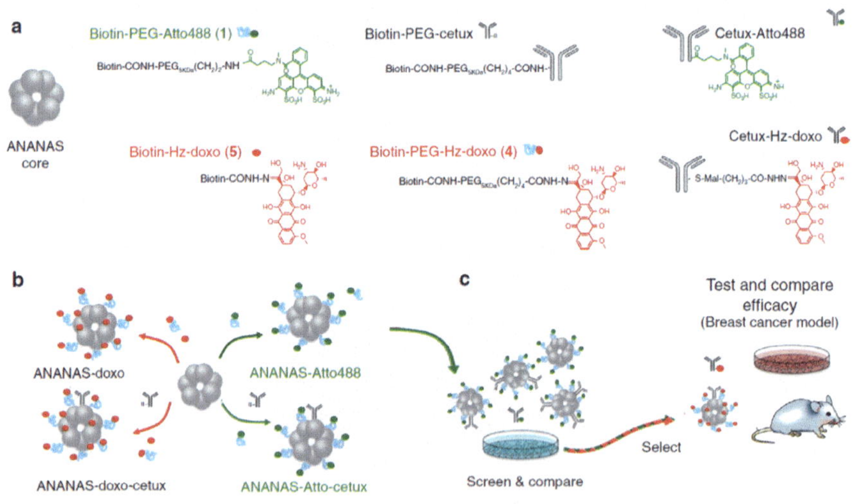

Fig. 2.9 ANANAS assembly and study design. **a** The building blocks for ANANAS functional assembly and the reference antibody conjugates synthesized; **b** ANANAS functional assembly and c in vitro and in vivo evaluations. Reprinted from Reference [8] with Creative Commons Attribution License

2.3 Welcome to Nano World

2.3.1 Nano World

How do you know about "nano?" As illustrated in Fig. 2.10, meter (m) and millimeter (mm $= 10^{-3}$ m) worlds are visible to naked eyes. Micron or micrometer (μm $=$ 10^{-6} m) is still visible but not precise. We need a microscope to see precise. Normal paper thickness is around 80 μm. Hair has the thickness of 60 $\sim$ 100 μm. Normally, animal cells are with the sizes of 6 $\sim$ 25 μm. Red blood cells are with 8 μm. Now, the next is the "nano" world. It is not "visible" because of the limit since the wavelength of visible light is 0.4 $\sim$ 0.7 μm $=$ 400 $\sim$ 700 nm. Butterfly is using this size. The periodical structure on the butterfly wing with 400 $\sim$ 700 nm pitches creates beautiful colors. Viruses are having typically 20 $\sim$ 120 nm size. A typical fullerene, C60, is with the size of 1 nm $= 1 \times 10^{-9}$ m. The atomic world is in nano ~ sub-nano world. The diameter of hydrogen atom is about 0.1 nm $= 1 \times 10^{-10}$m $=$ 1 Å. Proton's size can be understood as 1.78 fm $= 1.78 \times 10^{-15}$ m. Understood? In fact,

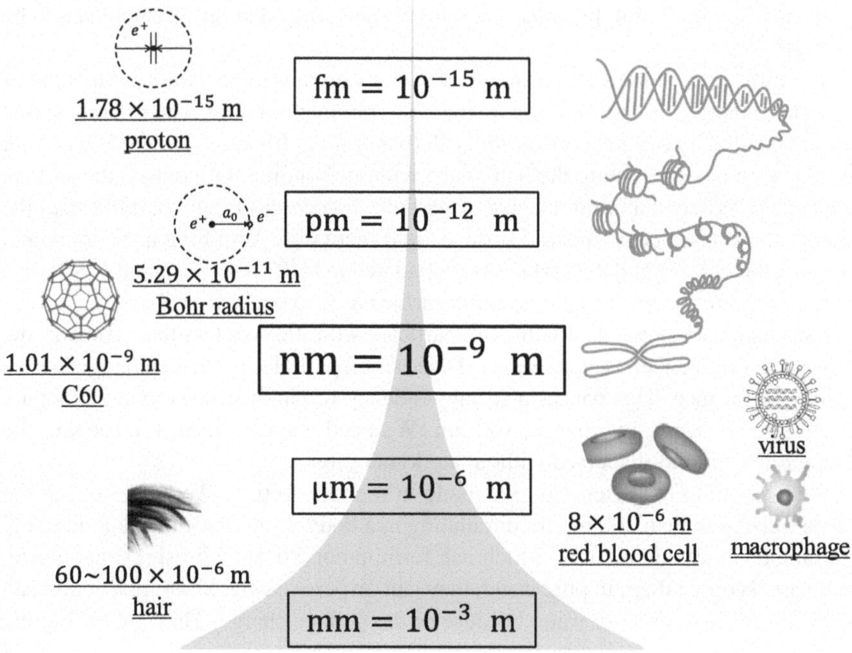

Fig. 2.10 "Nano" in scales

I am not sure if I am really understanding this. Anyway, the invisible world "nano" is there.

In 1959, Richard Feynman spoke "There's Plenty of Room at the Bottom." In 1986, K. Eric Drexler said "Engines of Creation: The Coming Era of Nanotechnology." Beside those, in 2000, economically most impacting speech was made by Bill Clinton as "National Nanotechnology Initiative." The speech strongly affected the governments of the world to results in huge investment to researches with the word "nano." A vast number of science and technologies for clarifying the "nano" world have been developed to date in the border of centuries.

2.3.2 What is Special in the "Nano" World?

Now what is special in the "nano" world? Let's think about gold atoms. The gold atoms can form clusters with certain number of atoms. In a case of a cluster with 2 nm diameter (with $r = 1$ nm), the number of atoms can be calculated to be 288 from the density. The surface area is $S = 4\pi r^2 = 12.6 \times 10^{-18}\,\text{m}^2$. The bonding length between two atoms is $d = 0.288\,\text{nm} = 2.88 \times 10^{-10}$ m. Assuming that one atom occupies an area with $d^2 = 8.29 \times 10^{-20}\text{m}^2$. We can now calculate the number of the surface atoms to be $S/d^2 = 152$. Surprisingly 53%, more than a half number, of atoms are on the surface! Let's compare with some visible case. In case of a gold particle with 2 mm diameter, the ratio of the surfaced atoms is calculated to be 53×10^{-6}%.

The difference of the surface atom and in-body atom is their stability. Any kind of condensed matters is stabilized by forming interatomic bonding. The in-body atoms are surrounded by two or more atoms with two or more bonding. On the other hand, the surface ones are having the half space without bonding. Obviously, the surface is unstable. When changing the size of particle, by decreasing the particle size, the ratio of the surface area (specific surface area) increases. Assuming a sphere with a radius r, the volume and the surface are calculated to be $V = 4\pi r^3/3$ and $S = 4\pi r^2$, respectively, and yields into the specific surface area to be $S/V = 3/r$.

Especially, when we deal with nanoparticles with the size less than 100 nm, the high energy of their surface is serious. Those small particles try to attach together for killing the surface. This causes a strong tendency of nanoparticles to agglomerate. The border is again, by chance, 100 nm. We need a special care for keeping the nanoparticles mono-dispersed without agglomeration.

Please remind the discussion we made in the last section. We have to keep the particle size less than 100 nm for circulating in a body. Iron, oxide or titanium oxide are famous inorganic matters, which can form nanoparticles with single nano-size. If you are keeping them in pure water, they can disperse easily. Mono-dispersion can be achieved easily in pure water because of the surface charge. They are having the same surface charge. Mostly in the case of ionic ceramics, the charge is positive, and in cases of metal or semiconductor nanoparticles, the charge is negative. Anyway, the particles are having the same charge for repulsion.

The surface charge can keep the number of particles mono-dispersed separately. However, think about the physiological condition. In water, 0.15 mol/L sodium chloride is contained. It is a very ionic condition. The surface charge of the nanoparticles is masked by the ions and loses the repulsion force. If you have a glass of water and dispersed nanoparticles in it, then you put a spoonful of salt in it, and eventually, the nanoparticles get together for forming a big agglomerate and sediments.

This fact means even if the particles are well dispersible in pure water, without having anything as a modifying agent, in ionic condition, they aggregate and settle down. We have to avoid this to happen in an animal body. How can we avoid? Also as discussed in the last section, there are a lot of kinds of proteins in a body with variety of the properties, hydrophobic, positively or negatively charged as shown in Fig. 2.11. In fact, as shown in Fig. 2.12 [9], proteins have various electrostatics. The proteins collect nanoparticles to stick onto them and form a big agglomerate to be a recognized by macrophages.

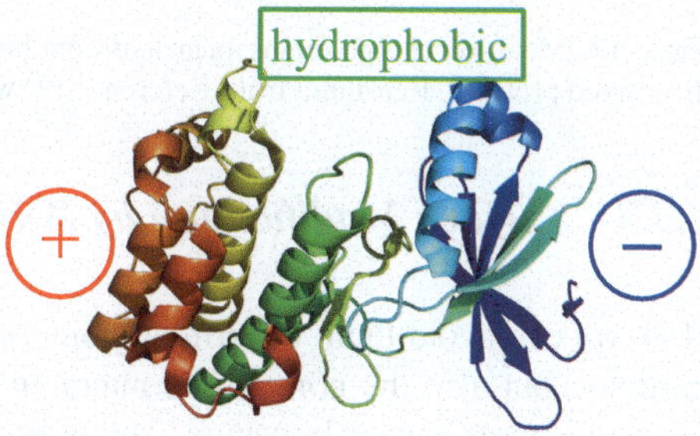

Fig. 2.11 Complex structure of protein. Parts of protein are positively or negatively charged or even hydrophobic

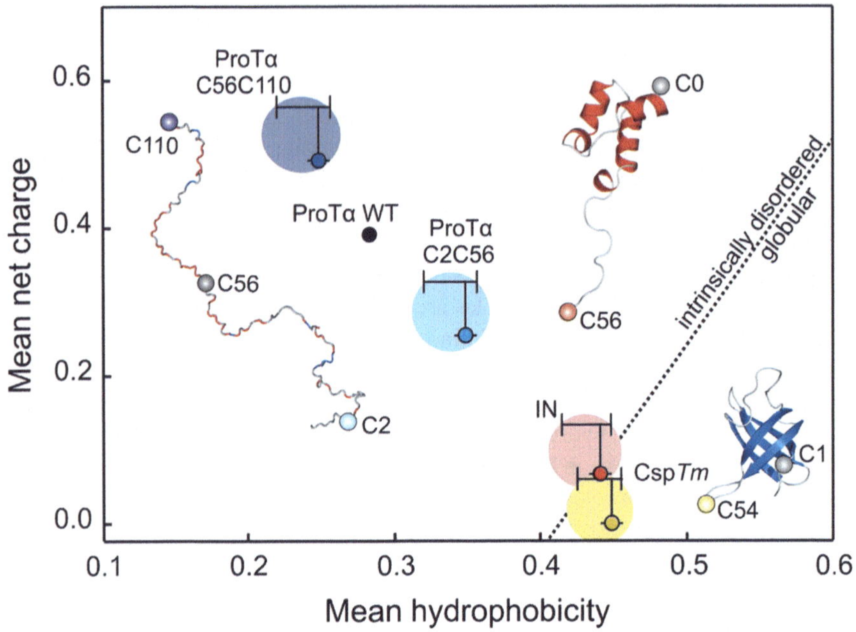

Fig. 2.12 Mean net charge versus mean hydrophobicity per residue of the globular and intrinsically disordered proteins. Reprinted from Reference [9] with permission from Natl. Acad. Sci. 2020

2.3.3 Surface Modification by Polymers

How we can avoid them to form agglomerate? Not only the interaction among the particles, but also the nonspecific interaction with the particles with non-targeting biological parts seriously occurs. The only presently known method to avoid these unfavorable attractions is the use of steric repulsion of polymers with the same property with the solvent as illustrated in Fig. 2.13.

The polymers are always moving, shrinking and stretching, at room temperature. The motion can avoid the particles to get together. This phenomenon is called steric repulsion. Also, very hydrophilic polymers, such as poly(ethylene glycol) (PEG), like water molecules so much that they are always wearing water molecules around. Thick water layer is formed on the surface of the nanoparticles with the polymers. The performance to avoid the agglomeration or nonspecific interaction depends on the property of the polymers such as polymer length (molecular weight). For more precise understanding of this phenomenon, we need to consider free energy change by that motion as described in literature [10].

Also, as discussed in the last section, maybe we can use an ionomer (polymers with charges) as long as it is charged negatively because the surface charge of the cells or organs is negative. If the ionomer has positive charges, it may be very toxic. We must be careful. For safety, the best solution is the electrically neutral property of

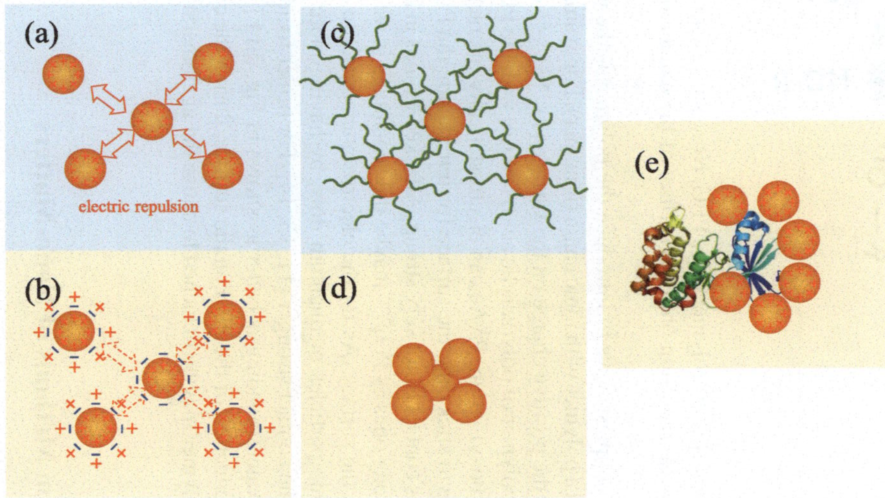

Fig. 2.13 Particles in aqueous solution with positive surface charge. **a** In pure water, they can be mono-dispersed because of the electric repulsion due to surface the same surface charge. **b** In salt solution, ions of salt will mask the surface charge to cancel the repulsive force to cause agglomeration as **d**. **e** Solution with protein, the charged parts on the protein attracts the charged particles to form a large aggregate. **c** Hydrophilic polymer on the surface of the particles keep the particles mono-dispersed by steric repulsion even with ions or proteins

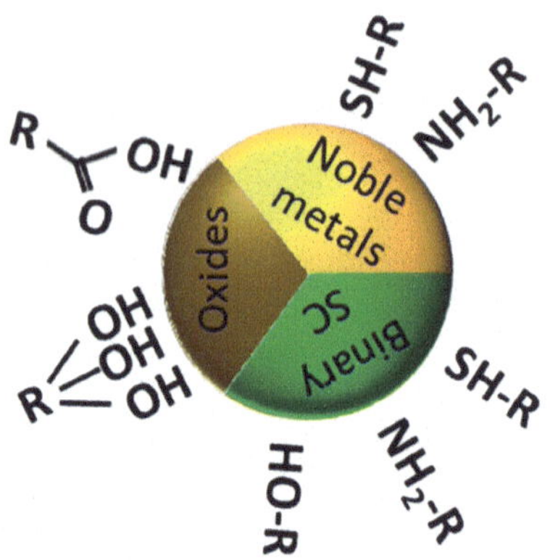

Fig. 2.14 Schematic representation of the most commonly used functional groups in surface ligands and their preferred affinity towards binding different NP materials. Reprinted from Reference [15] with Creative Commons Attribution License

the polymer. The most typically used polymer is the PEG. Now, we need to introduce the PEG for introducing steric repulsion onto the surface of the nanoparticle. How we can manage it?

One of the easiest and simplest ways to introduce the polymer is the use of co-block polymer of PEG and a polymer with ionic chain (ionomer). For example, some particles are charged with the positive charge on the surface [11, 12].

If we use the block co-polymer of poly(acrylic acid) (PAAc) with the PEG, as the ionic chain with many carboxyl group, PAAc can electrostatically interact with the surface of the nanoparticle to stick onto the surface for introducing the PEG on the surface. Unfortunately, this contact is not very strong. Attacked by ions in blood, not only Na^+ or Cl^-, some strong ones as PO_4^{3-}, maybe the attachment is weakened to remove the polymer from the surface. As for the electrostatic contact to introduce PEG to the surface of ionic particles or quantum dots, by having more than five charged group can make the contact stronger [11, 13, 14]. For firmer fixation of the PEG, the formation of covalent bonding is better as shown in Fig. 2.14 [15]. Even if we can confirm the introduction of the PEG on the surface, during the circulation in blood, gradually the polymer may be detached to cause the agglomeration of the nanoparticles.

2.4 Characterization Methods of Nano-Matters

2.4.1 Particle Size

2.4.1.1 Not See the Forest for the Trees

When you measure the particle size, the most important phrase is "Not See the Forest for the Trees." When you look at Fig. 2.15 as a picture of a forest, can you tell the tree size? Can you tell how is their crowdedness? If you see the picture D, you may say trees are small and scattered. If you see E, you say big tree is crowded. Now,

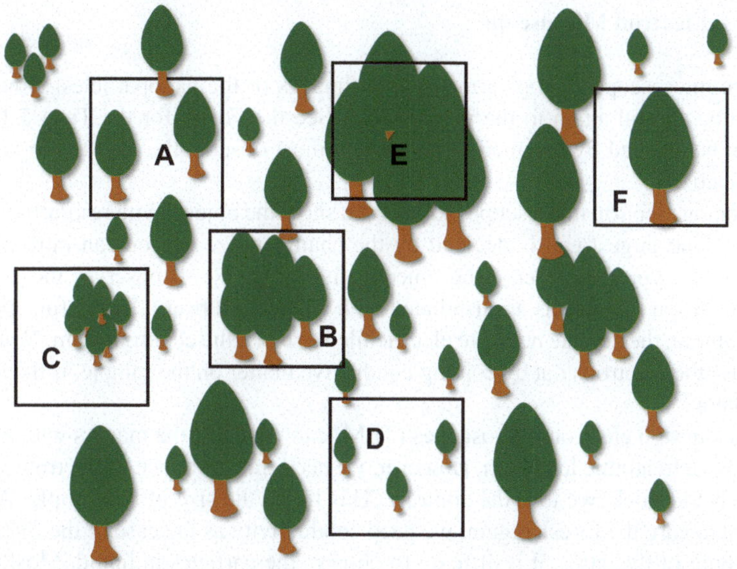

Fig. 2.15 Not See The Forest For The Trees

microscopic image seems to show the fact. However, the pictures are always trimmed. It may only focus on special case. For avoiding that, you always must apply various methods for deciding the particle size and dispersion state.

2.4.1.2 DLS: Dynamic Light Scattering

Dynamic light scattering (DLS) is now one of the most major methods for measuring particle size distribution. When you take a look at milk under a microscope, you can observe glitter of dots. The glitter moves and blinks. The glitter is scattering by the fat droplet in milk. You are observing Brownian motion. By analyzing the time correlation of the glitter, we can know the momentum of a droplet. This is how the DLS decides the particle size. From the temperature, you can know the kinetic energy. By combining the kinetic energy and the momentum, you can now know the mass of the droplet. If you know the density of the droplet material, the volume, in turn, the radius can be calculated. You have to know the viscosity of the dispersing liquid and the density of the particle material.

The merit of the DLS is that you can know the size distribution ranging from nm level to μm level. However, sedimentation of the particle may occur during the measurement. Also, if you have mixed sample of different particle sizes, you may miss the smaller particles information because larger one normally gives stronger scattering to mask the scattering light of from the smaller ones. Furthermore, if the concentration is too high, the scattering light cannot be detected fairly.

2.4.1.3 Electron Microscope

Electron microscopes can give most direct images of the nanoparticles. Now, you have to be careful again to the phrase, "Not See the Forest for the Trees." Easily you can be cheated by trimmed pictures. You must observe the forest with various magnifications.

Scanning electron microscopes (SEM) can show the objects with resolution down to 5 nm. The largest end is decided by the chamber size. If you can introduce to the chamber, you can observe the objects with large size. It observes the emitted electrons when the objects are irradiated with an electron beam. If electrons pile up to the sample, they create negative electric filed to resist the electron beam. You have to release the electrons out by coating conductive matter on the sample, if the object is insulator.

Transmission electron microscopes (TEM) can visualize the matters with a resolution down to atomic level, pm. However, we need transmittance to electrons. If the sample is too thick, we lose the contrast. This limits the size of the sample. Again, we need to care the forest. Again, we need conductivity as the case of the SEM.

For both of the cases, it is difficult to observe the particles in liquid. Mostly, the electron microscopes require vacuum conditions to efficiently fly the electrons. Also, during preparing the sample for observation, when you dry, it is easy to cause agglomeration. Also, the electron beam may damage the sample. Not only it is damaging the samples but also it may dirt the vacuum chamber with the decomposition products. Special care is required especially for the observation of organic matters.

2.4.1.4 Sedimentation (Turbidity)

By a simple and easy equipment, we can also measure the particle size distribution. A suspension of nanoparticles in a dispersion liquid is turbid because of the scattering due to the difference of refractive indices of the particles and the liquid. When the particles sediment down, the turbidity decreases. By measuring the transient transmittance of the suspension, we can observe the sedimentation velocity. Once the velocity is decided, we can calculate the diameter of the particles as discussed below. By observing the transient behavior of turbidity, we can know the particle size with distribution. Also, the measurement can give information of the dispersion stability.

According to Stokes' law, sedimentation velocity of a particle can be expressed by the following equation.

$$v = \frac{(\rho_S - \rho_L)g}{18\eta} d^2, \tag{2.1}$$

where $\rho_S[\mathrm{kg/m^3}]$ is the density of the solid particle, $\rho_L[\mathrm{kg/m^3}]$ is the density of the dispersion liquid, $g[\mathrm{m/s^2}]$ is the gravitational acceleration, $\eta[\mathrm{Pa \cdot s}]$ is the viscosity

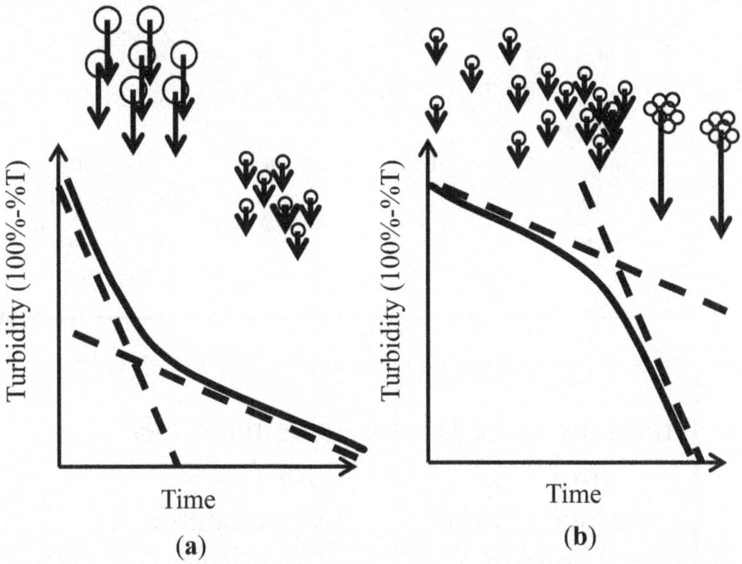

Fig. 2.16 Transient turbidity behavior and the situation of particles, **a** convex downward curve and **b** convex upward curve. **a** corresponds to case 2 and **b** does case 1 as explained in the text

of the dispersion liquid, and *d*[m] is the diameter of the particle. We can know the time of sedimentation from the transient transmittance measurement. For 3.2 the time into velocity, we need to know the distance. The distance can be decided with your system by measuring known standard sample.

Let's compare the two cases in Fig. 2.16. In (a), the curve is convex downward. In (b), it is convex upward. They are showing two different cases. Case 1 is that the particle size is single, but agglomeration occurs during the measurement to form larger particles with faster sedimentation. Case 2 is that two different particles with different sizes are mixed. Larger particle sediments are at first followed by the smaller particle sedimentation. Case 1 corresponds to (b) in the figure, and case 2 does to (a). By observing the curve, you can also know the dispersion stability, not only the size distribution. If the sedimentation is too fast or too slow, you can control them with the viscosity of the dispersion liquid. For increasing, possibly you can add some polymer in it. If too slow, you may increase temperature. In both cases, you must be careful not to have any changes to the particles.

2.4.2 *Surface Charge*

As discussed above, sometimes we need to know the surface charge of the particles. For attaching ionomer on the surface, we should know the surface charge. For checking the PEG installation, we can apply the surface charge characterization to

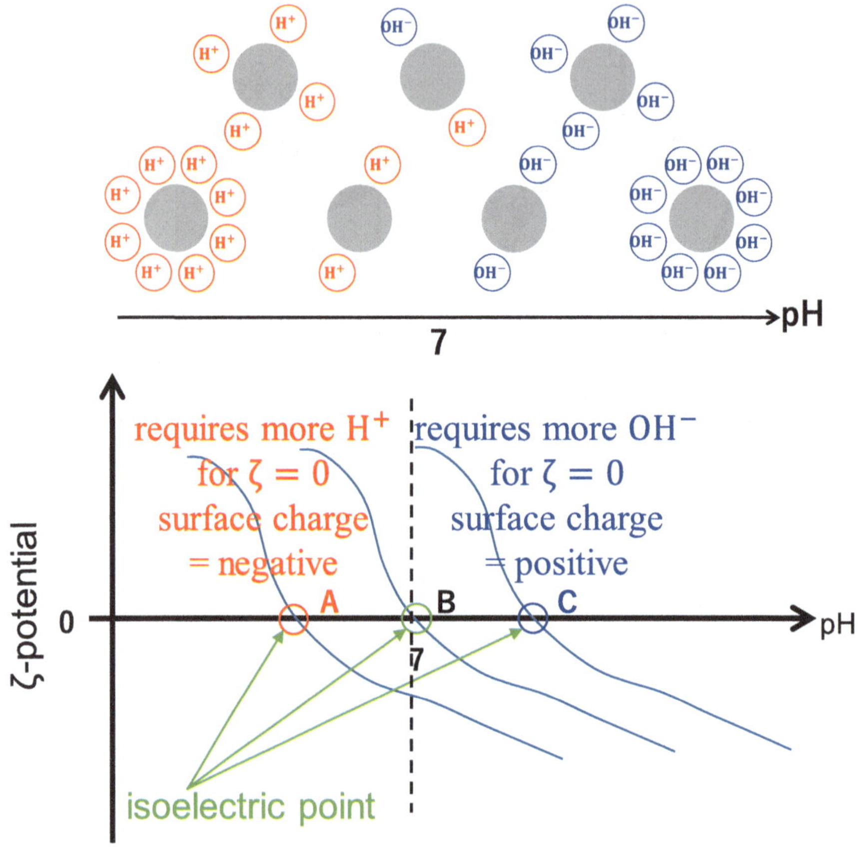

Fig. 2.17 Variation of ζ-potential by changing pH. Isoelectric point must be decided for judging the surface charge of the particles

show that neutral PEG is on the surface. Of course, for controlling the interaction, when you control the surface charge, you have to check how it is effective. The most popular way is the zeta (ζ) potential measurement. When you apply the measurement to the particles in an aqueous solution, always you must vary the pH in wide range for checking the net charge of the particle surface. As shown in Fig. 2.17.

2.4.3 Dispersion Stability

"Dispersibility" and "dispersion stability" are often confused. As discussed above, for guaranteeing circulation in animal body, the size-controlled nanoparticles must be dispersed well in advance. In any physiological condition, hydrophilic polymer must be installed well to keep the dispersion. Even in the case of drug delivery, the characters of the nanoparticles are well investigated in vitro. However, once coming into blood, it is very hard to trace what is going on in an animal body. Liposome is a very traditional drug carrier. Even the liposomes may be interacting or merged or destroyed during the circulation. For avoiding the surface property change, for keeping mono-dispersion, the surface polymer must be firmly attached on the surface of the nanostructure.

2.5 Summary

For applying the merits of fluorescence bioimaging, sensitivity, resolution, dynamicity and poly-chromatism for the near-infrared fluorescence bioimaging with transparency, we need to know "nano" world. In the world, higher specific surface energy makes things to have a strong tendency to agglomerate. Also, the size and mono-dispersion must be stably kept in biological system for controlling the distribution. In physiological condition, surface modification with hydrophilic polymer, such as poly(ethylene glycol), is inevitable since with strong ionicity and the existence of protein will make the particles larger to form aggregate. Once the size is larger than 100 nm, they cannot be circulated anymore in a body because of the trapping in RES. Also, endocytosis to take the particles into normal cells requires the size to be less than 100 nm. As for the excretion, if the size is less than 10 nm, there is more chance of renal excretion. Possibly, another route, bile excretion is useful to remove the probes from the body. The detailed matters and some applications of each phosphor are described in the following chapters.

References

1. Press release: The Nobel Prize in chemistry 2017. Available: https://www.nobelprize.org/prizes/chemistry/2017/press-release/
2. S. Nakamura, I. Grigoriev, T. Nogi et al., Dissecting the nanoscale distributions and functions of microtubule-end-binding proteins EB1 and ch-TOG in interphase HeLa cells. PLoS ONE **7**, e51442 (2012). https://doi.org/10.1371/journal.pone.0051442
3. H. Hama, H. Kurokawa, H. Kawano et al., Scale: a chemical approach for fluorescence imaging and reconstruction of transparent mouse brain. Nat. Neurosci. **14**, 1481 (2011). https://doi.org/10.1038/nn.2928
4. V. Haucke, On the endocytosis rollercoaster. Nature **517**, 446 (2015). https://doi.org/10.1038/nature14081
5. H. Moradian, T. Roch, A. Lendlein et al., mRNA transfection-induced activation of primary human monocytes and macrophages: dependence on carrier system and nucleotide modification. Sci. Rep. **10**, 4181 (2020). https://doi.org/10.1038/s41598-020-60506-4
6. J. Liu, M. Yu, C. Zhou et al., Renal clearable inorganic nanoparticles: a new frontier of bionanotechnology. Mater. Today **16**, 477 (2013). https://doi.org/10.1016/j.mattod.2013.11.003
7. I.I. Wistuba, A.F. Gazdar, Gallbladder cancer: lessons from a rare tumour. Nat. Rev. Cancer **4**, 695 (2004). https://doi.org/10.1038/nrc1429

8. F. Roncato, F. Rruga, E. Porcù et al., Improvement and extension of anti-EGFR targeting in breast cancer therapy by integration with the avidin-nucleic-acid-nano-assemblies. Nat. Commun. **9**, 4070 (2018). https://doi.org/10.1038/s41467-018-06602-6
9. S. Müller-Späth, A. Soranno, V. Hirschfeld et al., Charge interactions can dominate the dimensions of intrinsically disordered proteins. Proc. Natl. Acad. Sci. **107**, 14609 (2010). https://doi.org/10.1073/pnas.1001743107
10. Y. Nagasaki, Construction of a densely poly(ethylene glycol)-chain-tethered surface and its performance. Polym. J. **43**, 949 (2011). https://doi.org/10.1038/pj.2011.93
11. M. Kamimura, R. Saito, H. Hyodo et al., Over-1000 nm near-infrared fluorescence and SPECT/CT dual-modal in vivo imaging based on rare-earth doped ceramic nanophosphors. J. Photopolym. Sci. Technol. **29**, 525 (2016). https://doi.org/10.2494/photopolymer.29.525
12. E. Hemmer, N. Venkatachalam, H. Hyodo et al., The role of ph in PEG-*b*-PAAc modification of gadolinium oxide nanostructures for biomedical applications. Adv. Mater. Sci. Eng. **2012**, 748098 (2012). https://doi.org/10.1155/2012/748098
13. M. Kamimura, N. Kanayama, K. Tokuzen et al., Near-infrared (1550 nm) in vivo bioimaging based on rare-earth doped ceramic nanophosphors modified with PEG-*b*-poly(4-vinylbenzylphosphonate). Nanoscale **3**, 3705 (2011). https://doi.org/10.1039/c1nr10466g
14. X. Yuan, D. Fabregat, K. Yoshimoto et al., High PEGylation efficiency of pentaethylenehexamine-end poly(ethylene glycol) (mPEG-N6) for active-ester surface. Colloids Surf. B Biointerfaces **92**, 25 (2012). https://doi.org/10.1016/j.colsurfb.2011.11.013
15. L. Guerrini, R. A. Alvarez-Puebla, N. Pazos-Perez, Surface modifications of nanoparticles for stability in biological fluids. Materials (Basel), **11** (2018). https://doi.org/10.3390/ma11071154

Chapter 3
Organic Dyes for NIR Imaging

Masao Kamimura

Abstract Organic dyes are the most classical fluorescent materials for fluorescence bioimaging. Due to their low cost, small molecule size, and chemical reactivity, organic dyes are used as popular fluorescence probes. Recently, for deep tissue imaging, organic dyes which show near-infrared (NIR)-II (1050–1400 nm) emission is widely reported. Because of the small molecular size and biocompatibility, as compared to inorganic NIR-II fluorescent nanomaterials, these dyes have high potential for bioimaging applications. This chapter reviews the current progress of NIR-II fluorescent dyes-based bioimaging.

Keywords Organic dyes · Polymer micelles · Molecular imaging

3.1 Fluorescent Organic Dyes for Bioimaging

3.1.1 Traditional Visible Fluorescent Dyes

Organic dyes are the most traditional types of fluorescent materials for biological applications. In biological research, various fluorescent dyes, such as fluorescein, rhodamine, and cyanine-based fluorescent dyes are widely used as fluorescent labeling probes because of their low cost, small molecule size, and ability to attach biomolecules via chemical bonds. Although these dyes possess some drawbacks such as short Stokes shift and low photostability, organic fluorescent dyes are still popular materials. Therefore, organic dyes are essential fluorescent materials for fluorescence bioimaging [1–5].

For example, fluorescein derivatives, such as fluorescein-5-isothiocyanate (FITC), are amine-reactive fluorescent dyes and are widely used as "green" fluorescent labeling materials for biomolecules. Rhodamine derivatives, such as 5-carboxyrhodamine 6G, have been used as "red" fluorescent labeling materials. In

M. Kamimura (✉)
Department of Materials Science and Technology, Faculty of Industrial Science and Technology, Tokyo University of Science, 6-3-1 Niijuku, Katsushika, Tokyo 125-8585, Japan
e-mail: masaokamimura@rs.tus.ac.jp

K. Soga et al. (eds.), *Transparency in Biology*,
https://doi.org/10.1007/978-981-15-9627-8_3

addition, cyanine dyes are also major fluorescent dyes for bioimaging, due to their excellent photostability, high quantum yield, and water-soluble properties. These excitation and emission properties of cyanine-based dyes depend on the number of methine groups in the polyene chain. Thus, various cyanine-based dyes are synthesized and used for wide wavelength region, from red to near-infrared (NIR) (ca. 550–780 nm). For example, Cy3 (ex. 550 nm/em. 570 nm), Cy5 (ex. 640 nm/em. 670 nm), Cy5.5 (ex. 675 nm/em. 690 nm), and Cy7 (ex. 750 nm/em. 773 nm) are commercially available. Furthermore, Alexa dyes are also popular materials for fluorescence bioimaging in wide wavelength regions. The excitation wavelength of Alexa dyes covered from ultraviolet (UV) to NIR (350–750 nm) and thus are suitable for conventional excitation light sources. As compared to other organic dyes, Alexa dyes possess high stability against photobleaching and pH changes. Therefore, Alexa dyes are widely used in fluorescence bioimaging researches. These organic dyes are classical, but most used fluorescent labeling agents in biomedical field.

3.1.2 NIR Fluorescent Dyes

Although the various organic dyes are widely used in fluorescence bioimaging techniques, however, most commercially available organic dyes exhibit VIS or short-wavelength NIR emission (<800 nm). Therefore, for deeper tissue imaging based on over-1000 nm wavelength region (NIR-II: 1050–1400 nm, NIR-III: 1500–1700 nm) [6–24], inorganic nanophosphors such as semiconductor quantum dots (QDs), single-walled carbon nanotubes (SWCNTs), and rare earth-doped ceramic nanophosphors are mainly used as fluorescent probes (Fig. 3.1). Although these inorganic nanophosphors have excellent fluorescent properties, these probes have some drawbacks such as large probe size and low dispersion stability in biophysiological environment. Furthermore, renal clearance of these probes from the body is quite a serious problem. Therefore, the clinical application of these inorganic nanophosphors is limited.

To overcome such problems, NIR fluorescent organic dyes have been used for in vivo fluorescence bioimaging probes [17, 19, 25, 26]. These low molecular weight dyes can be cleared from the blood by a renal clearance. Thus, these dyes are promising candidates for clinical applications. For example, in the NIR-I (700–900 nm) wavelength region, indocyanine green (ICG) [24, 27–32] is already approved by US Food and Drug Administration (FDA) and is used as a fluorescence marker for clinical surgery.

Same as ICG, the ideal NIR-II fluorescent dye is expected to exhibit low toxicity and pharmacokinetics properties. As deep tissue NIR-II imaging agents, commercially available NIR dyes such as IR-26 or IR-1061 can be used as fluorescent dyes (Fig. 3.2). However, due to the hydrophobic structure and low NIR emission property in aqueous environment, these dyes are quite difficult to use as fluorescent probes under physiological conditions without conjugation with water-soluble polymers.

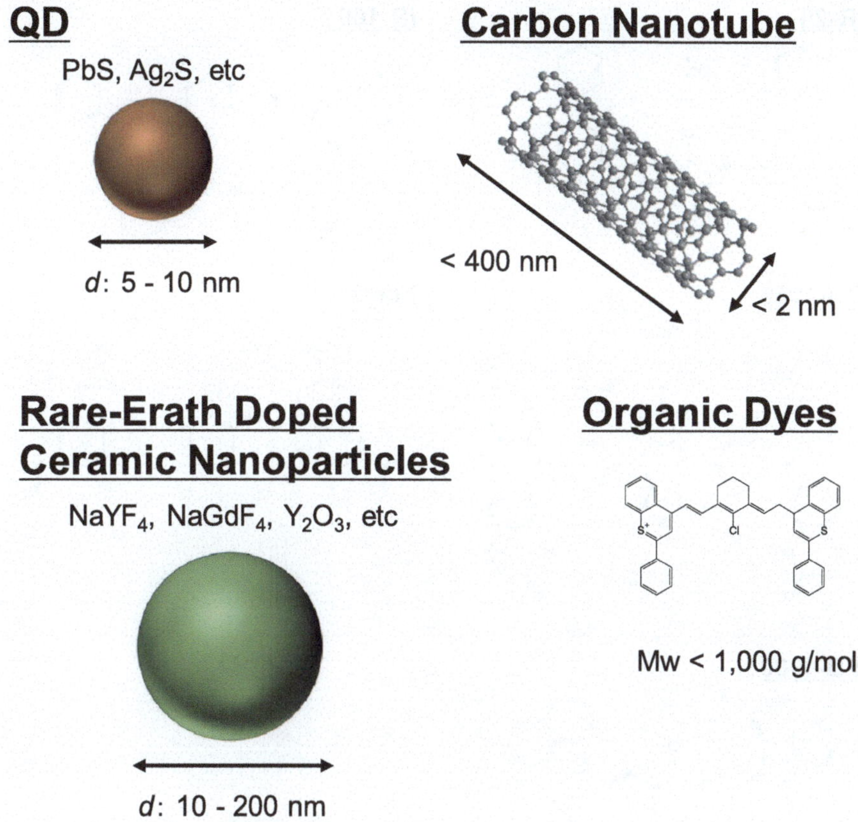

Fig. 3.1 Comparison of the size of various NIR nanophosphors

3.2 Organic Dyes for NIR-II Fluorescence Imaging

3.2.1 Commercially Available NIR-II Dye-Based Probes

In 2013, Prof. H. Dai and his co-workers first reported the NIR-II organic dye-based probes for in vivo imaging [33]. In this report, commercially available IR-1061 dye is incorporated in the amphiphilic polymer, poly(acrylic acid) and it is further modification with phospholipid-poly(ethylene glycol) (PEG). They demonstrated NIR-II imaging of mouse under 808 nm excitation. Since then, several groups reported commercially available IR-1061 dye-based probes for NIR-II imaging [33–36]. However, these IR-1061 based probes did not possess biodegradability and thus, these probes were difficult to remove completely from the body. To address this problem, Kamimura et al. reported IR-1061 dye loaded biodegradable polymer probes for NIR-II in vivo imaging. (Fig. 3.3) [34]. The polymer micelle is prepared using PEG possessing biodegradable poly(ε-caprolactone) block copolymer (PEG-*b*-PCL), and IR-1061 dye is incorporated into the hydrophobic and biodegradable

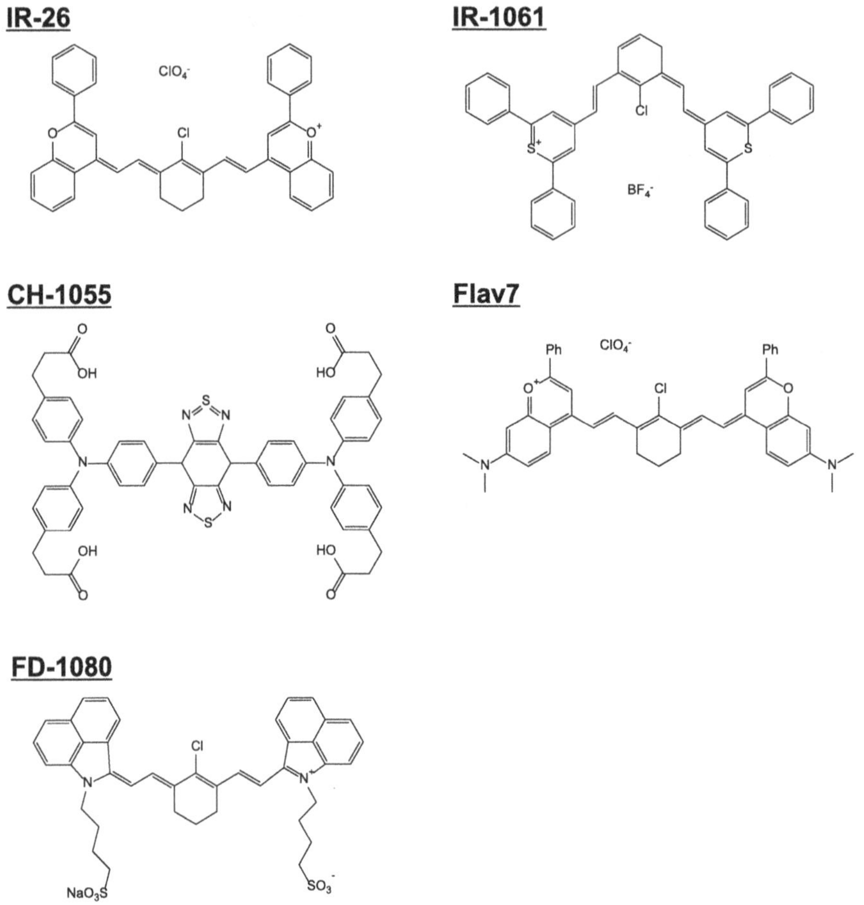

Fig. 3.2 Chemical structure of major NIR-II fluorescent organic dyes

PCL core of the micelles. The hydrophilic outer shell improves the biocompatibility and dispersion stability of the dye. Thus, the obtained probe shows high stability in aqueous environments and strong NIR emission. In addition, this probe also displays biodegradable property. Therefore, this IR-1061 dye loaded polymer micelle can be used as safety and high-performance NIR-II imaging agent (Fig. 3.4).

3.2.2 Novel Synthesized Dyes for NIR-II Imaging

Then, various NIR-II fluorescent organic dyes are newly synthesized and reported by many researchers [37–80]. For the design of NIR organic dyes, it is known that the energy band gap is greatly affected by the conjugation length of π-conjugated

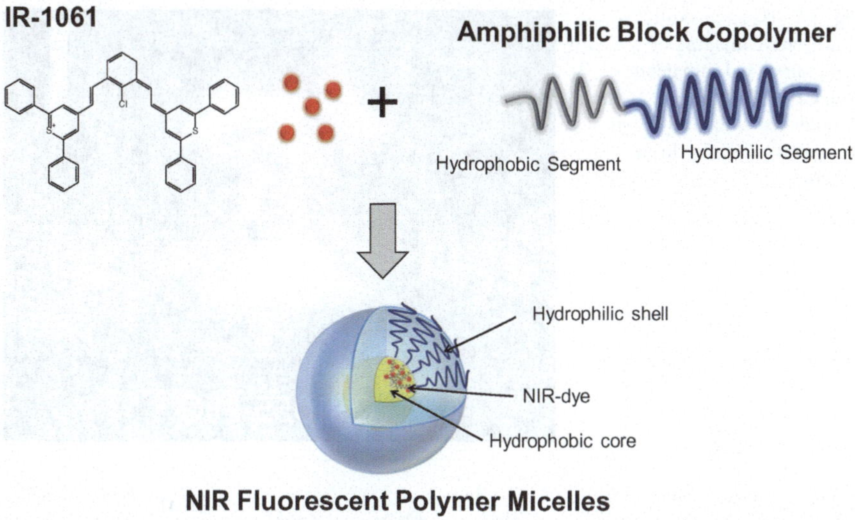

Fig. 3.3 Formation of NIR-II dye loaded polymer micelles

Fig. 3.4 NIR-II fluorescence in vivo imaging based on IR-1061 dye loaded biodegradable polymer micelles. Reproduced with permission from 33) © 2017 Springer Nature

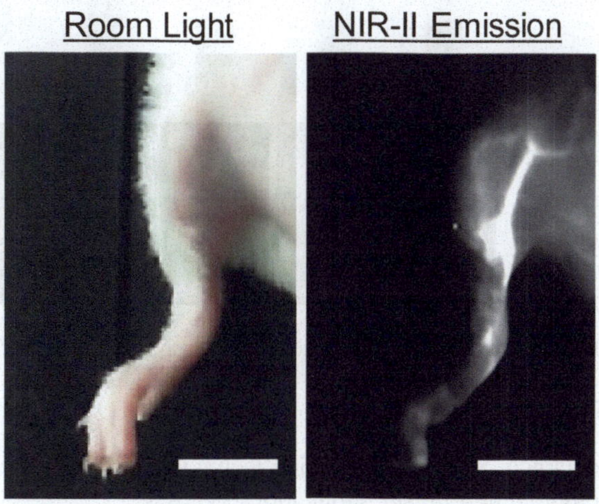

molecule and donor–acceptor (D-A) charge transfer. In 2016, H. Dai's group reported a synthetic 970-Da organic molecule dye with a donor–acceptor-donor (D-A-D) charge structure (CH1055) (Fig. 3.2) [38]. The CH1055 dye is conjugated with amphiphilic polymer, PEG (CH1055-PEG) and it shows about 1050 nm NIR-II emission in aqueous media (Fig. 3.5). Furthermore, they prepared dye-protein nanocomplexes that consist of CH1055 and bovine serum (Fig. 3.6) [39]. In this report, CH1055 is conjugated with taurine (CH-4 T) and then, formation of complex with bovine serum protein. After heating treatment, 50 times bright NIR-II probes are

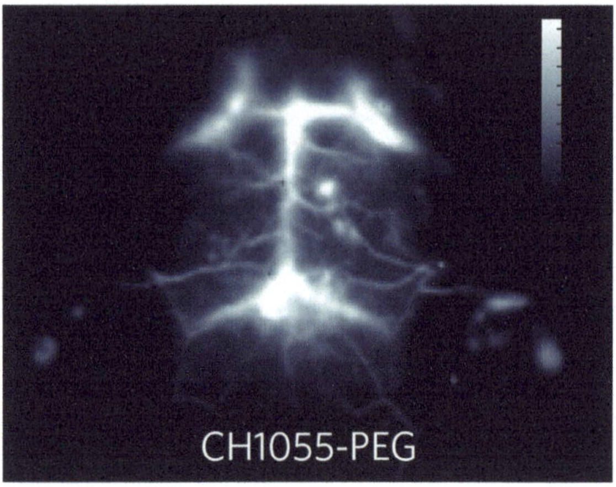

Fig. 3.5 Through-skull non-invasive imaging of brain tumor at 4 mm depth with CH1055-PEG. Reproduced with permission from 38) © 2015 Springer Nature

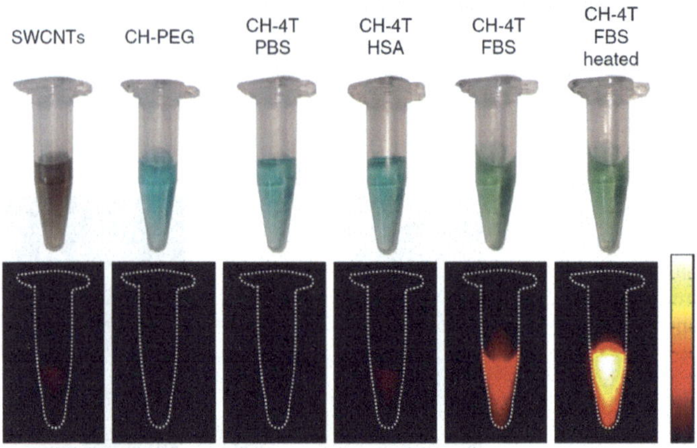

Fig. 3.6 Ultra-high quantum yield CH-4 T–protein complexes. Reproduced with permission from 39) © 2017 Springer Nature

obtained. Similar conjugation techniques based on NIR-II dye and protein are also reported in follower publications [81, 82].

Polymethine derivatives are also important dyes for NIR-II imaging. In 2017, E. M. Sletten and her co-workers reported flavylium polymethine based dye (Flav7) for NIR-II imaging (Fig. 3.2) [41]. In this report, 1050 nm fluorescence polymethine dye is synthesized and it shows about 13 times strong emission as compared to IR-26 dye. The synthesized Flav7 dye is incorporated into the core of phospholipid-PEG micelles and it visualizes whole body of nude mice. Furthermore, in 2018, other types of polymethine derivatives, FD-1080 are reported by Zhang's group (Fig. 3.2) [49]. The obtained FD-1080 is the first report of NIR-II dye which can be excited over 1000 nm wavelength region. The synthesized FD-1080 shows 1100 nm emission under 1064 nm excitation. The excitation wavelength of the FD-1080 is longer than previous reported NIR-II dyes, such as IR1061 and thus is prefer materials for deep tissue observation. Moreover, Zhang's group also reported the similar pentamethine cyanine dye, BTC-1070 and they performed lymphatic imaging [83].

3.2.3 NIR-I Fluorescent Dyes-Based NIR-II Imaging

Recently, application of NIR-I fluorescent dye, ICG for NIR-II fluorescence imaging are also reported [30–32]. As compared to other NIR-II fluorescent dyes, emission intensity of ICG in NIR-II wavelength region is weak, however, it can be used for NIR-II in vivo imaging applications. For example, in 2018, ICG based NIR-II imaging of blood and lymph vessels are first reported. Following this report, hind limb and intracranial vasculatures imaging of mice by using ICG loaded liposome fluorescent probe is also reported. Moreover, combination fluorescent probe based on ICG, IR-800, and IR12N3 are also reported for NIR-II imaging [84]. Because the ICG is FDA approved dye, ICG may open a new pathway of clinical application of NIR-II imaging.

3.3 Aggregation-Induced Emission Based Probes

3.3.1 Aggregation-Induced Emission for Fluorescence Imaging

Generally, aggregation of organic dyes in the media causes quenching of emission and thus, is a serious problem for biological applications. However, recently, an abnormal emission process of organic dyes, so-called "aggregation-induced emission (AIE)" is widely applied for fluorescence bioimaging, including NIR-II imaging [85–90]. In this emission process, emission intensity increases with increase of dye concentration (i.e. aggregation).

3.3.2 Aggregation-Induced Emission for NIR-II Imaging

For example, NIR AIE lumino-gens (AIEgens), TQ-BPN (Fig. 3.7), which shows 700–1200 nm are reported for imaging agent of vascular disease in brain region [87]. The TQ-BPN probe can also use for two-photon excitation process, 3D imaging of vascular and mouse brain is successfully performed. Furthermore, recently reported two-photon AIEgen, BTPETQ show NIR-I emission under NIR-II excitation, and the BTPETQ is used for brain and vascular imaging [91]. Moreover, another NIR

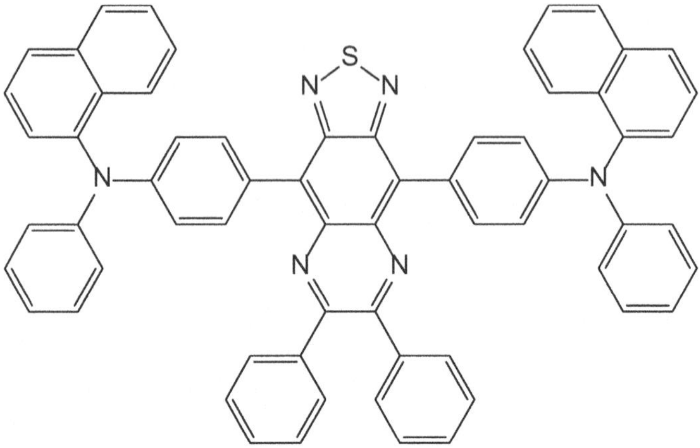

Fig. 3.7 Chemical structure of TQ-BPN

AIEgen, HLZ-BTED is also reported [92]. The reported HLZ-BTED show NIR-II (1034 nm) emission and it applied for in vivo tracking of breast cancer cells.

3.4 Fluorescent Polymer Based Probes

3.4.1 Fluorescent Polymers for Bioimaging

Recently, fluorescent polymer materials have also become an important probe for in vivo bioimaging. Compared to currently used low molecular fluorescent dyes, fluorescent polymers possess several advantages, such as sensitivity, robustness, and functionality. In addition, synthetic polymer chemistry can install various responsibility into the polymers, such as pH-, temperature-, or photo-sensitive properties. Therefore, polymer fluorescence probes are applied for stimuli-responsive imaging techniques. Same as low molecular dyes, most of current reported fluorescent polymer probes are mainly show VIS emission. However, recently, NIR fluorescent polymers are also developed and applied for deep tissue in vivo imaging.

3.4.2 NIR-II Fluorescent Polymers for in Vivo Imaging

Fluorescent polymer is also powerful tools for NIR-II in vivo imaging [91–97]. Because of the most of low molecular NIR-II fluorescent dyes possesses hydrophobic structures, conjugation with polymer surfactant are required for bioimaging applications. In the case of NIR-II fluorescent polymers, these polymers are designed

Fig. 3.8 Chemical structure of pDA polymer

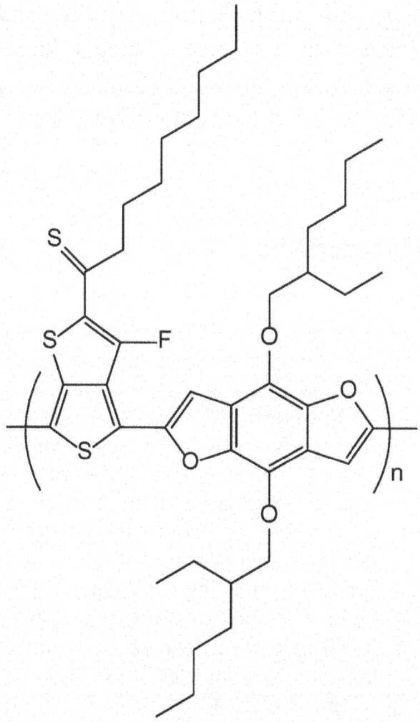

and synthesized for bioimaging applications and thus, these polymers are generally water-soluble and biocompatible materials. Therefore, these polymers can be used for bioimaging directly without further conjugation with biocompatible materials. Because of these NIR-II fluorescent polymers possess π-conjugated backbone and reactive side chains, these polymers show excellent fluorescence properties and functionality.

In 2014, H. Dai's group have first reported NIR-II fluorescent polymer probe, poly(benzo[1,2-b:3,4-b']difuran-alt-fluorothieno-[3,4-b]thiophen (pDA) [93] (Fig. 3.8). In this report, biocompatible polymer possesses D-A structure is synthesized and used for NIR-II imaging probe. The pDA polymer is synthesized via copolymerization of electron-donating and electron-withdrawing monomers. The pDA is further conjugation with phospholipid-PEG, and prepared PEGylated fluorescent polymer nanoparticles are used for blood flow imaging. In addition, other types of NIR-II fluorescent polymer, PDFT1032 is reported for imaging guide agent of tumor surgery [95]. The particle size of PEGylated PDFT1032 is ca. 70 nm, and it shows 1032 nm NIR-II emission under 809 nm NIR-I excitation. The obtained PEGylated PDFT1032 is successfully used for tumor and sentinel lymph node imaging.

3.5 Summary

In this chapter, the applications of organic fluorescent dyes for NIR-II imaging is reviewed. Compared to inorganic phosphors, organic dyes have several advantages, especially high biocompatibility and renal clearance from the body. Therefore, in the past 7 years, many researchers reported various NIR-II fluorescent dyes and polymers. In near future, it is anticipated that the clinical applications of NIR-II imaging is probably performed by using these organic fluorescent materials.

References

1. Z. Guo, S. Park, J. Yoon et al., Recent progress in the development of near-infrared fluorescent probes for bioimaging applications. Chem. Soc. Rev. **43** (2014). https://doi.org/10.1039/c3cs60271k
2. H. Kobayashi, M. Ogawa, R. Alford et al., New strategies for fluorescent probe design in medical diagnostic imaging. Chem. Rev. **110** (2010). https://doi.org/10.1021/cr900263j
3. A. Roda, M. Guardigli, R. Ziessel et al., Molecular luminescence imaging. Microchem. J. **85**, 5 (2007). https://doi.org/10.1016/j.microc.2006.04.010
4. Y. Urano, Novel live imaging techniques of cellular functions and in vivo tumors based on precise design of small molecule-based 'activatable' fluorescence probes. Curr. Opin. Chem. Biol. **16** (2012). https://doi.org/10.1016/j.cbpa.2012.10.023
5. L. Yuan, W. Lin, K. Zheng et al., Far-red to near infrared analyte-responsive fluorescent probes based on organic fluorophore platforms for fluorescence imaging. Chem. Soc. Rev. **42** (2013). https://doi.org/10.1039/c2cs35313j
6. S. Hameed, H. Chen, M. Irfan et al., Fluorescence guided sentinel lymph node mapping: from current molecular probes to future multimodal nanoprobes. Bioconjug. Chem. **30** (2019). https://doi.org/10.1021/acs.bioconjchem.8b00812
7. S. He, J. Song, J. Qu et al., Crucial breakthrough of second near-infrared biological window fluorophores: design and synthesis toward multimodal imaging and theranostics. Chem. Soc. Rev. **47** (2018). https://doi.org/10.1039/c8cs00234g
8. E. Hemmer, P. Acosta-Mora, J. Mendez-Ramos et al., Optical nanoprobes for biomedical applications: shining a light on upconverting and near-infrared emitting nanoparticles for imaging, thermal sensing, and photodynamic therapy. J. Mater. Chem. B, **5** (2017). https://doi.org/10.1039/c7tb00403f
9. G. Hong, A.L. Antaris, H. Dai, Near-infrared fluorophores for biomedical imaging. Nature Biomed. Eng. **1** (2017). https://doi.org/10.1038/s41551-016-0010
10. D. Jaque, C. Richard, B. Viana et al., Inorganic nanoparticles for optical bioimaging. Adv. Opt. Photonics **8**, 1 (2016). https://doi.org/10.1364/aop.8.000001
11. Y. Jiang, K. Pu, Multimodal biophotonics of semiconducting polymer nanoparticles. Acc. Chem. Res. **51** (2018). https://doi.org/10.1021/acs.accounts.8b00242
12. T. Jin, Critical review—recent progress in NIR fluorophores emitting over 1000 nm for bioimaging. ECS J. Solid State Sci. Technol. **8**, R9 (2019). https://doi.org/10.1149/2.0111901jss
13. Kenry, Y. Duan, B. Liu, Recent advances of optical imaging in the second near-infrared window. Adv. Mater. **30** (2018). https://doi.org/10.1002/adma.201802394
14. C. Li, Q. Wang, Challenges and opportunities for intravital near-infrared fluorescence imaging technology in the second transparency window. ACS Nano.**12** (2018). https://doi.org/10.1021/acsnano.8b07536
15. C. Li, Q. Wang, Advanced NIR-II fluorescence imaging technology for in vivo precision tumor theranostics. Adv. Ther. **2**, 1900053 (2019). https://doi.org/10.1002/adtp.201900053
16. J.-B. Li, H.-W. Liu, T. Fu et al., Recent progress in small-molecule near-IR probes for bioimaging. Trends Chem. **1**, 224 (2019). https://doi.org/10.1016/j.trechm.2019.03.002
17. P. Liu, X. Mu, X. D. Zhang et al., The near-infrared-ii fluorophores and advanced microscopy technologies development and application in bioimaging. Bioconjug. Chem. **31** (2020). https://doi.org/10.1021/acs.bioconjchem.9b00610

18. I. Martinić, S.V. Eliseeva, S. Petoud, Near-infrared emitting probes for biological imaging: organic fluorophores, quantum dots, fluorescent proteins, lanthanide(III) complexes and nanomaterials. J. Lumin. **189**, 19 (2017). https://doi.org/10.1016/j.jlumin.2016.09.058
19. P. Reineck, B.C. Gibson, Near-infrared fluorescent nanomaterials for bioimaging and sensing. Adv. Opt. Mater. **5**, 1600446 (2017). https://doi.org/10.1002/adom.201600446
20. S. Tsuboi, S. Yamada, Y. Nakane et al., Critical review—water-soluble near-infrared fluorophores emitting over 1000 nm and their application to in vivo imaging in the second optical window (1000–1400 nm). ECS J. Solid State Sci. Technol. **7**, R3093 (2017). https://doi.org/10.1149/2.0141801jss
21. H. Wan, H. Du, F. Wang et al., Molecular imaging in the second near-infrared window. Adv. Funct. Mater. **29** (2019). https://doi.org/10.1002/adfm.201900566
22. M. Zhao, B. Li, Y. Fan et al., In vivo assembly and disassembly of probes to improve near-infrared optical bioimaging. Adv. Healthc. Mater. **8** (2019). https://doi.org/10.1002/adhm.201801650
23. X. Zhao, S. He, M.C. Tan, Design of infrared-emitting rare earth doped nanoparticles and nanostructured composites. J. Mater. Chem. C **4**, 8349 (2016). https://doi.org/10.1039/c6tc02373h
24. S. Zhu, B.C. Yung, S. Chandra et al., Near-infrared-II (NIR-II) bioimaging via off-peak NIR-I fluorescence emission. Theranostics **8** (2018). https://doi.org/10.7150/thno.27995
25. S. Luo, E. Zhang, Y. Su et al., A review of NIR dyes in cancer targeting and imaging. Biomaterials**32** (2011). https://doi.org/10.1016/j.biomaterials.2011.06.024
26. J. Ripoll, B. Koberstein-Schwarz, V. Ntziachristos, Unleashing optics and optoacoustics for developmental biology. Trends Biotechnol. **33** (2015). https://doi.org/10.1016/j.tibtech.2015.08.002
27. A. Fernandez-Fernandez, R. Manchanda, T. Lei et al., Comparative study of the optical and heat generation properties of IR820 and indocyanine green. Mol. Imag. **11** (2012). https://doi.org/10.2310/7290.2011.00031
28. M.V. Marshall, J.C. Rasmussen, I.C. Tan et al., Near-infrared fluorescence imaging in humans with indocyanine green: a review and update. Open Surg. Oncol. J. **2** (2010). https://doi.org/10.2174/1876504101002010012
29. M. Ogawa, N. Kosaka, P.L. Choyke et al., In vivo molecular imaging of cancer with a quenching near-infrared fluorescent probe using conjugates of monoclonal antibodies and indocyanine green. Cancer Res. **69** (2009). https://doi.org/10.1158/0008-5472.CAN-08-3116
30. C. Zeng, W. Shang, K. Wang et al., Intraoperative identification of liver cancer microfoci using a targeted near-infrared fluorescent probe for imaging-guided surgery. Sci. Rep. **6** (2016). https://doi.org/10.1038/srep21959
31. Z. Starosolski, R. Bhavane, K.B. Ghaghada et al., Indocyanine green fluorescence in second near-infrared (NIR-II) window. PLoS One**12** (2017). https://doi.org/10.1371/journal.pone.0187563
32. Y. Suo, F. Wu, P. Xu et al., NIR-II fluorescence endoscopy for targeted imaging of colorectal cancer. Adv. Healthc. Mater. **8** (2019). https://doi.org/10.1002/adhm.201900974
33. Z. Tao, G. Hong, C. Shinji et al., Biological imaging using nanoparticles of small organic molecules with fluorescence emission at wavelengths longer than 1000 nm. Angew. Chem. Int. Ed. Engl. **52** (2013). https://doi.org/10.1002/anie.201307346.
34. M. Kamimura, S. Takahiro, M. Yoshida et al., Over-1000 nm near-infrared fluorescent biodegradable polymer nanoparticles for deep tissue in vivo imaging in the second biological window. Polymer J. **49** (2017). https://doi.org/10.1038/pj.2017.59
35. X. Xie, Y. Hu, C. Zhang et al., A targeted biocompatible organic nanoprobe for photoacoustic and near-infrared-II fluorescence imaging in living mice. RSC Adv. **9**, 301 (2019). https://doi.org/10.1039/c8ra08163h
36. Y. Tang, Y. Li, X. Hu et al., Dual "lock-and-key"-controlled nanoprobes for ultrahigh specific fluorescence imaging in the second near-infrared window. Adv. Mater. **30** (2018). https://doi.org/10.1002/adma.201801140

37. N. Alifu, A. Zebibula, J. Qi et al., Single-molecular near-infrared-II theranostic systems: ultrastable aggregation-induced emission nanoparticles for long-term tracing and efficient photothermal therapy. ACS Nano **12** (2018). https://doi.org/10.1021/acsnano.8b05937
38. A.L. Antaris, H. Chen, K. Cheng et al., A small-molecule dye for NIR-II imaging. Nat. Mater. **15** (2016). https://doi.org/10.1038/nmat4476
39. A.L. Antaris, H. Chen, S. Diao et al., A high quantum yield molecule-protein complex fluorophore for near-infrared II imaging. Nat. Commun. **8** (2017). https://doi.org/10.1038/ncomms15269
40. K. Cheng, H. Chen, C.H. Jenkins et al., Synthesis, characterization, and biomedical applications of a targeted dual-modal near-infrared-II fluorescence and photoacoustic imaging nanoprobe. ACS Nano **11** (2017). https://doi.org/10.1021/acsnano.7b05966
41. E.D. Cosco, J.R. Caram, O.T. Bruns et al., Flavylium polymethine fluorophores for near- and shortwave infrared imaging. Angew. Chem. Int. Ed. Engl. **56** (2017). https://doi.org/10.1002/anie.201706974
42. B. Ding, Y. Xiao, H. Zhou et al., Polymethine thiopyrylium fluorophores with absorption beyond 1000 nm for biological imaging in the second near-infrared subwindow. J. Med. Chem. **62** (2019). https://doi.org/10.1021/acs.jmedchem.8b01682
43. F. Ding, C. Li, Y. Xu et al., PEGylation regulates self-assembled small-molecule dye-based probes from single molecule to nanoparticle size for multifunctional nir-ii bioimaging. Adv. Healthc. Mater. **7** (2018). https://doi.org/10.1002/adhm.201800973
44. Y. Feng, S. Zhu, A.L. Antaris et al., Live imaging of follicle stimulating hormone receptors in gonads and bones using near infrared II fluorophore. Chem. Sci. **8** (2017). https://doi.org/10.1039/c6sc04897h
45. J. Huang, C. Xie, X. Zhang et al., Renal-clearable molecular semiconductor for second near-infrared fluorescence imaging of kidney dysfunction. Angew. Chem. Int. Ed. Engl. **58** (2019). https://doi.org/10.1002/anie.201909560
46. Y. Jiang, J. Li, X. Zhen et al., Dual-peak absorbing semiconducting copolymer nanoparticles for first and second near-infrared window photothermal therapy: a comparative study. Adv. Mater. **30** (2018). https://doi.org/10.1002/adma.201705980
47. S. Kurbegovic, K. Juhl, H. Chen et al., Molecular targeted NIR-II probe for image-guided brain tumor surgery. Bioconjug. Chem. **29** (2018). https://doi.org/10.1021/acs.bioconjchem.8b00669
48. Z. Lei, C. Sun, P. Pei et al., Stable, wavelength-tunable fluorescent dyes in the NIR-II region for in vivo high-contrast bioimaging and multiplexed biosensing. Angew. Chem. Int. Ed. Engl. **58** (2019). https://doi.org/10.1002/anie.201904182
49. B. Li, L. Lu, M. Zhao et al., An efficient 1064 nm NIR-II excitation fluorescent molecular dye for deep-tissue high-resolution dynamic bioimaging. Angew. Chem. Int. Ed. Engl. **57** (2018). https://doi.org/10.1002/anie.201801226
50. D. Li, C. Qu, Q. Liu et al., Monitoring the real-time circulatory system-related physiological and pathological processes in vivo using a multifunctional NIR-II probe. Adv. Func. Mater. **30**, 1906343 (2019). https://doi.org/10.1002/adfm.201906343
51. T. Li, C. Li, Z. Ruan et al., Polypeptide-conjugated second near-infrared organic fluorophore for image-guided photothermal therapy. ACS Nano **13** (2019). https://doi.org/10.1021/acsnano.9b00452
52. X. Lu, P. Yuan, W. Zhang et al., A highly water-soluble triblock conjugated polymer for in vivo NIR-II imaging and photothermal therapy of cancer. Polym. Chem. **9**, 3118 (2018). https://doi.org/10.1039/c8py00215k
53. Z. Ma, H. Wan, W. Wang et al., A theranostic agent for cancer therapy and imaging in the second near-infrared window. Nano Res. **12** (2019). https://doi.org/10.1007/s12274-018-2210-x
54. Y. Miao, C. Gu, B. Yu et al., Conjugated-polymer-based nanoparticles with efficient NIR-II fluorescent, photoacoustic and photothermal performance. Chembiochem **20** (2019). https://doi.org/10.1002/cbic.201900309
55. K. Qian, C. Qu, X. Ma et al., Tuning the near infrared II emitting wavelength of small molecule dyes by single atom alteration. Chem. Commun. (Camb) **56** (2020). https://doi.org/10.1039/c9cc08434g

56. Y. Shi, W. Yuan, Q. Liu et al., Development of polyene-bridged hybrid rhodamine fluorophores for high-resolution NIR-II imaging. ACS Mater. Lett. **1**, 418 (2019). https://doi.org/10.1021/acsmaterialslett.9b00265
57. K. Shou, C. Qu, Y. Sun et al., Multifunctional biomedical imaging in physiological and pathological conditions using a NIR-II probe. Adv. Funct. Mater. **27** (2017). https://doi.org/10.1002/adfm.201700995
58. C. Sun, B. Li, M. Zhao et al., J-aggregates of cyanine dye for NIR-II in vivo dynamic vascular imaging beyond 1500 nm. J. Am. Chem. Soc. **141** (2019). https://doi.org/10.1021/jacs.9b10043
59. Y. Sun, F. Ding, Z. Chen et al., Melanin-dot-mediated delivery of metallacycle for NIR-II/photoacoustic dual-modal imaging-guided chemo-photothermal synergistic therapy. Proc. Natl. Acad. Sci. U S A **116** (2019). https://doi.org/10.1073/pnas.1908761116
60. Y. Sun, F. Ding, Z. Zhou et al., Rhomboidal Pt(II) metallacycle-based NIR-II theranostic nanoprobe for tumor diagnosis and image-guided therapy. Proc. Natl. Acad. Sci. U S A **116** (2019). https://doi.org/10.1073/pnas.1817021116
61. Y. Sun, M. Ding, X. Zeng et al., Novel bright-emission small-molecule NIR-II fluorophores for in vivo tumor imaging and image-guided surgery. Chem. Sci. **8** (2017). https://doi.org/10.1039/c7sc00251c
62. Y. Sun, C. Qu, H. Chen et al., Novel benzo-bis(1,2,5-thiadiazole) fluorophores for in vivo NIR-II imaging of cancer. Chem. Sci. **7** (2016). https://doi.org/10.1039/c6sc01561a
63. Y. Tang, Y. Li, X. Lu et al., Bio-erasable intermolecular donor-acceptor interaction of organic semiconducting nanoprobes for activatable NIR-II fluorescence imaging. Adv. Func. Mater. **29**, 1807376 (2019). https://doi.org/10.1002/adfm.201807376
64. Y. Tang, Y. Li, Z. Wang et al., Organic semiconducting nanoprobe with redox-activatable NIR-II fluorescence for in vivo real-time monitoring of drug toxicity. Chem. Commun. (Camb), **55** (2018). https://doi.org/10.1039/c8cc08413k
65. H. Wan, H. Ma, S. Zhu et al., Developing a bright NIR-II fluorophore with fast renal excretion and its application in molecular imaging of immune checkpoint PD-L1. Adv. Funct. Mater. **28** (2018). https://doi.org/10.1002/adfm.201804956
66. H. Wan, J. Yue, S. Zhu et al., A bright organic NIR-II nanofluorophore for three-dimensional imaging into biological tissues. Nat. Commun. **9** (2018). https://doi.org/10.1038/s41467-018-03505-4
67. Q. Wang, Y. Dai, J. Xu et al., All-in-one phototheranostics: single laser triggers NIR-II fluorescence/photoacoustic imaging guided photothermal/photodynamic/chemo combination therapy. Adv. Func. Mater. **29**, 1901480 (2019). https://doi.org/10.1002/adfm.201901480
68. Q. Wang, B. Xia, J. Xu et al., Biocompatible small organic molecule phototheranostics for NIR-II fluorescence/photoacoustic imaging and simultaneous photodynamic/photothermal combination therapy. Mater. Chem. Front. **3**, 650 (2019). https://doi.org/10.1039/c9qm00036d
69. W. Wang, Z. Ma, S. Zhu et al., Molecular cancer imaging in the second near-infrared window using a renal-excreted NIR-II fluorophore-peptide probe. Adv. Mater. **30** (2018). https://doi.org/10.1002/adma.201800106
70. W. Wang, Q. Yang, Y. Du et al., Metabolic labeling of peptidoglycan with NIR-II dye enables in vivo imaging of gut microbiota. Angew. Chem. Int. Ed. Engl. **59** (2020). https://doi.org/10.1002/anie.201910555
71. G. Xu, Q. Yan, X. Lv et al., Imaging of colorectal cancers using activatable nanoprobes with second near-infrared window emission. Angew. Chem. Int. Ed. Engl. **57** (2018). https://doi.org/10.1002/anie.201712528
72. Q. Yang, Z. Hu, S. Zhu et al., Donor engineering for NIR-ii molecular fluorophores with enhanced fluorescent performance. J. Am. Chem. Soc. **140** (2018). https://doi.org/10.1021/jacs.7b10334
73. Q. Yang, Z. Ma, H. Wang et al., Rational design of molecular fluorophores for biological imaging in the NIR-II window. Adv. Mater. **29** (2017). https://doi.org/10.1002/adma.201605497

74. W. Yi, H. Zhou, A. Li et al., A NIR-II fluorescent probe for articular cartilage degeneration imaging and osteoarthritis detection. Biomater. Sci. **7** (2019). https://doi.org/10.1039/c8bm01440j
75. X. Zeng, L. Xue, D. Chen et al., A bright NIR-II fluorescent probe for breast carcinoma imaging and image-guided surgery. Chem. Commun. (Camb)**55** (2019). https://doi.org/10.1039/c9cc07694h
76. R. Zhang, Y. Xu, Y. Zhang et al., Rational design of a multifunctional molecular dye for dual-modal NIR-II/photoacoustic imaging and photothermal therapy. Chem. Sci. **10** (2019). https://doi.org/10.1039/c9sc03504d.
77. X.D. Zhang, H. Wang, A.L. Antaris et al., Traumatic brain injury imaging in the second near-infrared window with a molecular fluorophore. Adv. Mater. **28** (2016). https://doi.org/10.1002/adma.201600706
78. H. Zhou, W. Yi, A. Li et al., Specific small-molecule NIR-II fluorescence imaging of osteosarcoma and lung metastasis. Adv. Healthc. Mater.**9** (2020). https://doi.org/10.1002/adhm.201901224
79. S. Zhu, S. Herraiz, J. Yue et al., 3D NIR-II molecular imaging distinguishes targeted organs with high-performance NIR-II bioconjugates. Adv. Mater. **30** (2018). https://doi.org/10.1002/adma.201705799
80. S. Zhu, Q. Yang, A.L. Antaris et al, Molecular imaging of biological systems with a clickable dye in the broad 800- to 1,700-nm near-infrared window. Proc. Natl. Acad. Sci. U S A **114** (2017). https://doi.org/10.1073/pnas.1617990114
81. R. Tian, Q. Zeng, S. Zhu et al., Albumin-chaperoned cyanine dye yields superbright NIR-II fluorophore with enhanced pharmacokinetics. Sci. Adv. **5** (2019). https://doi.org/10.1126/sciadv.aaw0672
82. X. Zeng, Y. Xiao, J. Lin et al., Near-infrared II dye-protein complex for biomedical imaging and imaging-guided photothermal therapy. Adv. Healthc. Mater. **7** (2018). https://doi.org/10.1002/adhm.201800589
83. S. Wang, Y. Fan, D. Li et al., Anti-quenching NIR-II molecular fluorophores for in vivo high-contrast imaging and pH sensing. Nat. Commun. **10** (2019). https://doi.org/10.1038/s41467-019-09043-x
84. S. Zhu, Z. Hu, R. Tian et al., Repurposing cyanine NIR-I dyes accelerates clinical translation of near-infrared-II (NIR-II) bioimaging. Adv. Mater. (2018). https://doi.org/10.1002/adma.201802546
85. J. Du, S. Liu, P. Zhang et al., highly stable and bright NIR-II AIE dots for intraoperative identification of ureter. ACS Appl. Mater. Interfaces**12** (2020). https://doi.org/10.1021/acsami.9b22957
86. S. Liu, C. Chen, Y. Li et al., Constitutional isomerization enables bright NIR-II AIEgen for brain-inflammation imaging. Adv. Func. Mater. **30**, 1908125 (2019). https://doi.org/10.1002/adfm.201908125
87. J. Qi, C. Sun, A. Zebibula et al., Real-time and high-resolution bioimaging with bright aggregation-induced emission dots in short-wave infrared region. Adv. Mater. **30** (2018). https://doi.org/10.1002/adma.201706856
88. Z. Sheng, B. Guo, D. Hu et al., Bright aggregation-induced-emission dots for targeted synergetic NIR-II fluorescence and NIR-I photoacoustic imaging of orthotopic brain tumors. Adv. Mater. (2018). https://doi.org/10.1002/adma.201800766
89. W. Wu, Y. Yang, Y. Yang et al., molecular engineering of an organic NIR-II fluorophore with aggregation-induced emission characteristics for in vivo imaging. Small **15** (2019). https://doi.org/10.1002/smll.201805549
90. P. Xu, F. Kang, W. Yang et al., Molecular engineering of a high quantum yield NIR-II molecular fluorophore with aggregation-induced emission (AIE) characteristics for in vivo imaging. Nanoscale **12** (2020). https://doi.org/10.1039/c9nr09999a
91. S. Wang, J. Liu, C.C. Goh et al., NIR-II-excited intravital two-photon microscopy distinguishes deep cerebral and tumor vasculatures with an ultrabright NIR-I AIE luminogen. Adv. Mater. **31** (2019). https://doi.org/10.1002/adma.201904447

92. J. Lin, X. Zeng, Y. Xiao et al., Novel near-infrared II aggregation-induced emission dots for in vivo bioimaging. Chem. Sci. **10** (2019). https://doi.org/10.1039/c8sc04363a
93. G. Hong, Y. Zou, A.L. Antaris et al., Ultrafast fluorescence imaging in vivo with conjugated polymer fluorophores in the second near-infrared window. Nat. Commun. **5** (2014). https://doi.org/10.1038/ncomms5206
94. P. Koralli, A.D. Nega, L.E. Vagiaki et al., New conjugated polymer nanoparticles with high photoluminescence quantum yields for far-red and near infrared fluorescence bioimaging. Mater. Chem. Front. (2020). https://doi.org/10.1039/d0qm00195c
95. K. Shou, Y. Tang, H. Chen et al., Diketopyrrolopyrrole-based semiconducting polymer nanoparticles for in vivo second near-infrared window imaging and image-guided tumor surgery. Chem. Sci. **9** (2018). https://doi.org/10.1039/c8sc00206a
96. S. Wang, J. Liu, G. Feng et al., NIR-ii excitable conjugated polymer dots with bright NIR-I emission for deep in vivo two-photon brain imaging through intact skull. Adv. Func. Mater. **29**, 1808365 (2019). https://doi.org/10.1002/adfm.201808365
97. L. e. Zhang, Z. Zhang, C. Liu et al., NIR-II dye-labeled cylindrical polymer brushes for in vivo imaging. ACS Macro Lett. **8**, 1623 (2019). https://doi.org/10.1021/acsmacrolett.9b00815.

Chapter 4
Semiconductor Quantum Dots for NIR Bioimaging

Masao Kamimura

Abstract Quantum dots (QDs) are nanosized semiconductor crystals and are well known as bright nanophosphors. QDs show particle size-dependent emission wavelength due to their quantum size effect, and thus, these nanoparticles can be used for multicolor fluorescence imaging under single excitation light. Furthermore, QDs possess sharp fluorescence peak and show strong fluorescence signals for a long time. Therefore, in the past two decades, QDs became crucial tool as fluorescence bioimaging agents. This chapter reviews materials design of QDs for tissue penetrable near-infrared (NIR) fluorescence bioimaging.

Keywords Quantum dots · Nanoparticles · Surface modification · Quantum size effect · Semiconductor

4.1 Quantum Dots (QDs)

Quantum dots (QDs) are semiconductor nanocrystals, which show unique and special optical properties due to quantum confinement effect. Generally, particle size of QDs is ca.2–10 nm, and these QDs exhibit size-dependent emission spectra owing to their quantum size effect (Fig. 4.1) [1–3]. Therefore, QDs can be used to realize multicolor imaging under a single excitation light. In addition, QDs exhibit a sharp fluorescent peak and strong fluorescence signals for a long time. Traditionally, photobleaching is one of the most serious problems of fluorescence bioimaging. For example, emission intensity of organic fluorescent dyes will be decreased during long time irradiation of excitation light, due to "fading" of organic dye, which is caused by dissociation of molecular structure under strong excitation light irradiation. On the other hand, QDs show excellent photostability, which show no photobleaching and can be used for long time observation.

In 1998, Nie and Alivisatos individually reported the first demonstration of QD-based fluorescence bioimaging [4, 5]. Following these reports, QD-based bioimaging

M. Kamimura (✉)
Department of Materials Science and Technology, Faculty of Industrial Science and Technology, Tokyo University of Science, 6-3-1 Niijuku, Katsushika, Tokyo 125-8585, Japan
e-mail: masaokamimura@rs.tus.ac.jp

K. Soga et al. (eds.), *Transparency in Biology*,
https://doi.org/10.1007/978-981-15-9627-8_4

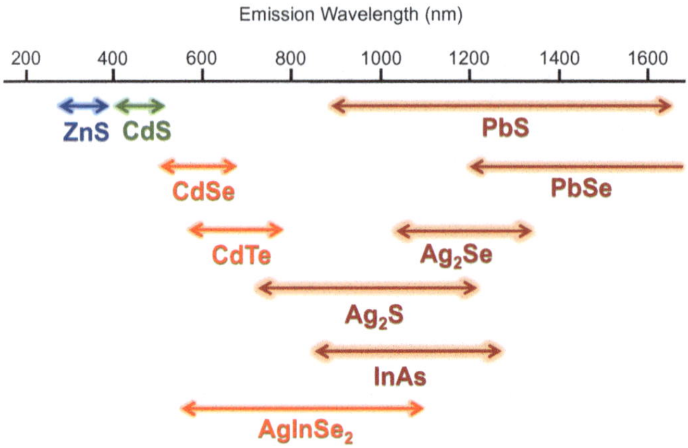

Fig. 4.1 Emission wavelengths of typical QDs

studies have been performed both in vitro and in vivo. Since then, in the past two decades, QDs have gained immense attention as fluorescence bioimaging probes. To improve the optical properties of QDs, such as emission intensity, fabrication of core–shell structure methods is widely reported. For example, core–shell-structured QDs such as CdSe/ZnS QDs are synthesized and used as highly fluorescent bioimaging probes. In addition, for the specific targeting of QDs against target part of the living body, researchers installed various ligand molecules such as antibody, protein, or sugar on the surface of QDs, and targeted in vivo delivery of these QDs was performed.

As just described, QDs work as excellent fluorescence bioimaging probes. However, because the QDs are formed from heavy metals (e.g., Pb, Cd, In, etc.), they exhibit cytotoxicity. Therefore, surface modification of QDs with biocompatible materials is quite important for biomedical applications. For example, poly(ethylene glycol) (PEG), which exhibits high biocompatibility, is widely used as a surface modification agent of QDs. The surface modification of QDs with biocompatible polymers such as PEG improves not only their biocompatibility but also their dispersion stability under physiological environment. In addition, the surface modification of QDs prevents non-specific adsorption of biomolecules, such as protein in the blood stream. Thus, most of QDs researches use PEG as surface modification agent of nanoparticles. In addition, several commercially available QDs for in vivo bioimaging also modified with PEG.

4.2 Near-Infrared Fluorescent QDs

Today, for deep tissue in vivo imaging, near-infrared (NIR) fluorescence phosphor-based imaging has become an important technique due to low scattering and absorption of various biomolecules and water in NIR region. Therefore, NIR imaging in these wavelength regions is one of the most promising ideas for successful in vivo imaging in deep tissue. Thus, both of excitation and emission wavelengths of QDs are shifting from visible (VIS) region to NIR (>700 nm) region [6–65]. For example, indium (In) or lead (Pb)-based QDs, such as InAs, PbSe, or PbS, are widely reported as NIR imaging probe. These QDs can be excited by NIR-I (800–980 nm) light and it shows NIR-II (1050–1400 nm) emission. Thus, these QDs are important candidate materials for NIR fluorescent probes for deep tissue imaging.

4.3 Lead Chalcogenide QDs for NIR-II Imaging

Lead chalcogenide, such as PbS or PbSe, is one of the major QDs for NIR-II fluorescence imaging. In 2007, Hyun et al. reported on NIR-II emission of PbS and PbSe QDs under NIR-I excitation [66]. This paper first reported in vitro NIR-II fluorescence imaging of living cells (endothelial human colon carcinoma HT29 cells). Then, in vivo imaging by using NIR-I or NIR-II emission of PbS or PbSe is widely reported. For example, Hu et al. reported PEGylated PbS QDs for NIR-I in vivo imaging [7]. In addition, Wan et al. also reported silica-PEG modified PbS QDs for NIR-I imaging [67]. Moreover, Shuhendler et al. reported surface modification of PbSe QDs with fatty ester and PEG and its application for in vivo imaging [68]. By using PEG and fatty ester modified PbSe QDs, they successfully performed in vivo imaging based on NIR-I (840 nm) emission of PbSe under NIR-I (710 nm) excitation light.

In the case of NIR-II region, Won et al. reported NIR-II imaging of mice by using PbSe QDs. In this work, they introduced tissue permeability of NIR light, including both NIR-I and II region, by using NIR-I fluorescent QDs (CdTeSe) and NIR-II fluorescent QDs (PbSe) [8]. In 2013, T. Jin and his co-workers reported GSH-modified PbS-based NIR-II in vivo imaging [10]. This work successfully prepared water dispersible PbS QDs and its application for in vivo tumor observation. They also reported PbS/CdS core–shell QDs for NIR-II imaging [17]. The PbS/CdS core–shell QDs were conjugated with Fe_3O_4 magnetic nanoparticles and applied for NIR-II and magnetic resonance (MR) multi-modal imaging. In addition, the PbS/CdS core–shell QDs were also used for NIR-II imaging of mouse head [18]. Jin's group also reported ligand installed NIR-II QDs for specific imaging [21]. In this report, recombinant protein (GST-EGFP-GB1)-coated PbS QDs were prepared. The GST-EGFP-GB1 protein is used as a template for growing PbS QDs in the aqueous phase. Furthermore, antibody recognizes the protein G B1 domain on the QDs surface. The obtained complexes showed nice NIR-II imaging properties. By using this probe, specific tumor imaging of mice was successfully performed. Moreover, they also reported thrombotic state in septic mouse brain [26].

To improve the luminescence properties of QDs, such as emission intensity or stability, construction of core–shell structure is major technique. In the case of lead chalcogenide, several groups reported on core–shell design of PbS QDs. For example, Dongling Ma and her co-workers developed core–shell structure of PbS core-based QDs, such as PbS/CdS/ZnS [37] and PbS/CdS [56] for various applications. They synthesized PbS/CdS/ZnS QDs and it showed excellent photostability [37]. The obtained PbS/CdS/ZnS QDs was extremely stable; it can disperse in phosphate buffer saline and show photoluminescence at least for 14 months. Furthermore, they also reported PbS/CdS-based probe for multimodal applications [56]. In this report, they synthesize brightness PbS/CdS QDs and conjugated with Fe_3O_4 magnetic nanoparticles. The obtained complex shows NIR-II emission for fluorescence imaging. Furthermore, Fe_3O_4 can be used for magnetic resonance (MR) imaging probe and heat source for hyperthermia under magnetic field. Thus, prepared nanocomplexes can be used for NIR-II and MR multi-modal imaging and thermal therapy of cancer in deep tissues.

The core–shell QDs are also modified with ligand molecules and applied for targeted imaging. Jeong et al. reported folic acid installed PbS/CdS core–shell QDs for NIR-II imaging [43]. Because of the folic acid is specifically bind to the tumor cells, the prepared QDs showed tumor accumulation properties. Yu et al. also reported antibody conjugated PbS/CdS QDs for targeted NIR-II imaging [58].

Moreover, core–shell QDs is also applied for diagnosis applications. Huang and his co-workers reported real-time imaging hemodynamics and vascular structure in mice with experimentally induced peripheral arterial disease (PAD) [69]. In this report, they used PbS/CdS QDs as NIR-II fluorescent probe for vascular contrast agent.

For deep tissue imaging, development of observation techniques is also important point. Recently, light-sheet microscopy is having much attention as 3D imaging technique for deep tissus. In 2018, Hongjie Dai's group reported light-sheet microscopy in the NIR-II wavelength region by using PbS/CdS QDs as fluorescent probe (Fig. 4.2) [70]. Light-sheet microscopy in NIR-II wavelength region is promising idea for non-invasive deep tissue 3D imaging of live mammals with high spatiotemporal resolution. This work reported the 3D imaging of mouse brain tissues, and this technique can be used for monitoring of the traumatic brain injury.

As describe in this section, PbS QDs were widely used as NIR-II fluorescent probes. However, because of Pb possesses strong toxicity, so far, it is difficult to use as safety fluorescent materials, including clinical applications.

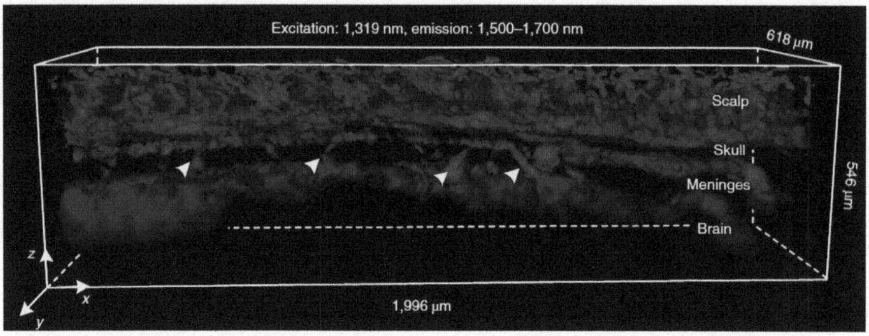

Fig. 4.2 NIR-II light-sheet imaging of blood vessels in mouse head region by using PbS/CdS QDs. Reproduced with permission from 70) © 2019 Springer Nature

4.4 Silver Chalcogenide QDs for NIR-II Imaging

In order to overcome toxicity problem of lead-based QDs such as PbS and PbSe, silver (Ag)-based QDs have attracted much attention. Ag is one of safety metals, and thus, it can be used for construction of low toxic QDs. In 2012, Hongjie Dai's group reported NIR-II in vivo imaging by using Ag_2S QDs [6]. In this work, PEG modified Ag_2S were developed and it showed NIR-II (1200 nm) emission under NIR-I (808 nm) excitation. Based on this material, tumor imaging of mouse was demonstrated. Following this report, various NIR-II imaging of Ag_2S QD-based probes were reported. Yan Zhang et al. reported biodistribution and pharmacokinetics of Ag_2S QDs [11]. In this paper, they studied long term in vivo distribution of PEG modified Ag_2S QDs. Chunyan Li et al. also reported real-time visualization of blood flow and angiogenesis by using NIR-II emission of Ag_2S QDs [14]. Furthermore, Jun Chen et al. studied synthesis of β-lactoglobulin capped Ag_2S QDs [24]. In this reported, β-lactoglobulin was used as capping agent of Ag_2S synthesis and obtained β-lactoglobulin capped Ag_2S QDs show high stability and strong NIR-II emission. Thus, successful in vivo imaging based on obtained QDs was also demonstrated.

Furthermore, other chalcogenides, such as selenium (Se) or tellurium (Te), were also used for synthesis of Ag-based QDs. For example, several groups reported Ag_2Se QDs for NIR-II imaging. Qiangbin Wang's group reported PEGylated Ag_2Se QDs for vascular and organ imaging [9]. In this work, the surface of Ag_2Se QDs was modified with PEG, and obtained probe was used for NIR-II in vivo imaging. Lianjiang Tan et al. also reported aqueous synthesis of multidentate polymer capped Ag_2Se QDs [16]. In this work, biocompatible polymer modified QDs were successfully obtained and it was applied for NIR-II in vivo imaging of live mouse. In addition, Huan Tang et al. reported biodistribution and clearance of Ag_2Se QDs [30]. They performed NIR-II imaging for analysis of biodistribution and clearance of Ag_2Se QDs in mice body. Moreover, Ag_2Te is also excellent probe for NIR-II imaging. For example, Min Yang et al. reported multivalent polymer capped Ag_2Te QDs for live cell imaging [22]. In 2019, Dai-Wen Pang's group reported cell membrane conjugated Ag_2Te QDs for homotypic tumor imaging [59]. Furthermore, in 2020, Zhi-Quan Tian's group applied Ag_2Te QDs for multimodal imaging of NIR-II and CT imaging [65].

Because of its low toxicity, Ag-based QDs are well used for in vivo tracking of cells, such as stem cell tracking studies. In 2013, Qiangbin Wang and his co-wokers have first reported in vivo tracking of transplanted mesenchymal stem cells (MSCs) by using NIR-II emission of Ag_2S QDs [12]. In this study, PEG modified Ag_2S QDs labeled human MSCs were transplanted in living mouse and fluorescence tracking was successfully performed. Following this work, in 2015, they also reported real-time visualization of MSCs tropism for cutaneous regeneration [19]. Furthermore, in 2017, they reported dynamic monitoring of fate of trans-planted stem cells by using NIR-II emission of Ag_2S QDs and endogenous red bioluminescence imaging by using firefly Luciferase [39]. Moreover, in 2018, they performed monitoring the location and osteogenic differentiation of transplanted hMSCs [50]. The demonstrated results in these works suggest that the Ag_2S QDs can be used for stem cell tracking and powerful tools for regenerative medicine.

Ag-based QDs were also applied for combination technique with other imaging method or therapy technology [20, 36, 41, 46]. For example, Qiangbin Wang and his co-workers reported Gd-DOTA conjugated Ag_2S QDs for multimodal imaging of NIR-II and MR [20]. Aihua Qu et al. reported AuNR, Pt, and Ag_2S QDs nanocomplexes for NIR-II fluorescence, CT, photoacoustic imaging and photothermal therapy [36].

Recently, Daniel Jaque's group reported several unique studies by using Ag_2S QDs. For example, in 2018, contactless brain thermal sensing through the scalp and skull in real time with a subdegree resolution was performed based on temperature sensitive NIR-II emission of Ag_2S QDs [40]. Furthermore, they reported Ag_2S nanodots for NIR-II imaging of damaged myocardium tissues after a myocardial infarction induced by either partial or global ischemia [51]. Moreover, in 2020, they reported 80-fold increment in the quantum yield of Ag_2S super dot possessing AgCl shell (Fig. 4.3) [63]. The obtained novel Ag_2S QD-based probe enables in vivo imaging, such as whole body, blood vessel, and biodistribution tracking under using ultra-low excitation power.

Due to the low toxicity of Ag_2S QDs, they are one of the most promising candidates for safety NIR-II in vivo imaging agent. Further improvement of the performance of Ag_2S QDs, such as emission properties, dispersion suitability under physiological environment, or targeting ability facilitates its application for imaging studies, including medical trials.

4.5 Indium-Based QDs for NIR-II Imaging

Recently, new types of NIR-II QDs, In-based QDs are also reported for in vivo imaging. In 2016, Kameyama et al. reported synthesize of rod-shaped $AgInTe_2$ QDs and its application for NIR-II imaging [27]. In this work, $AgInTe_2$ QDs were incorporated into the liposome particles and used as fluorescent probe. The injected QDs

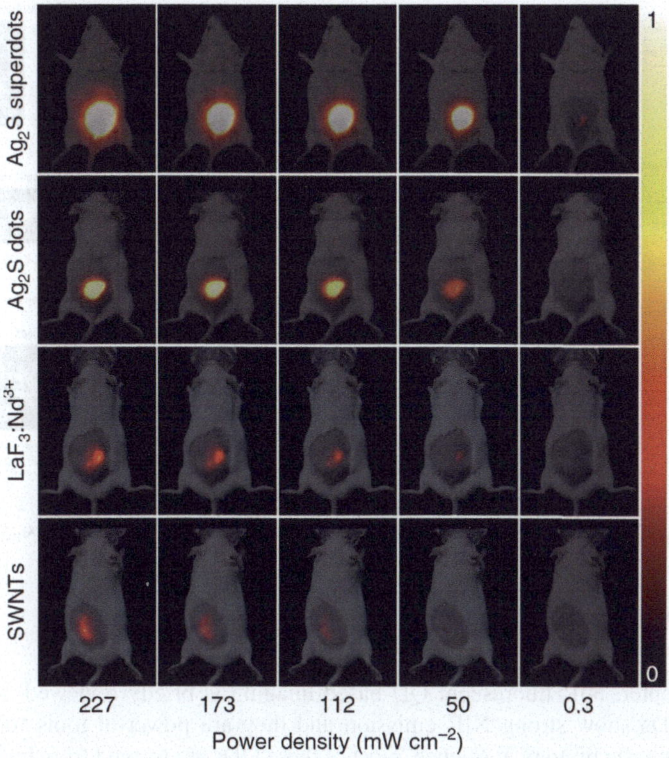

Fig. 4.3 NIR-II fluorescence images of mice subcutaneously injected with colloidal aqueous dispersions containing Ag_2S superdots, commercial Ag_2S QDs, single-walled carbon nanotubes (SWNTs), and LaF_3:Nd nanoparticles (NPs). Reproduced with permission from 63) © 2020 Springer Nature

loaded liposome showed NIR-II emission and organs of live mouse were visualized. Furthermore, Bawendi and his co-workers reported InAs/CdSe/CdS tri-layered QDs based in vivo imaging [25]. This study shows synthesis of bright NIR-II QDs, "core–shell-shell" InAs/CdSe/CdS QDs, and visualization of blood vessel of mice were performed. In addition, they further reported various InAs QD-based probes (Fig. 4.4) [32]. In this work, they synthesized various QDs probes possessing InAs core and CdS, CdSe, or ZnSe shell structures. These core–shell or core–shell-shell QDs are incorporated into several kinds of nanostructures, such as phospholipid micelles, nanosome, and composite nanoparticles.

In short, In-based QDs are also attractive materials for NIR-II imaging. Core–shell or core–shell-shell structure improves emission properties, such as wavelength, emission intensity, or photostability. Further modification with organic materials (liposome, polymer micelles, etc.) is also work for improvement of biocompatibility of QDs. Thus, In-based QDs have promising values for high performance in vivo imaging probes.

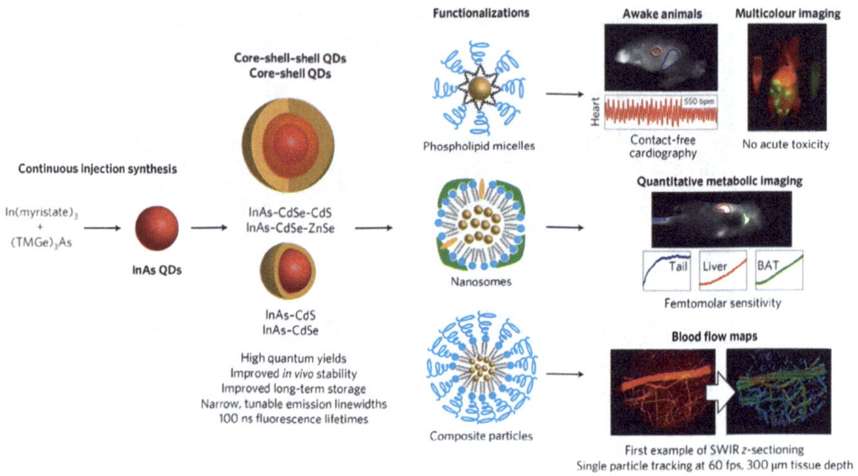

Fig. 4.4 A schematic overview of the synthesis of core–shell and core–shell–shell InAs QD-based probes for NIR-II imaging. Reproduced with permission from 32) © 2017 Springer Nature

4.6 Summary

In this chapter, NIR fluorescent QD-based imaging is briefly reviewed. Although various QDs show strong NIR emission and they are powerful tools for in vivo imaging in basic biological science, most of these QDs are formed from highly toxic elements. Therefore, clinical application of QDs is still challenging. Thus, to develop the QD-based probes for human clinical use, non-toxic material-based QDs and/or complete surface design, such as surface modification with biocompatible polymer or construction of core–shell structure with safety materials, are required. As compared to other fluorescent nanoparticles, small particle size of QDs is strong point for in vivo imaging. In addition, decoration of QDs surface with ligand molecules leads to targeted imaging. Overall, current QDs reached such a high degree of perfection as fluorescence bioimaging probe. Further update of materials design and surface modification may be able to achieve human clinical applications in near future.

References

1. H. Yukawa, Y. Baba, In vivo fluorescence imaging and the diagnosis of stem cells using quantum dots for regenerative medicine. Anal. Chem. **89** (2017). https://doi.org/10.1021/acs.analchem.6b04763
2. H. Yukawa, Y. Baba, <i>In Vivo</i> imaging technology of transplanted stem cells using quantum dots for regenerative medicine. Anal. Sci. **34** (2018). https://doi.org/10.2116/analsci.17R005
3. H. Lu, G.M. Carroll, N.R. Neale, et al., Infrared quantum dots: progress, challenges, and opportunities. ACS Nano **13** (2019). https://doi.org/10.1021/acsnano.8b09815

4. M. Bruchez, M. Moronne, P. Gin, et al., Semiconductor nanocrystals as fluorescent biological labels. Science **281** (1998). https://doi.org/10.1126/science.281.5385.2013
5. W.C.W. Chan, S. Nie, Quantum dot bioconjugates for ultrasensitive nonisotopic detection. Science **281** (1998). https://doi.org/10.1126/science.281.5385.2016
6. G. Hong, J.T. Robinson, Y. Zhang, et al., In vivo fluorescence imaging with Ag_2S quantum dots in the second near-infrared region. Angew. Chem. Int. Ed. Engl. **51** (2012). https://doi.org/10.1002/anie.201206059
7. R. Hu, W.C. Law, G. Lin, et al., PEGylated phospholipid micelle-encapsulated near-infrared PbS quantum dots for in vitro and in vivo bioimaging. *Theranostics* **2** (2012). https://doi.org/10.7150/thno.4275
8. N. Won, S. Jeong, K. Kim, et al., Imaging depths of near-infrared quantum dots in first and second optical windows. Molecular Imaging **11** (2012). 7290.2011.00057, https://doi.org/10.2310/7290.2011.00057
9. B. Dong, C. Li, G. Chen et al., Facile synthesis of highly photoluminescent Ag_2Se quantum dots as a new fluorescent probe in the second near-infrared window for in vivo imaging. Chem. Mater. **25**, 2503 (2013). https://doi.org/10.1021/cm400812v
10. Y. Nakane, Y. Tsukasaki, T. Sakata, et al., Aqueous synthesis of glutathione-coated PbS quantum dots with tunable emission for non-invasive fluorescence imaging in the second near-infrared biological window (1000–1400 nm). Chem. Commun. (Camb) **49** (2013). https://doi.org/10.1039/c3cc44000a
11. Y. Zhang, Y. Zhang, G. Hong, et al., Biodistribution, pharmacokinetics and toxicology of Ag_2S near-infrared quantum dots in mice. Biomaterials **34** (2013). https://doi.org/10.1016/j.biomaterials.2013.01.089
12. G. Chen, F. Tian, Y. Zhang et al., Tracking of transplanted human mesenchymal stem cells in living mice using near-infrared Ag_2S quantum dots. Adv. Func. Mater. **24**, 2481 (2014). https://doi.org/10.1002/adfm.201303263
13. M. Corricelli, N. Depalo, E. Di Carlo, et al., Biotin-decorated silica coated PbS nanocrystals emitting in the second biological near infrared window for bioimaging. Nanoscale **6** (2014). https://doi.org/10.1039/c4nr01025f
14. C. Li, Y. Zhang, M. Wang, et al., In vivo real-time visualization of tissue blood flow and angiogenesis using Ag_2S quantum dots in the NIR-II window. Biomaterials **35** (2014). https://doi.org/10.1016/j.biomaterials.2013.10.010
15. L. Tan, A. Wan, H. Li, Synthesis of near-infrared quantum dots in cultured cancer cells. ACS Appl. Mater. Interfaces **6** (2014). https://doi.org/10.1021/am404534v
16. L. Tan, A. Wan, T. Zhao, et al., Aqueous synthesis of multidentate-polymer-capping Ag_2Se quantum dots with bright photoluminescence tunable in a second near-infrared biological window. ACS Appl. Mater. Interfaces **6** (2014). https://doi.org/10.1021/am5015088
17. Y. Tsukasaki, A. Komatsuzaki, Y. Mori, et al., A short-wavelength infrared emitting multimodal probe for non-invasive visualization of phagocyte cell migration in living mice. Chem. Commun. (Camb) **50** (2014). https://doi.org/10.1039/c4cc06542e
18. Y. Tsukasaki, M. Morimatsu, G. Nishimura et al., Synthesis and optical properties of emission-tunable PbS/CdS core–shell quantum dots for in vivo fluorescence imaging in the second near-infrared window. RSC Adv. **4**, 41164 (2014). https://doi.org/10.1039/c4ra06098a
19. G. Chen, F. Tian, C. Li, et al., In vivo real-time visualization of mesenchymal stem cells tropism for cutaneous regeneration using NIR-II fluorescence imaging. Biomaterials **53** (2015). https://doi.org/10.1016/j.biomaterials.2015.02.090
20. C. Li, L. Cao, Y. Zhang, et al., Preoperative detection and intraoperative visualization of brain tumors for more precise surgery: a new dual-modality MRI and NIR nanoprobe. Small **11** (2015). https://doi.org/10.1002/smll.201500997
21. A. Sasaki, Y. Tsukasaki, A. Komatsuzaki, et al., Recombinant protein (EGFP-Protein G)-coated PbS quantum dots for in vitro and in vivo dual fluorescence (visible and second-NIR) imaging of breast tumors. Nanoscale **7** (2015). https://doi.org/10.1039/c4nr06480a
22. M. Yang, R. Gui, H. Jin, et al., Ag_2Te quantum dots with compact surface coatings of multivalent polymers: ambient one-pot aqueous synthesis and the second near-infrared bioimaging. Colloids Surf. B Biointerfaces **126** (2015). https://doi.org/10.1016/j.colsurfb.2014.11.030

23. J. Chen, Y. Kong, S. Feng et al., Recycled synthesis of whey-protein-capped lead sulfide quantum dots as the second near-infrared reporter for bioimaging application. ACS Sustain. Chem. Eng. **4**, 2932 (2016). https://doi.org/10.1021/acssuschemeng.6b00490
24. J. Chen, Y. Kong, Y. Wo, et al., Facile synthesis of beta-lactoglobulin capped Ag_2S quantum dots for in vivo imaging in the second near-infrared biological window. J. Mater. Chem. B **4** (2016). https://doi.org/10.1039/c6tb01186a
25. D. Franke, D.K. Harris, O. Chen, et al., Continuous injection synthesis of indium arsenide quantum dots emissive in the short-wavelength infrared. Nat. Commun. **7** (2016). https://doi.org/10.1038/ncomms12749
26. Y. Imamura, S. Yamada, S. Tsuboi, et al., Near-infrared emitting PbS quantum dots for in vivo fluorescence imaging of the thrombotic state in septic mouse brain. Molecules **21** (2016). https://doi.org/10.3390/molecules21081080
27. T. Kameyama, Y. Ishigami, H. Yukawa, et al., Crystal phase-controlled synthesis of rod-shaped $AgInTe_2$ nanocrystals for in vivo imaging in the near-infrared wavelength region. Nanoscale **8** (2016). https://doi.org/10.1039/c5nr07532g
28. Y. Kong, J. Chen, H. Fang, et al., Highly fluorescent ribonuclease-A-encapsulated lead sulfide quantum dots for ultrasensitive fluorescence in vivo imaging in the second near-infrared window. Chem. Mater. **28** (2016). https://doi.org/10.1021/acs.chemmater.6b00208
29. D.H. Ortgies, L. de la Cueva, B. Del Rosal, et al., In vivo deep tissue fluorescence and magnetic imaging employing hybrid nanostructures. ACS Appl. Mater. Interfaces **8** (2016). https://doi.org/10.1021/acsami.5b10617
30. H. Tang, S.T. Yang, Y.F. Yang, et al., Blood clearance, distribution, transformation, excretion, and toxicity of near-infrared quantum dots Ag_2Se in mice. ACS Appl. Mater. Interfaces **8** (2016). https://doi.org/10.1021/acsami.6b05057
31. H. Zhao, G. Sirigu, A. Parisini, et al., Dual emission in asymmetric “giant” PbS/CdS/CdS core/shell/shell quantum dots. Nanoscale **8** (2016). https://doi.org/10.1039/c5nr08881j
32. O.T. Bruns, T.S. Bischof, D.K. Harris, et al., Next-generation in vivo optical imaging with short-wave infrared quantum dots. Nat. Biomed. Eng. **1** (2017). https://doi.org/10.1038/s41551-017-0056
33. N. Depalo, V. De Leo, M. Corricelli, et al., Lipid-based systems loaded with PbS nanocrystals: near infrared emitting trackable nanovectors. J. Mater. Chem. B **5** (2017). https://doi.org/10.1039/c6tb02590k
34. E.G. Durmusoglu, M.M. Yildizhan, M.A. Gulgun et al., Production of small, stable PbS/CdS quantum dots via room temperature cation exchange followed by a low temperature annealing processes. J. Phys. Chem. C **121**, 25520 (2017). https://doi.org/10.1021/acs.jpcc.7b06153
35. S. Jeong, J. Song, W. Lee, et al., Cancer-microenvironment-sensitive activatable quantum dot probe in the second near-infrared window. Nano Lett. **17** (2017). https://doi.org/10.1021/acs.nanolett.6b04261
36. A. Qu, L. Xu, M. Sun et al., Photoactive hybrid AuNR-Pt@Ag_2S core-satellite nanostructures for near-infrared quantitive cell imaging. Adv. Func. Mater. **27**, 1703408 (2017). https://doi.org/10.1002/adfm.201703408
37. F. Ren, B. del Rosal, S.Y. An et al., Development and investigation of ultrastable PbS/CdS/ZnS quantum dots for near-infrared tumor imaging. Part. Part. Syst. Charact. **34**, 1600242 (2017). https://doi.org/10.1002/ppsc.201600242
38. J.A. Carr, M. Aellen, D.Franke, et al., Absorption by water increases fluorescence image contrast of biological tissue in the shortwave infrared. Proc. Natl. Acad. Sci. USA **115** (2018). https://doi.org/10.1073/pnas.1803210115
39. G. Chen, S. Lin, D. Huang, et al., Revealing the fate of transplanted stem cells in vivo with a novel optical imaging strategy. Small **14** (2018). https://doi.org/10.1002/smll.201702679
40. B. del Rosal, D. Ruiz, I. Chaves-Coira et al., In vivo contactless brain nanothermometry. Adv. Func. Mater. **28**, 1806088 (2018). https://doi.org/10.1002/adfm.201806088
41. X. Hao, C. Li, Y. Zhang, et al., Programmable chemotherapy and immunotherapy against breast cancer guided by multiplexed fluorescence imaging in the second near-infrared window. Adv. Mater. **30** (2018). https://doi.org/10.1002/adma.20180443

42. J. He, L. Ai, X. Liu, et al., Plasmonic CuS nanodisk assembly based composite nanocapsules for NIR-laser-driven synergistic chemo-photothermal cancer therapy. J. Mater. Chem. B **6** (2018). https://doi.org/10.1039/c7tb02772a
43. S. Jeong, Y. Jung, S. Bok, et al., Multiplexed in vivo imaging using size-controlled quantum dots in the second near-infrared window. Adv. Healthc Mater. **7** (2018). https://doi.org/10.1002/adhm.201800695
44. F. Zamberlan, L. Turyanska, A. Patane, et al., Stable DHLA-PEG capped PbS quantum dots: from synthesis to near-infrared biomedical imaging. J. Mater. Chem. B **6** (2018). https://doi.org/10.1039/c7tb02912h
45. M. Zhang, J. Yue, R. Cui, et al., Bright quantum dots emitting at approximately 1,600 nm in the NIR-IIb window for deep tissue fluorescence imaging. Proc. Natl. Acad. Sci. USA **115** (2018). https://doi.org/10.1073/pnas.1806153115
46. D.H. Zhao, J. Yang, R.X. Xia, et al., High quantum yield Ag_2S quantum dot@polypeptide-engineered hybrid nanogels for targeted second near-infrared fluorescence/photoacoustic imaging and photothermal therapy. Chem. Commun. (Camb) **54** (2018). https://doi.org/10.1039/c7cc09266k
47. Y. Zou, H. Jin, F. Sun et al., Design and synthesis of a lead sulfide based nanotheranostic agent for computer tomography/magnetic resonance dual-mode-bioimaging-guided photothermal therapy. ACS Appl. Nano Mater. **1**, 2294 (2018). https://doi.org/10.1021/acsanm.8b00359
48. Z. Feng, Y. Yang, J. Zhang et al., In vivo and in situ real-time fluorescence imaging of peripheral nerves in the NIR-II window. Nano Res. **12**, 3059 (2019). https://doi.org/10.1007/s12274-019-2552-z
49. C. Huang, Z. Sun, H. Cui, et al., InSe nanosheets for efficient NIR-II-responsive drug release. ACS Appl. Mater. Interfaces **11** (2019). https://doi.org/10.1021/acsami.9b06170
50. D. Huang, S. Lin, Q. Wang et al., An NIR-II fluorescence/dual bioluminescence multiplexed imaging for in vivo visualizing the location, survival, and differentiation of transplanted stem cells. Adv. Func. Mater. **29**, 1806546 (2019). https://doi.org/10.1002/adfm.201806546
51. D.H. Ortgies, Á.L. García-Villalón, M. Granado et al., Infrared fluorescence imaging of infarcted hearts with Ag_2S nanodots. Nano Res. **12**, 749 (2019). https://doi.org/10.1007/s12274-019-2280-4
52. G. Rotko, J. Cichos, E. Wysokinska, et al., Towards biocompatible NIR-II nanoprobes - transfer of hydrophobic Ag_2S quantum dots to aqueous solutions using phase transfer catalysed hydrolysis of poly(maleic anhydride-alt-1-octadecene). Colloids Surf. B Biointerfaces **181** (2019). https://doi.org/10.1016/j.colsurfb.2019.04.068
53. M. Wang, N. Chen, Three-dimensional cellular imaging in thick biological tissue with confocal detection of one-photon fluorescence in the near-infrared II window. J. Biophotonics **12** (2019). https://doi.org/10.1002/jbio.201800459
54. M. Wen, J. Ouyang, C. Wei, et al., Artificial enzyme catalyzed cascade reactions: antitumor immunotherapy reinforced by NIR-II light. Angew. Chem. Int. Ed. Engl. **58** (2019). https://doi.org/10.1002/anie.201909729
55. Q. Wen, Y. Zhang, C. Li, et al., NIR-II fluorescent self-assembled peptide nanochain for ultrasensitive detection of peritoneal metastasis. Angew. Chem. Int. Ed. Engl. **58** (2019). https://doi.org/10.1002/anie.201905643
56. F. Yang, A. Skripka, M.S. Tabatabaei, et al., Multifunctional self-assembled supernanoparticles for deep-tissue bimodal imaging and amplified dual-mode heating treatment. ACS Nano **13** (2019). https://doi.org/10.1021/acsnano.8b06563
57. Y. Yang, J. Chen, X. Shang, et al., Visualizing the fate of intra-articular injected mesenchymal stem cells in vivo in the second near-infrared window for the effective treatment of supraspinatus tendon tears. Adv. Sci. (Weinh) **6** (2019). https://doi.org/10.1002/advs.201901018
58. G.T. Yu, M.Y. Luo, H. Li, et al., Molecular targeting nanoprobes with non-overlap emission in the second near-infrared window for in vivo two-color colocalization of immune cells. ACS Nano **13** (2019). https://doi.org/10.1021/acsnano.9b05038
59. J.J. Zhang, Y. Lin, H. Zhou, et al., Cell membrane-camouflaged NIR II fluorescent Ag_2Te quantum dots-based nanobioprobes for enhanced in vivo homotypic tumor imaging. Adv. Health Mater. **8** (2019). https://doi.org/10.1002/adhm.201900341

60. D.H. Zhao, X.Q. Yang, X.L. Hou, et al., In situ aqueous synthesis of genetically engineered polypeptide-capped Ag_2S quantum dots for second near-infrared fluorescence/photoacoustic imaging and photothermal therapy. J. Mater. Chem. B **7** (2019). https://doi.org/10.1039/c8tb03043j
61. C. Li, W. Li, H. Liu, et al., An activatable NIR-II nanoprobe for in vivo early real-time diagnosis of traumatic brain injury. Angew. Chem. Int. Ed. Engl. **59** (2020). https://doi.org/10.1002/anie.201911803
62. A. Ortega-Rodriguez, Y. Shen, I. Zabala Gutierrez, et al., 10-fold quantum yield improvement of Ag_2S nanoparticles by fine compositional tuning. ACS Appl. Mater. Interfaces **12** (2020). https://doi.org/10.1021/acsami.9b22827
63. H.D.A. Santos, I. Zabala Gutierrez, Y. Shen, et al., Ultrafast photochemistry produces superbright short-wave infrared dots for low-dose in vivo imaging. Nat. Commun. **11** (2020). https://doi.org/10.1038/s41467-020-16333-2
64. X. Shi, S. Chen, M.-Y. Luo et al., Zn-doping enhances the photoluminescence and stability of PbS quantum dots for in vivo high-resolution imaging in the NIR-II window. Nano Res **13**, 2239 (2020). https://doi.org/10.1007/s12274-020-2843-4
65. M.-X. Yu, J.-J. Ma, J.-M. Wang et al., Ag_2Te quantum dots as contrast agents for near-infrared fluorescence and computed tomography imaging. ACS Appl. Nano Mater. **3**, 6071 (2020). https://doi.org/10.1021/acsanm.0c01274
66. B.-R. Hyun, Chen, D. A. Rey, et al., Near-infrared fluorescence imaging with water-soluble lead salt quantum dots. J. Phys. Chem. B **111** (2007). https://doi.org/10.1021/jp068455j
67. H.-Y. Yang, Y.-W. Zhao, Z.-Y. Zhang, et al., One-pot synthesis of water-dispersible Ag_2S quantum dots with bright fluorescent emission in the second near-infrared window. Nanotechnology **24** (2013). https://doi.org/10.1088/0957-4484/24/5/055706
68. I. Moreels, Y. Justo, B. De Geyter, et al., Size-tunable, bright, and stable PbS quantum dots: a surface chemistry study. ACS Nano **5** (2011). https://doi.org/10.1021/nn103050w
69. Z. Ma, M. Zhang, J. Yue, et al., Near-infrared IIb fluorescence imaging of vascular regeneration with dynamic tissue perfusion measurement and high spatial resolution. Adv. Funct. Mater. **28** (2018). https://doi.org/10.1002/adfm.201803417
70. F. Wang, H. Wan, Z. Ma, et al., Light-sheet microscopy in the near-infrared II window. Nature Methods **16** (2019). https://doi.org/10.1038/s41592-019-0398-7

Chapter 5
Carbon Nanotubes—Potential of Use for Deep Bioimaging

Atsuto Onoda and Masakazu Umezawa

Abstract Single-walled carbon nanotubes (CNTs) are a major over 1000-nm near-infrared fluorophore that enables in vivo deep imaging in the "second biological window." Because of the unique properties with their tube-shaped morphology, the CNTs show extremely unique in vivo behaviors including the migration and retention in the organs and time-dependent change in the cellular uptake followed by intracellular distribution. This chapter reviews the in vivo deep imaging results using CNTs and unique in vivo behavior and biological effects of CNTs comparatively with their types with different properties. The information is important for future safe and further functional biomedical use of CNTs.

Keywords Blood vessel imaging · Chemical functionalization · Polymer modification · Tissue distribution · Frustrated phagocytosis · Safer by design

5.1 Single-Walled Carbon Nanotubes (SWCNTs) as an Over 1000-nm Near-Infrared (OTN-NIR) Fluorophore

Carbon nanotubes (CNTs) are an allotrope of carbon, intermediate between fullerene cages and flat graphene with diameters typically in the range of nanometers. The CNTs have been indispensable materials for the development of current science and technology since they were first reported by Dr. Sumio Iijima in 1991 [1] and constantly manufactured in the 1990s [2–4]. Unique properties and functions of CNTs have enabled development of new products and novel techniques that are

A. Onoda (✉)
Department of Toxicology and Health Science, Faculty of Pharmaceutical Sciences, Sanyo-Onoda City University, 1-1-1 University Street, Sanyo-Onoda, Yamaguchi 756-0884, Japan
e-mail: 3B13624@alumni.tus.ac.jp

M. Umezawa
Department of Materials Science and Technology, Faculty of Industrial Science and Technology, Tokyo University of Science, 6-3-1 Niijuku, Katsushika, Tokyo 125-8585, Japan
e-mail: masa-ume@rs.tus.ac.jp

K. Soga et al. (eds.), *Transparency in Biology*,
https://doi.org/10.1007/978-981-15-9627-8_5

impossible with conventional materials. Especially, further biomedical applications of CNTs are currently expected such as bioimaging and drug delivery; therefore, lots of studies have been conducted to regulate the properties of CNTs for their further applications [5, 6]. In the CNTs, single-walled CNTs (SWCNTs) are a major fluorophore that emits over-thousand-nanometer (OTN) near-infrared (NIR) [7] that enables in vivo deep imaging in the "second biological window" [8]. SWCNTs are either of metallic or semiconductive depending on their own rolling pattern of graphene sheet that composes the carbon tubes (Fig. 5.1). The rolling pattern of the graphene lattice, the "chirality," influences the electronic band structure which largely determines the optical properties (light absorption and emission spectra) of CNTs [7]. While the metallic CNTs are not excitonically luminescent because the holes generated by excitation are immediately filled by many other electrons available in the metal, semiconductive ones emit NIR light upon photoexcitation followed by phonon-assisted energy transfer [7]. The visible colors of multi-walled CNTs (MWCNTs), consisting of nested SWCNTs, and SWCNTs with mixed chirality are almost black because they contain various absorption bands. In contrast, singly separated SWCNTs that can be prepared by gradient ultracentrifugation [9] (Fig. 5.2) or a designed single-surfactant multi-column gel chromatography [10] (Fig. 5.3) are colored. Oxygen doping to SWCNT enhances its OTN-NIR emission intensity by modifying the perturbations to the ground and the lowest bright exciton states, and thus shifting its major emission bands from its absorption band by oxygenation

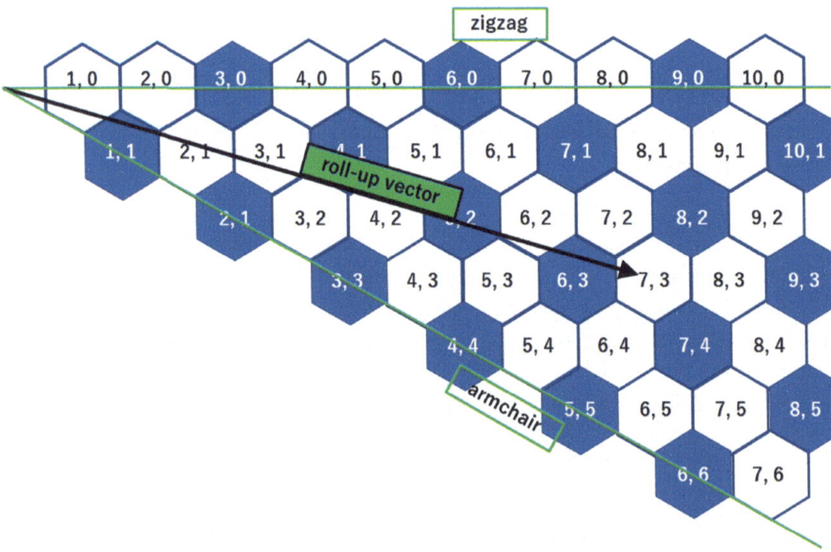

Fig. 5.1 Graphene sheet segment showing indexed lattice points composing carbon nanotubes (CNTs). CNTs designated (n, m) are obtained by rolling the sheet form (0, 0) to (n, m) along a roll-up vector. Especially, the CNT is metallic if $|m - n|$ is a multiple of three (blue hexagones) and is semiconductive otherwise (white)

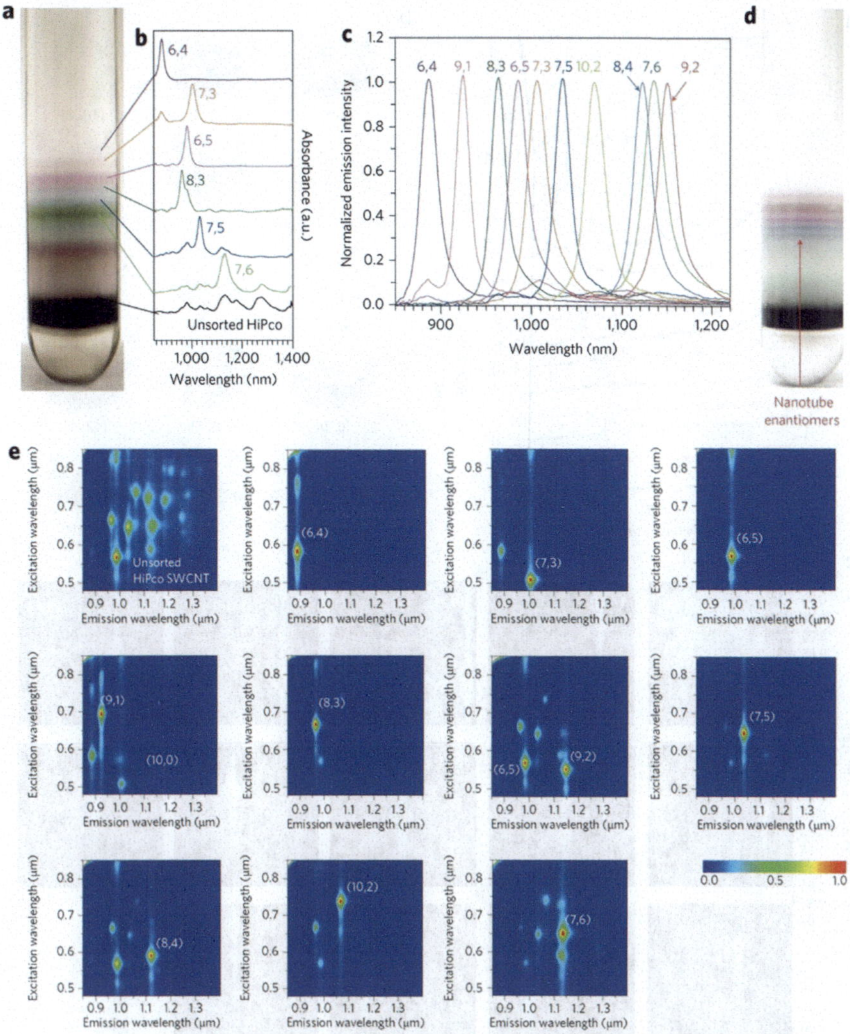

Fig. 5.2 Single-chirality single-walled carbon nanotubes (SWCNTs) sorted by single-step nonlinear density-gradient ultracentrifugation (DGU) and their optical properties. a Image of a centrifuge tube containing high-pressure carbon monoxide (HiPco) SWCNTs sorted by one 18-h nonlinear DGU. The distinct colored bands are layers enriched in different SWCNT species. **b** Near-infrared absorbance spectra of the marked colored layers, labeled with the main (n, m) component. **c** Photoluminescence spectra of separated fractions excited for each (n, m) component. **d** Image of a centrifuge tube showing resolved pairs of enantiomer bands sorted from HiPco SWCNTs by nonlinear DGU with a single surfactant (dosium cholate). **e** Photoluminescence contour maps plotted as a function of excitation and emission wavelengths. The unsorted sample shows numerous semi-conducting species, whereas each of sorted fractions shows significant enrichment of an individual semiconducting species. Reprinted from Reference [9] with permission from Springer Nature

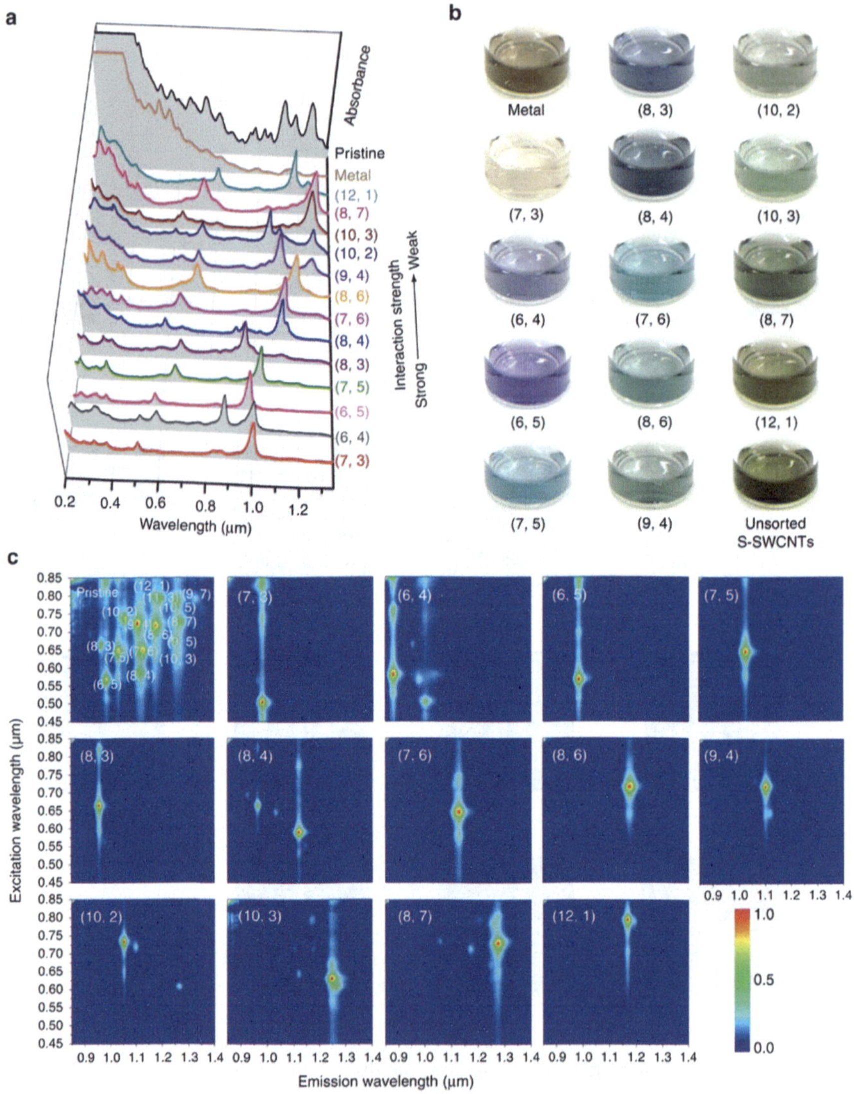

Fig. 5.3 Separated single-chirality and metallic single-walled carbon nanotubes (SWCNTs). **a** Optical absorbance spectra of 13 (n, m) semiconducting species and those of metallic and pristine HiPco mixtures. Chirality separation of SWCNTs was done by using single-surfactant multi-column gel chromatography. The spectra are ranked according to interaction strength between the (n, m) species and the gel. **b** Photographs of various (n, m) species showing their distinct colors. **c** Photoluminescence contour maps of separated (n, m) fractions. Each of single chirality of semiconducting species is highly enriched in each fraction. Reprinted from Reference [10] with permission from Springer Nature

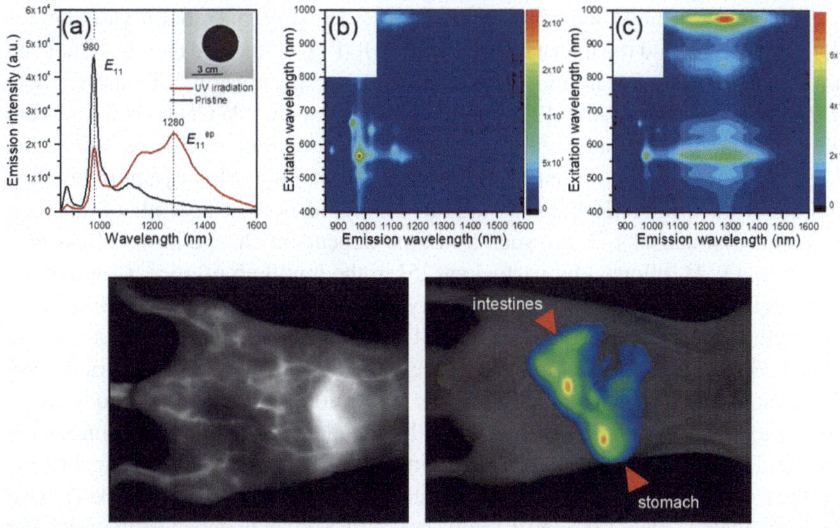

Fig. 5.4 Spectral changes and demonstrative in vivo imaging results of SWCNTs after oxygenation via exposure to ultraviolet (UV)-ozone process. a Photoluminescence spectra of SWCNTs after UV irradiation (red) and pristine SWCNTs (black) excited by 570-nm light. (inset) Image of treated SWCNT on a thin film. **b, c** 2D excitation-emission contour plots of **b** pristine and **c** oxygenated SWCNTs. Bottom pictures are OTN-NIR fluorescence images of mouse after (left) intravenous injection and (right) oral administration of oxygenated SWCNTs. The pictures were recorded by a portable in vivo NIR imaging system SAI-1000 (Shimadzu Co., Kyoto, Japan) [16]. Reprinted from Reference [12] with permission with the terms of the Creative Commons CC BY license

[11]. In addition to the ether-type oxygen adduct of SWCNT, epoxide-type oxygen adduct that can be prepared easily by a solid-state oxygenation approach was also reported with demonstrative data of its application to in vivo deep imaging in mice [12] (Fig. 5.4). Because the emission of SWCNTs is quenched by bundling via aggregation, surface modification is needed to make the CNTs monodispersed. Use of oligonucleotide [13] and poly(ethylene glycol) (PEG) conjugated with phospholipid [12, 14, 15] has been reported for preparation of OTN-NIR emissive monodispersed SWCNT suspensions. Development of monodispersed and highly emissive probes has realized their use of imaging of the biological structures and dynamic phenomena in small animals.

5.2 In Vivo Deep Imaging and Biosensing by OTN-NIR Fluorescence Imaging with SWCNTs

The intrinsic OTN-NIR fluorescence of monodispersed SWCNTs has realized deep imaging of biological structures and the dynamics of the change in blood circulation.

For example, the whole-body blood vessels [12, 15] (Fig. 5.4), brain vasculatures [17] (Fig. 5.5a, b), and the internal organs [18–20] (Fig. 5.5c) were clearly visualized without opening the skin in mice. OTN-NIR fluorescence imaging of cancer model [19, 21, 22] and its metastasis [23] was also reported using SWCNTs as fluorescent probes. One of the most advantages of the use of fluorescence for deep bioimaging is high temporal resolution that allows video-rate imaging. This dynamic imaging technique with OTN-NIR light is suitable for liquid flow at a depth of subcutaneous and submucosal 1–2 cm, such as a time-dependent change in the blood flow via ischemia [24] followed by perfusion [25] in the hindlimb of mice. Conjugation of SWCNTs with polymer, poly(2-methacryloyloxy-ethyl phosphorylcholine-co-n-butyl methacrylate), succeeded imaging of adipose tissues [26]. Antigen-specific cancer recognition using OTN-NIR fluorescence of SWCNTs was also reported using modification with a mouse/human chimeric monoclonal anti-CD20 antibody rituximab (Rituxan) for B-cell lymphoma and with a humanized, recombinant monoclonal anti-HER2 antibody trastuzumab (Herceptin) for HER2-positive breast cancer cells [27]; thus, the protocols for surface modification of SWCNTs with antibodies have been investigated. Also, OTN-NIR fluorescence of SWCNTs depends on molecular

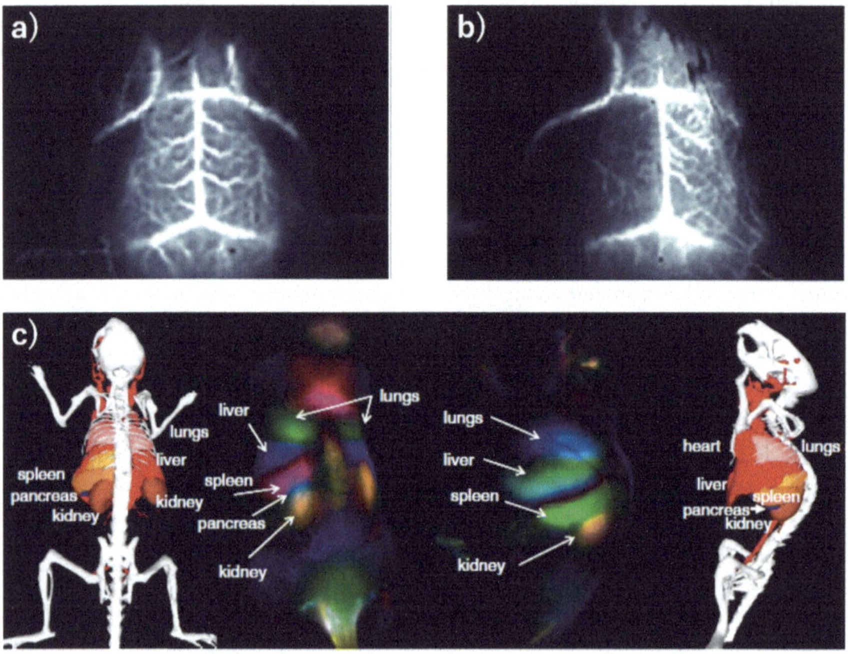

Fig. 5.5 In vivo imaging of deep tissues by using SWCNTs. a, b Brain blood vessels of **a** a normal mouse and **b** a mouse with a stroke model induced by middle cerebral artery occlusion to the left cerebral hemisphere. **c** Dynamic contrast-enhanced imaging of internal organs with SWCNTs through principle component analysis (PCA). PCA was performed for images taken over the first 130 s following injection. Adapted from **a, b** Reference [17] with permission from Springer Nature, and **c** Reference [20] with permission from National Academy of Science

adsorption and is useful for biosensing. For example, the fluorescence of SWCNTs conjugated with PEG and a specific oligonucleotide can be used for in vivo biosensing of nitric oxide and inflammation in organs [28]. For better application of SWCNTs as bioimaging probes, not only their optical properties but also their unique in vivo behavior should be noted.

5.3 Various Physicochemical Properties of CNTs

For better use of the CNTs in various field including medical biology, their unique in vivo behaviors, such as cellular uptake followed by intracellular distribution and the migration into and retention in biological organs, and their interaction with the organs should be comprehensively understood. Comprehending and controlling the in vivo behavior and biological effects of CNTs has been an important issue to enhance their function and to minimize their undesired side effects [29]; thus, the relation between the physicochemical properties of CNTs and biological responses to them merits discussion for better biomedical use of CNTs. The physicochemical properties that may be associated with the in vivo behavior and biological effects of CNTs are their size, nanoscaled structure related to mechanical property, constituent elements, and surface modification as summarized as follows.

5.3.1 Size (Length) and Mechanical Properties

First, CNTs are the nanomaterials, which have extremely small sizes. The primary size of CNTs, as well as their agglomerate/aggregate size (secondary size), influences the cellular uptake and biological response to the nanomaterials [29, 30]. Second, the in vivo behavior and biological effects of CNTs are also dependent on their shape derived from the nanoscaled structure, such as aspect ratio (tube diameter and length) and their mechanical property owing to layer number (single-walled or multi-walled) and other factors. In vitro studies have demonstrated that human peripheral mononuclear cells cannot completely phagocytose MWCNTs (50 μm length) because the MWCNTs were too large to be fully internalized. This frustrated phagocytosis was associated with induction of oxidative stress and production of a representative inflammation marker. In contrast, short MWCNTs (10 μm length) were fully phagocytosed by the mononuclear cells and induced no increase in the oxidative stress marker and inflammation marker [31]. Similar observations have been clearly reported in animal experiments [32]. The evidence highlights that the length and aspect ratio of CNTs strongly affects the cellular uptake and thereby impacting on the cellular responses including inflammation. The layer number of CNTs appears to be also associated with their mechanical property like stiffness which may be also related to their in vivo behavior [33].

5.3.2 Constituent Elements and Surface Modification

Third, the constituent elements directly influence the physicochemical properties of CNTs. CNTs are almost entirely composed by elemental carbon in graphite form and contained very small amounts of oxygen and hydrogen atoms on its surface [34]. If the CNTs are doped by oxygen, the adsorption of biomolecules on their surface may be changed. Thereby, the doped CNTs exhibit different in vivo behaviors and optical properties. Finally, the surface modification of CNTs is important because their surface interacts with various substances including biomolecules. Surface chemistry of nanomaterials influences their surface charge, reactivity, defects, and agglomeration/aggregation [35]. Previous studies have shown the monodispersion and reduced toxicity of CNTs by their surface modification with PEG [36]. Active targeting of cancer cells by conjugating of SWCNTs with folic acid is also reported [37]. Compared with constituent atoms, size, and shape of nanomaterials, the surface character can be flexibly regulated by surface modification and functionalization with coating molecules such as PEG. Hence, the surface modification is challenged as effective approaches to functionalize the CNTs with oligonucleotides [28], antibodies [27, 38] or polymers for bioimaging and biosensing using the intrinsic NIR fluorescence [26, 39–41] and to minimize their toxicity [42]. Controlling the nanoparticle–protein interaction such as adsorption and intercalation of proteins on PEG cloud of particle surface has been a challenging topic. A recent study reported that the combination uses of PEG with different length (carboxy-PEG_{2000} and methoxy-PEG_{550}) for surface modification of polymer nanoparticles can control the adsorption of proteins and largely suppress nanoparticle-mediated complement activation [43].

5.4 Unique In Vivo Behavior of CNTs

5.4.1 Tissue Distribution

Nanomaterials generally show unique biodistribution owing to their small size and surface character [44]. High barrier permeability of nanomaterials particularly allows their translocation to various organs that micro-sized objects cannot translocate [45–48]. For example, nanoparticles with a diameter less than 200 nm can pass through blood-placental barriers [48] and reach the fetal lipid-rich organs (brain, liver, and testis) [49, 50]. The translocated nanoparticles from pregnant mice to their offspring are likely to regulate the fetal development and following growth such as behavioral neural function [51], persistence activation of brain perivascular glial cells (astrocytes) [52, 53], inheritance liver defects [54], serum testosterone concentration [55], and sperm production of the testis [56, 57]. To understand the biodistribution of administered nanomaterials including CNTs is important for predicting their dynamic in vivo behavior as well as the biological responses to nanomaterials. Moreover, the dynamic behavior following the distribution such as the retention in the blood circulation and each organ and the clearance from the organs and the whole body should be visualized. The behavior of CNTs following inhalation has been a majorly discussed point with the perspectives of public and occupational hygiene [58]. Inhaled long MWCNTs (30 mg/m^3 for 6 h) reach the subpleural tissue, which is famous as the target of nanoparticles with high aspect ratio, as deposited in the subpleural wall and within subpleural macrophages in mice [59]. The behavior of CNTs in the chest wall is also dependent on their shape and mechanical properties (Fig. 5.6).

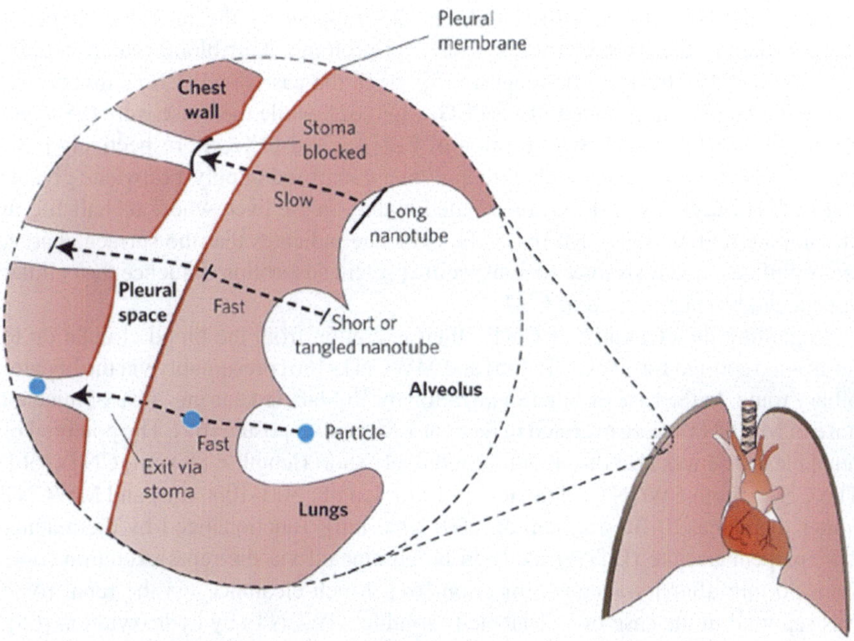

Fig. 5.6 Movement of particles through the lung. The behavior of nanoparticles including CNTs in the lungs and chest cavity, surrounded by thin membranes called pleura (red lines), is shown. Between the two membranes is the pleural space, which contains fluid that lubricates the movement of lungs. Nano-sized particles (blue circles) and short or tangled CNTs (black lines) can move rapidly through to the pleural space and drain easily into the lymphatic system through pores in the chest wall (3–8 μm diameter), whereas long, fibrous CNTs cannot. Long CNTs that lodge at the drainage points and block them can cause inflammation, leading to pathological changes over time. Reprinted from Reference [60] with permission from Springer Nature

5.4.2 Blood Retention and In Vivo Behavior Toward Clearance

With attention to the biological tissue-barrier permeability, the method to control the blood retention and following in vivo behavior should be understood for nanomaterials including CNTs. Surface modification of CNTs is a major method to control their in vivo behavior. For example, mild cationic surface charge altered the rate of clearance from organs and increased retention time a little in the blood circulation, where their blood retention half-life is 3–3.5 h [61]. In general, functionalization of SWCNTs with PEG reduces the trapping by the reticuloendothelial system (mainly the liver and the spleen) and prolongs their blood retention half-life. The half-life likely to be longest (15.3 h) in the case of SWCNTs functionalized with diamine-terminated PEG (PEG_{1500N}) [62], while the those were 0.5 h and 2 h for SWCNTs coated by adsorption of PEG_{2000} and PEG_{5400}, respectively [63]. The SWCNTs dispersed by a block-copolymer composed of poly(propylene glycol) and PEG (PEG-*b*-PPG-*b*-PEG) distributed mainly in the liver, where its half-life in the blood circulation was 1 h [64]. The evidence indicates that the surface charge, hydrophilicity, and resistance to nonspecific protein adsorption influence the cellular uptake and blood retention of CNTs.

Regarding the clearance of CNTs, their excretion from the blood circulation to feces was reported for SWCNTs [65] and MWCNTs [66] presumably via the hepatobiliary route. In the case of functionalization by ^{14}C-labeled taurine, 74% of injected dose of MWCNTs were excreted in feces at 12-h post-exposure [67]. The potential of renal clearance was also discussed for relatively short (length < 500 nm) CNTs [66]. The CNTs [both SWCNT (diameter: ≈ 1 nm, length: 300–1000 nm) and MWCNT (outer diameter: 20–30 nm, length: 500–2000 nm)] functionalized by diethylenetriaminepentaacetate (DTPA) likely to be eliminated via the renal excretion route from 30 min after intravenous injection [61]. Much clearance via the renal route was reported in the case of ^{125}I-labeled "soluble" SWCNTs by hydroxylation [68]. Generally, nanoparticles with hydrodynamic diameter of < 6 nm can be cleared by the renal excretion; however, the adsorption of biomolecules on the particle surface (e.g., adsorption of albumin on anionic materials) may suppress the excretion via the renal filtration [69].

5.4.3 Interaction with Cells and Intracellular Behavior

Recently, the unique time-dependent change in monodispersed SWCNTs in cultured cells [14] and tissues [70] was found by imaging of their OTN-NIR fluorescence. In the cellular study, delayed concentrating of monodispersed SWCNTs in cells was suggested by Raman and OTN-NIR fluorescence microscopy. Interestingly, almost no OTN-NIR fluorescence derived from the CNT was observed in the cells after a 24-h incubation of cultured cells with phospholipid-PEG-adsorbed SWCNTs. The emission was observed after three-day incubation and increased in the later stage up to the seventh day, even though the cells were cultured in fresh medium without the CNTs since the second day [14] (Fig. 5.7). Time-dependent change in the distribution of Raman signal of the CNTs was similar, suggesting that the lack of OTN-NIR fluorescence at the first 24 h was not due to quenching of the SWCNT-PEG. The evidence of this unique intracellular behavior of SWCNTs is supported by a previous Raman microscopic observation of well-dispersed SWCNT conjugated with collagen in cultured stem cells [71]. They observed an increase in intracellular Raman signal of the collagen-SWCNT after its three-day incubation with the cells. On the other hand, most of previous studies have reported intracellular CNT signals from cultured cells at quite earlier stages (1–4 h) after adding the CNTs to the culture medium. In these cases, CNT samples contained MWCNTs or cationic modified SWCNTs [72]. Different mechanical and physiological properties of MWCNTs from SWCNTs

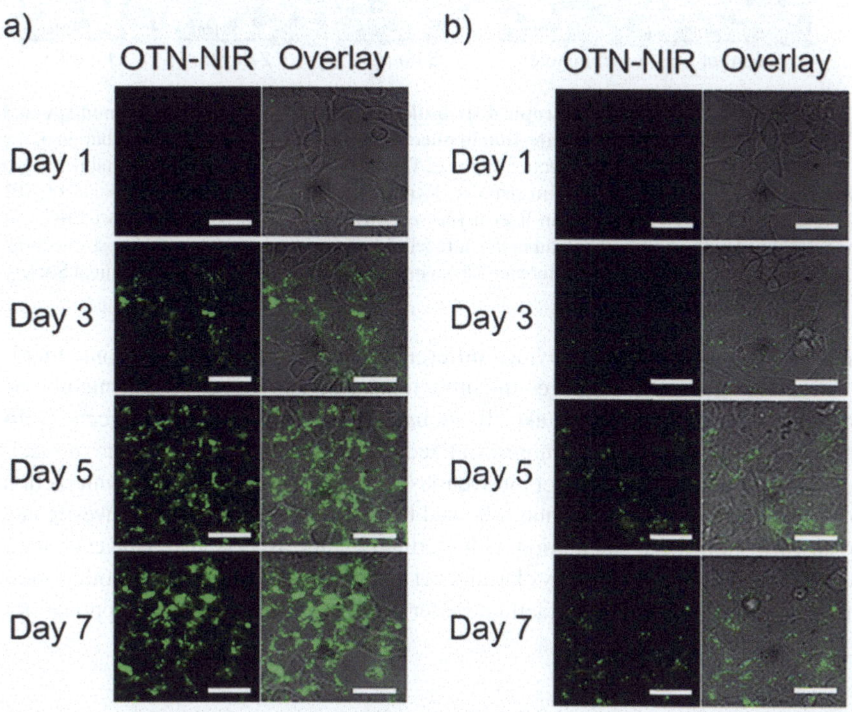

Fig. 5.7 Time-dependent change in the OTN-NIR fluorescence signal (under 980-nm excitation) of a cultured Colon-26 treated with SWCNT-PEG. a After the cultured Colon-26 cells were incubated with the SWCNTs at day 0, the culture media were exchanged every 48 h (at day 2, 4, and 6) with new medium containing SWCNTs. **b** As a comparison, the medium was exchanged every 48 h with new one without the SWCNTs after incubation of the cells with the SWCNT on day 0–2. Green color indicates the OTN-NIR fluorescence of the SWCNT-PEG. Scale bars represent 30 μm. Reprinted from Reference [14] with permission from American Chemical Society

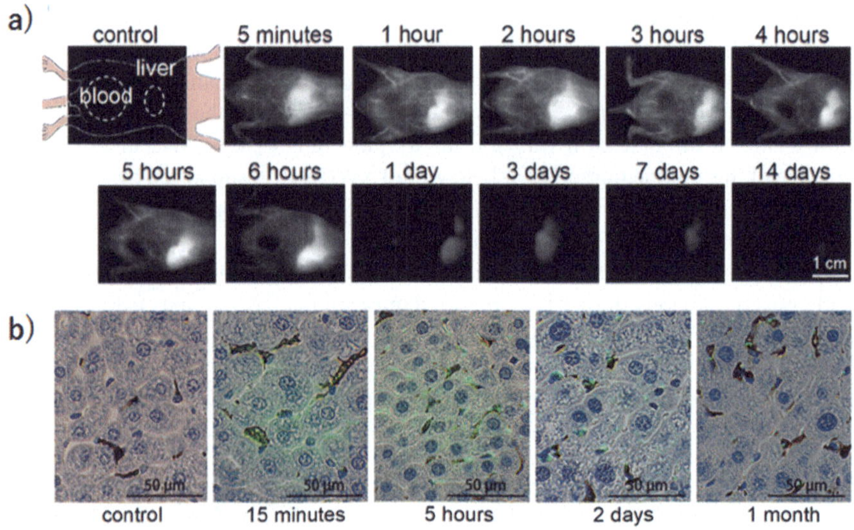

Fig. 5.8 Macroscopic and microscopic distribution analysis of oxygen-doped monodispersed SWCNT-PEG after intravenous injection in mice. **a** Dynamic change in tissue distribution of the SWCNT-PEG after intravenous injection in mice. The images were recorded by a portable in vivo NIR imaging system SAI-1000 (Shimadzu Co., Kyoto, Japan) [16]. **b** Microscopic images of NIR fluorescence of the SWCNT-PEG in liver tissue sections. The well-dispersed SWCNT-PEG was more localized in the sinus rather than in the intracellular area of the liver at two-day post-injection. Reprinted with permission from Reference [70] with permission from American Chemical Society

such as stiffness may have provided different interaction with cells. Cationic modification also generally enhances the attraction of nanomaterials to cell membrane followed by their cellular uptake. In mouse studies, OTN-NIR fluorescent well-dispersed SWCNT-PEG was more localized in the sinus rather than in the intracellular area of the liver at 48 h after intravenous injection [70] (Fig. 5.8). The interaction of SWCNTs with cell membrane followed by cellular uptake has been investigated by in vitro studies and calculation with thermodynamic aspects [73]. However, their unique interactions are not fully clarified yet and should be better understood. Understanding these properties will facilitate safer use and further biomedical applications of the SWCNTs.

5.5 Unique Biological Effects of CNTs

In addition to the in vivo behavior of CNTs, the evidence on their unique biological effects is also valuable for designing more functionalized products with improved biocompatibility. An in vitro study evaluating the biological responses of five types of carbon nanomaterials on human fibroblast cells demonstrated a hierarchy of cytotoxicity based on the observation of cell adhesion and apoptosis as follows: SWCNTs (diameter: 2 nm, length: 500 nm) > carbon black nanoparticles > activated carbons > MWCNTs (diameter: 50 nm, length: 5 μm) [74]. This report discussed that the difference in the cytotoxicity between SWCNT and MWCNT of this case may be caused by the relative surface areas of the CNTs; however, the potential of difference in their extra- and intracellular behaviors should be also noted.

The unique biological effects of CNTs have been also investigated in in vivo studies. For instance, CNT inhalation induced overexpression of circulating immunomodulatory cytokines including transforming growth factor-β in lung, thereby affecting the T-cell maturation and proliferation in the spleen via changing in the abnormal expression of prostaglandin E2 and interleukin-10 [75] (Fig. 5.9). In short, biomolecules secreted from lung upon CNT exposure suppressed the immune function in the spleen. Also, the cross-talk between lung and secondary organs is an important issue of inhalation toxicity of nanomaterials caused by abnormal systemic circulation of biomolecules (Fig. 5.10). Expression levels of genes encoding key proteins related to inflammatory responses and oxidative stress in lung and plasma were particularly enhanced by MWCNT and SWCNT comparing with carbon black nanoparticles [76]. The evidence suggested that it is worth monitoring the levels of circulating biomolecules to clarify the biological effects of CNT administration. Thus, much attention should be paid to the "far-reaching effects" that mean the

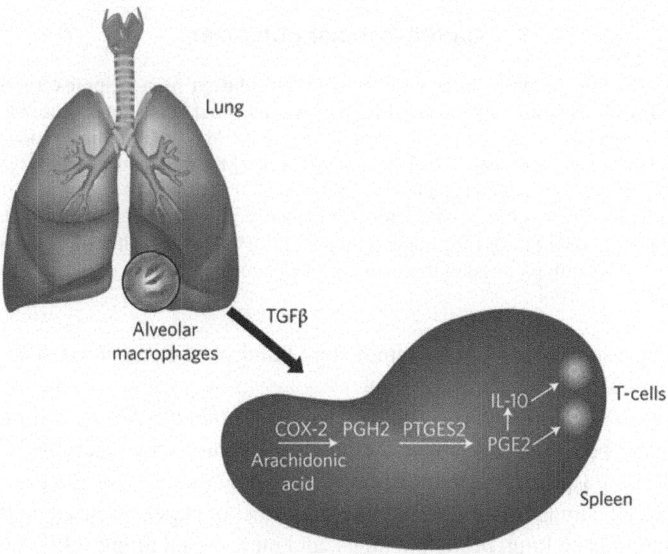

Fig. 5.9 Schematic of the proposed mechanism by which inhaled multi-walled carbon nanotubes induce systemic immune suppression. Transforming growth factor beta (TGFβ) released by the lung upon MWCNT inhalation likely to result in systemic immunosuppression. Further attention should be paid to the effects of CNTs on not only their administered and primary targeted sites but also "far-reaching effects" on distant organs [77]. Reprinted from Reference [75] with permission from Springer Nature

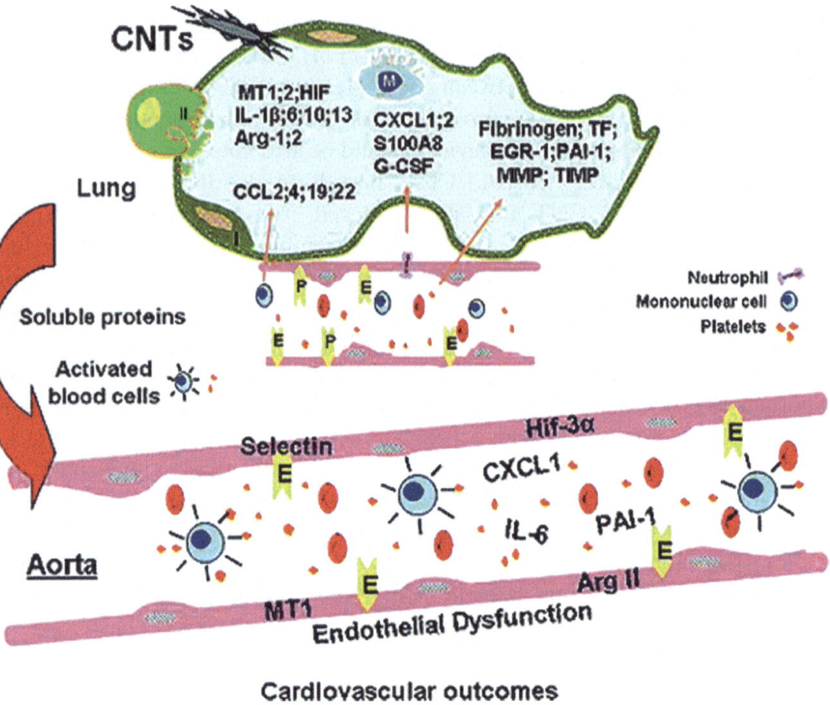

Fig. 5.10 Cross-talk between lung and systemic circulation as a response to respiratory administration of CNTs. SWCNTs and MWCNTs administered in the lung induce an acute lung and extra-pulmonary effect, which is more pronounced in the MWCNT administration. Abbreviations include (I) alveolar type I cell; (II) alveolar type II cell; (M) alveolar macrophage; (MT) metallothionein; (HIF) hypoxia inducible factor; (IL) interleukin; (Arg) arginase; (G-CSF) granulocyte-colony stimulating factor; (TF) tissue factor; (EGR) early growth response; (PAI) plasminogen activator inhibitor; (MMP) matrix metalloproteinase; (TIMP) tissue inhibitor of metalloproteinase. Reprinted from [76] with permission from American Chemical Society

influence on distant organs from administered and primary targeted sites of CNTs [77].

Not only the acute responses, but the late-onset manifestations following CNT exposure have been also investigated primarily because they resemble asbestos in shape with high aspect ratio that displayed carcinogenicity [78]. Because asbestos mainly induced damages in pulmonary organs, most of papers focused on the effects of these materials on lung, bronchial tubes, and mesothelial lining following respiratory administration. A potential of toxic effects of CNTs similar to those of asbestos has been concerned [32, 79–81]. However, comparison of the biological effects of CNTs and asbestos is not simple because they can induce different biologically mechanistic pathways. Although both MWCNTs and asbestos suppressed cellular survival rate and proliferation, MWCNTs dysregulate the pathways including oxidative phosphorylation and mitochondrial dysfunction, while asbestos disturbed chronic inflammation pathways [82]. Thus, the biological responses of CNTs should be discussed separately from those of asbestos.

Moreover, the types of CNTs and their dispersion state should be paid attention to understand their biological effects. Highly dispersed SWCNTs produced only slight fibrosis without inflammation in the large airways, where granuloma was induced with severe fibrosis, inflammation, and macrophage activation related to frustrated phagocytosis by agglomerated SWCNTs and asbestos at 30 days after exposure [83]. The cellular response to highly dispersed and agglomerated SWCNTs is likely dependent on their effects on phagolysosome and the difference in gradual clearance over time. Less cytotoxicity of highly dispersed SWCNTs than agglomerated SWCNTs is also shown by in vitro data [84]. In addition to SWCNTs, the cellular responses of MWCNTs also change depending on their dispersion states. Whereas the straight MWCNTs induced the frustrated phagocytosis [32] (Fig. 5.11) related to produc-

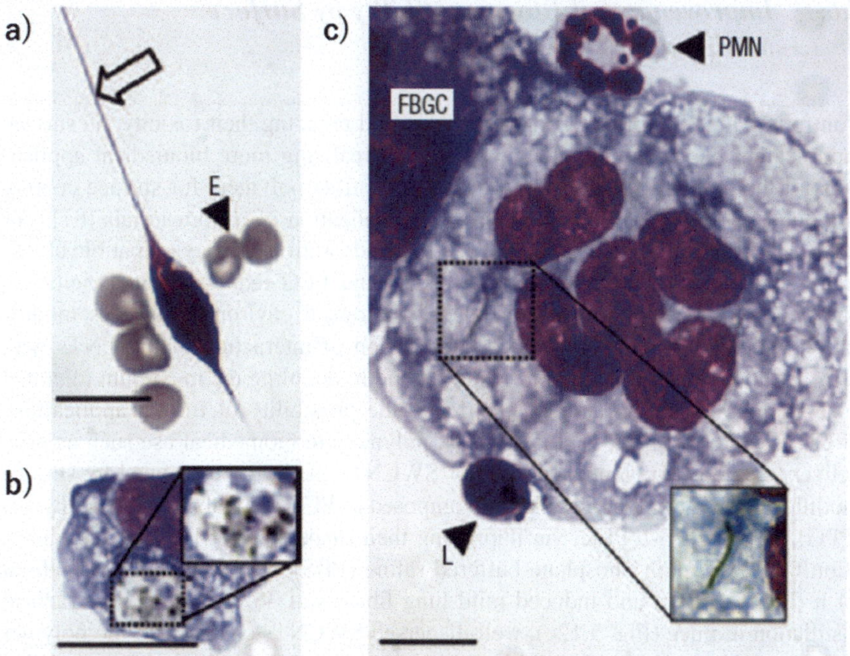

Fig. 5.11 Effect of the length of MWCNTs on phagocytosis by peritoneal macrophages. **a** Histological features show incorporation of long MWCNTs (arrow; length > 20 μm) resulting in frustrated phagocytosis. **b** In contrast, shorter MWCNTs (1–5 μm length) can be readily phagocytosed. **c** Foreign body giant cells, fused macrophages generated in response to the presence of large foreign bodies, are also present after administration of the long MWCNTs. Abbreviations: E: erythrocytes, F: foreign body giant cells, L: lymphocyte, PMN: polymorphonuclear leukocyte). Reprinted from Reference [32] with permission from Springer Nature

tion of the reactive oxygen species and inflammation markers in the mononuclear cells, entangled MWCNTs with aggregate forms were easily phagocyted by the cells resulting in little oxidative stress or inflammation [31]. Further, cellular uptake of thick or aggregative MWCNTs is different from that of asbestos. While asbestos injures mesothelial cells by internalization into the cells and severe inflammation, aggregative and thick MWCNTs did not penetrate mesothelial cells and induced only mild inflammation and macrophage activation [85]. These observations indicated that, as well as the in vivo behavior, the biological effects of CNTs are dependent on their shape (associated with the mechanical properties) and dispersion/aggregation states that can be controlled by surface modification as reviewed in the next section.

5.6 Control of Biological Response to CNTs by Their Surface Modification

5.6.1 Improvement of Biocompatibility by Surface Modification

Controlling the biological response to CNTs and reducing their toxicity via surface modification has been a challenging issue for realizing more biomedical applications of the material [86]. The use of biocompatible polymers for surface coating is a major choice of safer design and functionalization of nanomaterials [87]. For example, surface modification of MWCNTs by adsorption of a biocompatible block-copolymer composed of poly(lactide) (PLA) and PEG reduced their toxicity via improving their dispersion in aqueous physiological environments. This modification with PEG-*b*-PLA resulted in suppression of interaction of the CNTs with cells and long-term inflammatory response and in doubling the maximum tolerated dose in mice [36]. The authors mentioned the possibility of further applications of CNTs coated with the multifunctional polymers to biomedical use such as drug delivery. Similarly, potential toxicity of SWCNTs likely to be reduced by surface modification with a block-copolymer composed of PEG and poly(propylene glycol) (PPG), PEG-*b*-PPG-*b*-PEG, via improving their dispersion [83]. While agglomeration of SWCNTs in phosphate-buffered saline (PBS) deposited in the bronchi at 24 h (Fig. 5.12a, c) and induced mild lung fibrosis at 30 days after intratracheal instillation in mice (Fig. 5.12e), well-dispersed SWCNTs coated with the polymer distributed whole tissue of the lung with no adverse histopathological features after the instillation (Fig. 5.12b, d, f) and were removed possibly by macrophages within 90 days (Fig. 5.12g–i). Inflammatory responses of the airway were also suppressed by coating SWCNTs with the polymer prior to the instillation [83]. The use of α-tocopheryl polyethylene glycol succinate was also proposed for surface modification of MWCNT to improve their dispersity states and biocompatibility [88].

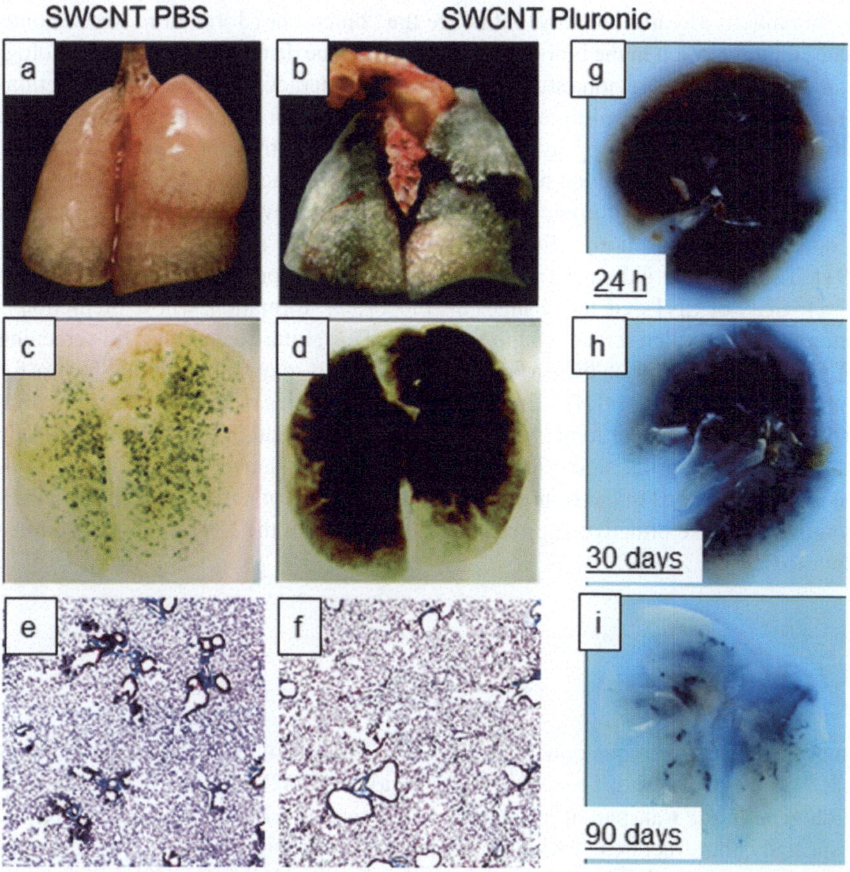

Fig. 5.12 Intrapulmonary distribution clearance of SWCNTs dispersed with and without polymer (PEG-*b*-PPA-*b*-PEG) coating and their effect. a, **b** Photographs of lungs inflated to 4% paraformaldehyde immediately after harvest at 24 h after intratracheal instillation of SWCNTs (**a**) aggregated in phosphate-buffered saline (PBS) or **b** dispersed by coating with the polymer (Pluronic). **c**, **d** Photographs of lungs after optically tissue clearing by salicylic acid and ethanol showing the distribution of the SWCNTs. **e**, **f** Photomicrographs of Trichrome stained sections of lungs collected at 30 days after the instillation. **g**–**i** Paraffin blocks containing fixed embedded lungs that were collected at **g** 24 h, **h** 30 days, and **i** 90 days after the instillation of SWCNTs dispersed with the polymer (Pluronic). Adapted from [83] with permission from American Chemical Society

5.6.2 Contribution of Protein Adsorption Forming "Biocorona"

Modification of the CNT surface properties also alters the profile of in vivo protein adsorption, also called "biocorona" formation, which may also change their in vivo behavior and biological effects. The properties such as surface charge and hydrophilicity (hydrophobicity) influence the "biocorona" formation, which dominates their retention in the blood circulation and tissue distribution as well as biological effects [89]. A previous study comparing the interaction of native and functionalized CNTs with serum proteins suggested that the corona formation via protein adsorption is influenced by hydrogen bonding and electrostatic interactions rather than hydrophobic interaction and pi-stacking [90]. Surface coating of SWCNTs by a nonionic surfactant, Pluronic F127, did not only improve their dispersion but also reduce protein adsorption, cellular uptake, inflammatory response, and cytotoxicity [91]. The evidence indicates that the surface modification can control the function and toxicity of CNTs via the changes in "biocorona" formation, whose prediction is attempted by in silico screening approach [92]. Thus, future techniques for screening of molecular behavior on the surface of nanomaterials may find important factors for enhancing the safety of CNTs. In the case of possibilities of the application to drug deliverly, the balance of toxicity of the carrier (CNTs) and the suppression efficacy of using CNTs on the side effects of native drugs by delivery targeting [88] will be further discussed in the future. Hence, the challenge of surface modification will continue to make effective approaches to functionalize the CNTs and to minimize their toxicity.

5.7 Summary

In this chapter, the unique optical properties, in vivo behavior, and biological effects of CNTs were reviewed. The OTN-NIR emission of SWCNTs is useful for observing biological tissue transparent to look into deep tissues without physical sectioning. Especially, the structures of whole-body blood vessels, brain vasculatures, gastrointestinal tract, and other internal organs can be visualized by OTN-NIR fluorescence imaging with SWCNTs as a contrast agent. Dynamic imaging is also realized by using the intrinsic fluorescence of SWCNTs. While further biomedical applications of CNTs are expected, their unique in vivo behavior and effects including toxicity have been concerned, as first mentioned by similarity in shape to asbestos composed of nanostructured long and thin fibrous crystals. There are actually lots of things that can be learnt from asbestos; however, the biological responses to asbestos and CNTs should be discussed carefully because they have different chemical compositions and physicochemical properties and induce different biologically mechanistic pathways. In addition to the lysosomal instability during frustrated phagocytosis of CNTs, abnormal intracellular behavior of monodispersed SWCNTs was also found recently. Frustrated phagocytosis was observed as an effect of long MWCNTs but not by shorter (1–5 μm length) MWCNTs. The in vivo behavior including blood retention, tissue distribution, and clearance of CNTs and their biological effects including toxicity are dependent on their shape, mechanical properties, and surface properties that can be controlled by chemical modification. We hope the knowledge reviewed here will facilitate valuable, functional, and safe biomedical application of CNTs with unique physicochemical properties.

References

1. S. Iijima, Helical microtubules of graphitic carbon. Nature **354**, 56 (1991). https://doi.org/10.1038/354056a0
2. A. Thess, R. Lee, P. Nikolaev et al., Crystalline Ropes of Metallic Carbon Nanotubes. Science **276**, 483 (1996). https://doi.org/10.1126/science.273.5274.483
3. W.Z. Li, S.S. Xie, L.X. Qian et al., Large-scale synthesis of aligned carbon nanotubes. Science **274**, 1701 (1996). https://doi.org/10.1126/science.274.5293.1701
4. T.W. Ebbesen, P.M. Ajayan, Large-scale synthesis of carbon nanotubes. Nature **358**, 220 (1992). https://doi.org/10.1038/358220a0
5. C. Farrera, F. Torres Andon, N. Feliu, Carbon nanotubes as optical sensors in biomedicine. ACS Nano. **11**, 10637 (2017). https://doi.org/10.1021/acsnano.7b06701
6. E. Hemmer, A. Benayas, F. Légaré et al., Exploiting the biological windows: current perspectives on fluorescent bioprobes emitting above 1000 nm. Nanoscale Horiz **1**, 168 (2016). https://doi.org/10.1039/C5NH00073D
7. S.M. Bachilo, C. Kittrell, R.H. Hauge et al., Structure-assigned optical spectra of single-walled carbon nanotubes. Science **298**, 2361 (2002). https://doi.org/10.1126/science.1078727
8. A.M. Smith, M.C. Mancini, S. Nie, Bioimaging: second window for in vivo imaging. Nat. Nanotechnol. **4**, 710 (2009). https://doi.org/10.1038/nnano.2009.326
9. S. Ghosh, S.M. Bachilo, R.B. Weisman, Advanced sorting of single-walled carbon nanotubes by nonlinear density-gradient ultracentrifugation. Nat. Nanotechnol. **5**, 443 (2010). https://doi.org/10.1038/nnano.2010.68
10. H. Liu, D. Nishide, T. Tanaka et al., Large-scale single-chirality separation of single-wall carbon nanotubes by simple gel chromatography. Nat. Commun. **2**, 309 (2011). https://doi.org/10.1038/ncomms1313
11. S. Ghosh, R.A. Simonette, K.M. Beckingham et al., Oxygen doping modifies near-infrared band gaps in fluorescent single-walled carbon nanotubes. Science **330**, 1656 (2010). https://doi.org/10.1126/science.1196382
12. Y. Iizumi, M. Yudasaka, J. Kim et al., Oxygen-doped carbon nanotubes for near-infrared fluorescent labels and imaging probes. Sci. Rep. **8**, 6272 (2018). https://doi.org/10.1038/s41598-018-24399-8
13. D.A. Heller, S. Baik, T.E. Eurell et al., Single-walled carbon nanotube spectroscopy in live cells: towards long-term labels and optical sensors. Adv. Mater. **17**, 2793 (2005). https://doi.org/10.1002/adma.200500477
14. S. Sekiyama, M. Umezawa, Y. Iizumi et al., Delayed increase in near-infrared fluorescence in cultured murine cancer cells labeled with oxygen-doped single-walled carbon nanotubes. Langmuir **35**, 831 (2019). https://doi.org/10.1021/acs.langmuir.8b03789
15. K. Welsher, Z. Liu, S.P. Sherlock et al., A route to brightly fluorescent carbon nanotubes for near-infrared imaging in mice. Nat. Nanotechnol. **4**, 773 (2009). https://doi.org/10.1038/nnano.2009.294
16. D. Jaque, C. Richard, B. Viana et al., Inorganic nanoparticles for optical bioimaging. Adv. Opt. Photonics **8**, 1 (2016). https://doi.org/10.1364/AOP.8.000001
17. G. Hong, S. Diao, J. Chang et al., Through-skull fluorescence imaging of the brain in a new near-infrared window. Nat. Photonics **8**, 723 (2014). https://doi.org/10.1038/nphoton.2014.166
18. X. Dang, L. Gu, J. Qi, et al., Layer-by-layer assembled fluorescent probes in the second near-infrared window for systemic delivery and detection of ovarian cancer. Proc. Natl. Acad. Sci. USA. **113**, 5179 (2016). https://doi.org/10.1073/pnas.1521175113
19. J.T. Robinson, G. Hong, Y. Liang et al., In vivo fluorescence imaging in the second near-infrared window with long circulating carbon nanotubes capable of ultrahigh tumor uptake. J. Am. Chem. Soc. **134**, 10664 (2012). https://doi.org/10.1021/ja303737a
20. K. Welsher, S.P. Sherlock, H. Dai, Deep-tissue anatomical imaging of mice using carbon nanotube fluorophores in the second near-infrared window. Proc. Natl. Acad. Sci. USA. **108**, 8943 (2011). https://doi.org/10.1073/pnas.1014501108

21. J.T. Robinson, S.M. Tabakman, Y. Liang et al., Ultrasmall reduced graphene oxide with high near-infrared absorbance for photothermal therapy. J. Am. Chem. Soc. **133**, 6825 (2011). https://doi.org/10.1021/ja2010175
22. H. Yi, D. Ghosh, M.H. Ham et al., M13 phage-functionalized single-walled carbon nanotubes as nanoprobes for second near-infrared window fluorescence imaging of targeted tumors. Nano Lett. **12**, 1176 (2012). https://doi.org/10.1021/nl2031663
23. C. Liang, S. Diao, C. Wang et al., Tumor metastasis inhibition by imaging-guided photothermal therapy with single-walled carbon nanotubes. Adv. Mater. **26**, 5646 (2014). https://doi.org/10.1002/adma.201401825
24. G. Hong, J.C. Lee, J.T. Robinson et al., Multifunctional in vivo vascular imaging using near-infrared II fluorescence. Nat. Med. **18**, 1841 (2012). https://doi.org/10.1038/nm.2995
25. G. Hong, J.C. Lee, A. Jha et al., Near-infrared II fluorescence for imaging hindlimb vessel regeneration with dynamic tissue perfusion measurement. Circ. Cardiovasc. Imaging **7**, 517 (2014). https://doi.org/10.1161/CIRCIMAGING.113.000305
26. M. Yudasaka, Y. Yomogida, M. Zhang et al., Near-infrared photoluminescent carbon nanotubes for imaging of brown fat. Sci. Rep. **7**, 44760 (2017). https://doi.org/10.1038/srep44760
27. K. Welsher, Z. Liu, D. Daranciang et al., Selective probing and imaging of cells with single walled carbon nanotubes as near-infrared fluorescent molecules. Nano Lett. **8**, 586 (2008). https://doi.org/10.1021/nl072949q
28. N.M. Iverson, P.W. Barone, M. Shandell et al., In vivo biosensing via tissue-localizable near-infrared-fluorescent single-walled carbon nanotubes. Nat. Nanotechnol. **8**, 873 (2013). https://doi.org/10.1038/nnano.2013.222
29. N. Saito, H. Haniu, Y. Usui et al., Safe clinical use of carbon nanotubes as innovative biomaterials. Chem. Rev. **114**, 6040 (2014). https://doi.org/10.1021/cr400341h
30. E. Durantie, D. Vanhecke, L. Rodriguez-Lorenzo et al., Biodistribution of single and aggregated gold nanoparticles exposed to the human lung epithelial tissue barrier at the air-liquid interface. Part. Fibre Toxicol. **14**, 49 (2017). https://doi.org/10.1186/s12989-017-0231-3
31. D.M. Brown, I.A. Kinloch, U. Bangert et al., An in vitro study of the potential of carbon nanotubes and nanofibres to induce inflammatory mediators and frustrated phagocytosis. Carbon **45**, 1743 (2007). https://doi.org/10.1016/j.carbon.2007.05.011
32. C.A. Poland, R. Duffin, I. Kinloch et al., Carbon nanotubes introduced into the abdominal cavity of mice show asbestos-like pathogenicity in a pilot study. Nat. Nanotechnol. **3**, 423 (2008). https://doi.org/10.1038/nnano.2008.111
33. K. Soga, M. Kamimura, K. Okubo et al., Near-infrared biomedical imaging for transparency. Journal of the Imaging Society of Japan **58**, 602 (2019). https://doi.org/10.11370/isj.58.602
34. L. Ma-Hock, V. Strauss, S. Treumann et al., Comparative inhalation toxicity of multi-wall carbon nanotubes, graphene, graphite nanoplatelets and low surface carbon black. Part. Fibre Toxicol. **10**, 23 (2013). https://doi.org/10.1186/1743-8977-10-23
35. K. Yan, F. Zhao, S. Li et al., Low-toxic and safe nanomaterials by surface-chemical design, carbon nanotubes, fullerenes, metallofullerenes, and graphenes. Nanoscale **3**, 362 (2011). https://doi.org/10.1039/c0nr00647e
36. T.L. Moore, J.E. Pitzer, R. Podila et al., Multifunctional polymer-coated carbon nanotubes for safe drug delivery. Part. Part. Syst. Charact. **30**, 365 (2013). https://doi.org/10.1002/ppsc.201200145
37. J.J. Castillo, T. Rindzevicius, L.V. Novoa et al., Non-covalent conjugates of single-walled carbon nanotubes and folic acid for interaction with cells over-expressing folate receptors. J. Mater. Chem. B **1**, 1475 (2013). https://doi.org/10.1039/C2TB00434H
38. Y. Iizumi, T. Okazaki, Y. Ikehara et al., Immunoassay with single-walled carbon nanotubes as near-infrared fluorescent labels. ACS Appl. Mater. Interfaces. **5**, 7665 (2013). https://doi.org/10.1021/am401702q
39. J.H. Kim, D.A. Heller, H. Jin et al., The rational design of nitric oxide selectivity in single-walled carbon nanotube near-infrared fluorescence sensors for biological detection. Nat. Chem. **1**, 473 (2009). https://doi.org/10.1038/nchem.332

40. J. Zhang, M.P. Landry, P.W. Barone et al., Molecular recognition using corona phase complexes made of synthetic polymers adsorbed on carbon nanotubes. Nat. Nanotechnol. **8**, 959 (2013). https://doi.org/10.1038/nnano.2013.236
41. S. Kruss, M.P. Landry, E.V. Ende et al., Neurotransmitter detection using corona phase molecular recognition on fluorescent single-walled carbon nanotube sensors. J. Am. Chem. Soc. **136**, 713 (2014). https://doi.org/10.1021/ja410433b
42. Q. Zhang, H. Zhou, B. Yan, Reducing nanotube cytotoxicity using a nano-combinatorial library approach. Methods Mol. Biol. **625**, 95 (2010). https://doi.org/10.1007/978-1-60761-579-8_9
43. M. Pannuzzo, S. Esposito, L.P. Wu et al., Overcoming nanoparticle-mediated complement activation by surface PEG pairing. Nano Lett. **20**, 4312 (2020). https://doi.org/10.1021/acs.nanolett.0c01011
44. W.G. Kreyling, M. Semmler-Behnke, J. Seitz et al., Size dependence of the translocation of inhaled iridium and carbon nanoparticle aggregates from the lung of rats to the blood and secondary target organs. Inhal. Toxicol. **21**(Suppl 1), 55 (2009). https://doi.org/10.1080/08958370902942517
45. W.G. Kreyling, M. Semmler, F. Erbe et al., Translocation of ultrafine insoluble iridium particles from lung epithelium to extrapulmonary organs is size dependent but very low. J. Toxicol. Environ. Health A **65**, 1513 (2002). https://doi.org/10.1080/00984100290071649
46. G. Oberdörster, Z. Sharp, V. Atudorei et al., Extrapulmonary translocation of ultrafine carbon particles following whole-body inhalation exposure of rats. J. Toxicol. Environ. Health A **65**, 1531 (2002). https://doi.org/10.1080/00984100290071658
47. H.S. Choi, Y. Ashitate, J.H. Lee et al., Rapid translocation of nanoparticles from the lung airspaces to the body. Nat. Biotechnol. **28**, 1300 (2010). https://doi.org/10.1038/nbt.1696
48. P. Wick, A. Malek, P. Manser et al., Barrier capacity of human placenta for nanosized materials. Environ. Health Perspect. **118**, 432 (2010). https://doi.org/10.1289/ehp.0901200
49. K. Takeda, K. Suzuki, A. Ishihara et al., Nanoparticles transferred from pregnant mice to their offspring can damage the genital and cranial nerve systems. J. Health Sci. **55**, 95 (2009). https://doi.org/10.1248/jhs.55.95
50. K. Yamashita, Y. Yoshioka, K. Higashisaka et al., Silica and titanium dioxide nanoparticles cause pregnancy complications in mice. Nat. Nanotechnol. **6**, 321 (2011). https://doi.org/10.1038/nnano.2011.41
51. T. Notter, L. Aengenheister, U. Weber-Stadlbauer et al., Prenatal exposure to TiO_2 nanoparticles in mice causes behavioral deficits with relevance to autism spectrum disorder and beyond. Transl. Psychiatry **8**, 193 (2018). https://doi.org/10.1038/s41398-018-0251-2
52. A. Onoda, K. Takeda, M. Umezawa, Dose-dependent induction of astrocyte activation and reactive astrogliosis in mouse brain following maternal exposure to carbon black nanoparticle. Part. Fibre. Toxicol. **14**, 4 (2017). https://doi.org/10.1186/s12989-017-0184-6
53. M. Umezawa, A. Onoda, I. Korshunova et al., Maternal inhalation of carbon black nanoparticles induces neurodevelopmental changes in mouse offspring. Part. Fibre. Toxicol. **15**, 36 (2018). https://doi.org/10.1186/s12989-018-0272-2
54. Y. Hao, J. Liu, Y. Feng et al., Molecular evidence of offspring liver dysfunction after maternal exposure to zinc oxide nanoparticles. Toxicol. Appl. Pharmacol. **329**, 318 (2017). https://doi.org/10.1016/j.taap.2017.06.021
55. N. Bara, M. Eshwarmoorthy, K. Subaharan K, et al., Mesoporous silica nanoparticle is comparatively safer than zinc oxide nanoparticle which can cause profound steroidogenic effects on pregnant mice and male offspring exposed in utero. Toxicol. Ind. Health. **34**, 507 (2018). https://doi.org/10.1177/0748233718757641
56. Z.O. Kyjovska, A.M.Z. Boisen, P. Jackson et al., Daily sperm production: application in studies of prenatal exposure to nanoparticles in mice. Reprod. Toxicol. **36**, 88 (2013). https://doi.org/10.1016/j.reprotox.2012.12.005
57. S. Yoshida, K. Hiyoshi, S. Oshio et al., Effects of fetal exposure to carbon nanoparticles on reproductive function in male offspring. Fertil. Steril. **93**, 1695 (2010). https://doi.org/10.1016/j.fertnstert.2009.03.094

58. M. Riediker, D. Zink, W. Kreyling et al., Particle toxicology and health—where are we? Part. Fibre. Toxicol. **16**, 19 (2019). https://doi.org/10.1186/s12989-019-0302-8
59. J.P. Ryman-Rasmussen, M.F. Cesta, A.R. Brody et al., Inhaled Carbon Nanotubes Reach the Subpleural Tissue in Mice. Nat. Nanotechnol. **4**, 747 (2009). https://doi.org/10.1038/nnano.2009.305
60. K. Donaldson, C.A. Poland, New insights into nanotubes. Nat. Nanotechnol. **4**, 708 (2009). https://doi.org/10.1038/nnano.2009.327
61. R. Singh, D. Pantarotto, L. Lacerda. et al., Tissue biodistribution and blood clearance rates of intravenously administered carbon nanotube radiotracers. Proc. Natl. Sci. USA, **103**, 3357 (2006). https://doi.org/10.1073/pnas.0509009103
62. S.T. Yang, K.A.S. Fernando, J.H. Liu et al., Covalently PEGylated carbon nanotubes with stealth character in vivo. Small **4**, 940 (2008). https://doi.org/10.1002/smll.200700714
63. Z. Liu, W. Cai, L. He et al., In vivo biodistribution and highly efficient tumour targeting of carbon nanotubes in mice. Nat. Nanotechnol. **2**, 47 (2007). https://doi.org/10.1038/nnano.2006.170
64. P. Cherukuri, C.J. Gannon, T.K. Leeuw, et al., Mammalian pharmacokinetics of carbon nanotubes using intrinsic near-infrared fluorescence. Proc. Natl. Acad. Sci. USA, **103**, 18882 (2006). https://doi.org/10.1073/pnas.0609265103
65. Z. Liu, C. Davis, W. Cai, et al., Circulation and long-term fate of functionalized, biocompatible single-walled carbon nanotubes in mice probed by Raman spectroscopy. Proc. Natl. Acad. Sci. USA. **105**, 1410 (2008). https://doi.org/10.1073/pnas.0707654105
66. S. Jain, V.S. Thakare, M. Das et al., Toxicity of multiwalled carbon nanotubes with end defects critically depends on their functionalization density. Chem. Res. Toxicol. **24**, 2028 (2011). https://doi.org/10.1021/tx2003728
67. X. Deng, G. Jia, H. Wang et al., Translocation and fate of multi-walled carbon nanotubes in vivo. Carbon **45**, 1419 (2007). https://doi.org/10.1016/j.carbon.2007.03.035
68. H. Wang, J. Wang, X. Deng et al., Biodistribution of carbon single-wall carbon nanotubes in mice. J. Nanosci. Nanotechnol. **4**, 1019 (2004). https://doi.org/10.1166/jnn.2004.146
69. J.E. Riviere, Pharmacokinetics of nanomaterials: an overview of carbon nanotubes, fullerenes and quantum dots. Wiley Interdiscip. Rev. Nanomed. Nanobiotechnol. **1**, 26 (2009). https://doi.org/10.1002/wnan.24
70. T. Takeuchi, Y. Iizumi, M. Yudasaka et al., Characterization and biodistribution analysis of oxygen-doped single-walled carbon nanotubes used as in vivo fluorescence imaging probes. Bioconjugate Chem. **30**, 1323 (2019). https://doi.org/10.1021/acs.bioconjchem.9b00088
71. H. Mao, N. Kawazoe, G. Chen, Uptake and intracellular distribution of collagen-functionalized single-walled carbon nanotubes. Biomaterials **34**, 2472 (2013). https://doi.org/10.1016/j.biomaterials.2013.01.002
72. K. Kostarelos, L. Lacerda, G. Pastorin et al., Cellular uptake of functionalized carbon nanotubes is independent of functional group and cell type. Nat. Nanotechnol. **2**, 108 (2007). https://doi.org/10.1038/nnano.2006.209
73. H. Jin, D.A. Heller, R. Sharma et al., Size-dependent cellular uptake and expulsion of single-walled carbon nanotubes: single particle tracking and a generic uptake model for nanoparticles. ACS Nano **3**, 149 (2009). https://doi.org/10.1021/nn800532m
74. F. Tian, D. Cui, H. Schwarz et al., Cytotoxicity of single-wall carbon nanotubes on human fibroblasts. Toxicol. In Vitro **20**, 1202 (2006). https://doi.org/10.1016/j.tiv.2006.03.008
75. L.A. Mitchell, F.T. Lauer, S.W. Burchiel et al., Mechanisms for how inhaled multiwalled carbon nanotubes suppress systemic immune function in mice. Nat. Nanotechnol. **4**, 451 (2009). https://doi.org/10.1038/nnano.2009.151
76. A. Erdely, T. Hulderman, R. Salmen, et al., Cross-talk between lung and systemic circulation during carbon nanotube respiratory exposure. Potential biomarkers. Nano Lett. **9**, 36 (2009). https://doi.org/10.1021/nl801828z
77. I. Lynch, Far-reaching effects from carbon nanotubes. Nat. Nanotechnol. **14**, 639 (2019). https://doi.org/10.1038/s41565-019-0477-z

78. V.C. Sanchez, J.R. Pietruska, N.R. Miselis et al., Biopersistence and potential adverse health impacts of fibrous nanomaterials: what have we learned from asbestos? Wiley Interdiscip. Rev. Nanomed. Nanobiotechnol. **1**, 511 (2009). https://doi.org/10.1002/wnan.41
79. A. Takagi, A. Hirose, T. Nishimura et al., Induction of mesothelioma in p53$^{+/-}$ mouse by intraperitoneal application of multi-wall carbon nanotube. J. Toxicol. Sci. **33**, 105 (2008). https://doi.org/10.2131/jts.33.105
80. J. Muller, F. Huaux, N. Moreau et al., Respiratory toxicity of multi-wall carbon nanotubes. Toxicol. Appl. Pharmacol. **207**, 221 (2005). https://doi.org/10.1016/j.taap.2005.01.008
81. H. Nagai, T. Toyokuni, Biopersistent fiber-induced inflammation and carcinogenesis: lessons learned from asbestos toward safety of fibrous nanomaterials. Arch. Biochem. Biophys. **502**, 1 (2010). https://doi.org/10.1016/j.abb.2010.06.015
82. J.M. Dymacek, B.N. Snyder-Talkington, R. Raese et al., Similar and differential canonical pathways and biological processes associated with multiwalled carbon nanotube and asbestos-induced pulmonary fibrosis: a 1-year postexposure study. Int. J. Toxicol. **37**, 276 (2018). https://doi.org/10.1177/1091581818779038
83. G.M. Mutlu, G.R. Budinger, A.A. Green et al., Biocompatible nanoscale dispersion of single-walled carbon nanotubes minimizes in vivo pulmonary toxicity. Nano Lett. **10**, 1664 (2010). https://doi.org/10.1021/nl9042483
84. P. Wick, P. Manser, L.K. Limbach et al., The degree and kind of agglomeration affect carbon nanotube cytotoxicity. Toxicol. Lett. **168**, 121 (2007). https://doi.org/10.1016/j.toxlet.2006.08.019
85. H. Nagai, Y. Okazaki, S.H. Chew, et al., Diameter and rigidity of multiwalled carbon nanotubes are critical factors in mesothelial injury and carcinogenesis. Proc. Natl. Acad. Sci. USA. **108**, E1330 (2011). https://doi.org/10.1073/pnas.1110013108
86. K.K. Jain, Advances in use of functionalized carbon nanotubes for drug design and discovery. Expert Opin. Drug Discov. **7**, 1029 (2012). https://doi.org/10.1517/17460441.2012.722078
87. M. Kamimura, N. Kanayama, K. Tokuzen et al., Near-infrared (1550 nm) in vivo bioimaging based on rare-earth doped ceramic nanophosphors modified with PEG-b-poly (4-vinylbenzylphosphonate). Nanoscale **3**, 3705 (2011). https://doi.org/10.1039/C1NR10466G
88. R.P. Singh, G. Sharma, Sonali, et al., Effects of transferrin conjugated multi-walled carbon nanotubes in lung cancer delivery. Mater. Sci. Eng. C Mater. Biol. Appl., **67**, 313 (2016). https://doi.org/10.1016/j.msec.2016.05.013
89. Y. Teow, P.V. Asharani, M.P. Hande et al., Health impact and safety of engineered nanomaterials. Chem. Commun. (Camb.) **47**, 7025 (2011). https://doi.org/10.1039/c0cc05271j
90. J.H. Shannahan, J.M. Brown, R. Chen et al., Comparison of nanotube-protein corona composition in cell culture media. Small **9**, 2171 (2013). https://doi.org/10.1002/smll.201202243
91. D. Dutta, S.K. Sundaram, J.G. Teeguarden et al., Adsorbed proteins influence the biological activity and molecular targeting of nanomaterials. Toxicol. Sci. **100**, 303 (2007). https://doi.org/10.1093/toxsci/kfm217
92. H. Zhou, Q. Mu, N. Gao et al., A Nano-combinatorial library strategy for the discovery of nanotubes with reduced protein-binding, cytotoxicity, and immune response. Nano Lett. **8**, 859 (2008). https://doi.org/10.1021/nl0730155

Chapter 6
Rare-Earth-Doped Ceramic Nanoparticles for Transparency in the Biomedical Field

Karina Nigoghossian

Abstract This chapter brings a comprehensive overview on rare-earth-doped ceramic nanoparticles (RED-CNPs) for biomedical applications, which have attracted much attention over the past two decades. Their appealing luminescent features are related to intraconfigurational 4f–4f electronic transitions, such as narrow luminescence peaks, high resistance to photobleaching, and long lifetimes. In this sense, RED-CNPs offer multiple applications in the near-infrared (NIR) range, such as time-gated imaging, magnetic resonance imaging (MRI), nanothermometry, and NIR-activated photodynamic therapy (PDT). Moreover, RED-CNPs are promising for building multifunctional nanosystems by integrating into a single platform luminophores, therapeutic agents, and targeting biomolecules. First, a general introduction is given on how RED-CNPs emerged in the nanomedicine field. A theoretical background on RE electronic transitions leads to the main issues for achieving efficient photoluminescence processes. Insights for developing RED-CNPs probes are provided in terms of material design and synthetic methods. Finally, potentialities for biomedical applications are highlighted.

Keywords Rare-earth ions · Ceramic nanoparticles · Luminescence · Nanomedicine · Bioimaging

6.1 Introduction

RED-CNPs offer transparency to biomedical imaging in the NIR range. The unique photophysical properties of trivalent rare-earth (RE) ions make these elements very attractive for designing luminescent probes. The RE valence electrons in 4f orbitals are shielded by $5s^2$ and $5p^6$ filled orbitals from changes in the surroundings. Thus, the

K. Nigoghossian (✉)
Department of Materials Science and Technology, Faculty of Industrial Science and Technology, Tokyo University of Science, 6-3-1 Niijuku, Katsushika-ku, Tokyo 125-8585, Japan
e-mail: karina.nig@gmail.com

ICGM, Univ. Montpellier, CNRS, ENSCM, UM, Bât 17, cc1701, Place Eugène Bataillon, 34095 Montpellier, France

K. Soga et al. (eds.), *Transparency in Biology*,
https://doi.org/10.1007/978-981-15-9627-8_6

dynamic and static electric fields caused by local atoms induce weak effects on the shielded electrons. As a result of such shielding effect, 4f–4f transitions possess long lifetimes, narrow emission peaks, and high resistance to photobleaching. The large number of discrete energy levels within 4f orbitals are responsible for transitions lines covering a broad wavelength range from the ultraviolet to the infrared region. The presence of diverse energy levels and narrow excitation and emission lines enables tuning the excitation and emission wavelengths by changing the activator ions and their relative concentrations [1, 2]. The possibility to discriminate the sharp spectral lines among different REs enables the rational design of a multiplexed biological labeling under a single excitation wavelength. The imaging capability can be therefore extended to simultaneous tracking of multiple molecular targets, allowing the classification and differentiation of complex human diseases [3]. The long radiative lifetimes (μs–ms) eliminate the autofluorescence effect, *i.e.,* simultaneous emission (ns) from both analyte and surroundings, and thus improve sensitivity of detection [4].

In particular cases, for certain RE dopants at suitable concentrations, successive levels have similar energy gap and form a ladder-like energy level diagram. Such metastable states have appropriate lifetime to store energy and enable the sequential absorption of two or more pump photons. Consecutive higher energy levels are stepwise populated. A subsequent emission of a single photon occurs, with higher energy compared with excitation photons individually. This nonlinear phenomenon is known as upconversion luminescence (UCL). There are several UCL mechanisms involving multiple photons that may lead to anti-Stokes emission. Among these mechanisms, the stepwise one, known as energy transfer upconversion (ETU), is considered the most efficient, as described by Auzel in 2004 [5]. Although anti-Stokes emission can also be observed in other luminescent materials, such as organic dyes [6–8] and quantum dots (QDs) [9, 10], a costly high-power pulsed laser is necessary to provide high photon density, and thus promote the simultaneous absorption of two-photos. In the case of RE ions, UCL is a multi-step excitation process, achievable by using a low power continuous-wave (CW)-NIR laser, due to characteristic long lifetimes of metastable levels.

The advances in nanotechnology have enabled the miniaturization of RE-based luminophores to sub-micron [11, 12] or nanoscale levels (<50 nm) [13]. The synthetic protocols of RED-CNPs have emerged in the early 2000s, along with the proposal of a novel class of bright and photostable probes for biomedical applications [14–16]. The early reports on RED-CNPs as bioprobes for biological labeling and imaging were elaborated to operate in the NIR-to-visible regime to overcome the issues related to conventional UV–visible excited fluorescent contrast agents [17]. As already introduced in Part I of this book, the use of NIR excitation enables deeper tissue penetration. The lower energy of NIR with respect to UV–visible light results in lower photo-damage, and a sensitive detection without the usual interference from autofluorescence background. Since then, the upconverting RED-CNPs has gained great attention for the development of multiplexed biosensing [18], diagnosis [19], and therapeutics [20, 21].

Despite the excellent penetration depth of 980-nm excitation laser, up to 1 cm in biological tissues, the detection of emitted visible light is limited to less than 2.5 mm [22]. NIR-emitting RED-CNPs were neglected in the early 2000s [23]. At that time, the lack of cameras for detection of NIR light of output signal limited the implementation of NIR-to-NIR bioimaging methods. Around 2005, InGaAs camera capable to detect light in the region over-1000-nm (OTN) became commercially available [23]. Since then, the nanoparticle design was rationally adjusted to shift the operational range to the NIR-to-NIR region, as well as development of devices for OTN-NIR fluorescence in vivo bioimaging system [23]. The use of NIR light for both pump and output, has opened the possibility to improve the observation depth from less than a millimeter to several centimeters. Furthermore, the downshifted OTN-NIR luminescence is much more efficient when compared to the UCL, because it involves a single photon for excitation, while the UCL requires multiple steps for excitation. Besides being less suitable for bioimaging, the NIR-to-visible UCL is still useful for delivering visible light to deep tissue layers to locally activate photosensitizers agents in PDT processes [24], and to trigger drug release [25]. Simultaneous imaging and PDT treatment can be performed by a single NIR-excited nanoplatform based on RED-CNPs emission at different wavelength ranges. The aim of this chapter is reviewing the contribution of RED-CNPs to NIR biophotonics.

6.2 Theoretical Background

6.2.1 Properties of 4f–4f Transitions and Design Guideline

RED-CNPs consist of a crystalline host doped with RE ions. In some cases, luminescence by direct excitation of RE dopant is not efficient because of the shielding effect on partially filled 4f orbitals. To overcome such low absorption coefficient, a RE ion with larger absorptivity is co-doped in the host lattice. This sensitizer ion maximizes the light absorption and transfers the excitation energy to the activator ion via resonant energy levels. The excited activator ion may undergo two alternative pathways to return to the ground state, whether in a non-radiative process by releasing the excitation energy as heat or go through a radiative pathway, which results in luminescence [26]. Such processes of heat release and light emission are competitive. In terms of quantum mechanics, these phenomena are related to electronic transitions between energy levels for emitting phonons and photons. Therefore, the design of RED-CNPs is basically aimed at emitting light instead of heat.

For RE ions, the non-radiative relaxation between electronic states is described as multiphonon emission. The phonon energy is related to the thermal vibrations of atoms surrounding optically active ions, which provide a field of phonons. The multiphonon relaxation rate (W_{MPR}) can be approximated as:

$$W_{MPR} = B\exp\left(-\frac{\Delta E}{\hbar\omega}\right),$$

where $\hbar\omega$ is the effective phonon energy of the host material, and ΔE is the energy separation between the excited energy level and the next lower one [27]. Essentially, this equation is showing that the rate of multiphonon relaxation is sensitive to the number of phonons required to bridge the energy gap. The closer the levels are in energy, the lower is the number of phonons necessary for relaxing the upper level to the lower-lying one, and higher are chances of releasing energy to the lattice non-radiatively as phonons. Note that, for NIR luminescence, the non-radiative deactivation has a greater impact in comparison with visible or UV transitions, because of the shorter energy distance between levels.

The B parameter is related to the electron–phonon interaction, *i.e.*, strength of interaction between crystal lattice vibrations and dopants. The B contribution to non-radiative process is not much discussed, since it is expected to be small for RE ions because of the shielding effect on 4f electrons. However, for emerging probes based on NIR-emitting RED-CNPs, the electron–phonon coupling becomes a relevant issue. The strong coupling is related to the small energy gap between the lattice vibrational modes and NIR transitions involved in photoexcitation and light emission, and thereby low efficiency of luminescence is expected [28]. Furthermore, other effects may be observed as a consequence of local ligand field in terms of spectral position, shape, and intensity. Changings in the local environment can increase the ΔE between electronic levels and shift the center of gravity of a transition line. Also, the lower the coupling strength between the electron and the vibrating surroundings can result in narrower bandwidths and longer lifetimes.

6.2.2 Host Lattice

The phonon concept introduced above is crucial to designing RED-CNPs-based bioprobes. This approach is clearly illustrated by the works of Jaque et al. [23] and Soga et al. [29]. The authors quantitatively compared a range of inorganic hosts and RE dopants from values found in literature. The possibility for high quantum efficiency is indicated by a W_{MPR} value lower than the typical radiative transition rate (W_{rad}) for REs (~10^3–10^4 s^{-1}). The phonon frequency ($\hbar\omega$) parameter is typically estimated from infrared, Raman, and vibronic spectra [23, 30]. A more comprehensive comparison in terms of dopant selection was demonstrated by Jaque et al. [23]. The discussion, aimed at designing NIR-emitting RED-CNPs, considered 12 different NIR transitions for 2 crystalline hosts (Y_2O_3 and YPO_4). Among the 24 candidates, only 10 were expected to be a suitable luminophore ($W_{MPR} < W_{rad}$). In the work of Soga et al. [29], literature data were compiled to compare the W_{MPR} of several combinations of RE ions and hosts (crystals and glasses). The chloride-based crystalline host ($LaCl_3$, $\hbar\omega = 240$ cm^{-1}) displayed dramatically lower contribution to W_{MPR}, of several orders of magnitude. After chloride, the second-lowest W_{MPR} value was exhibited by the fluoride-based crystalline host (LaF_3, $\hbar\omega = 310\ cm^{-1}$). Indeed, heavy halide-based hosts, such as chlorides, bromides, and iodides, offer low-phonon energies (<300 cm^{-1}). However, such materials are highly hygroscopic, which hampers their choice for designing RED-CNPs [31]. The lower emission efficiency of oxides and phosphates is a consequence of their quite high phonon energies (500–1400 cm^{-1}) [29].

The fluorides, such as $NaYF_4$ and $NaGdF_4$, combine the suitable features related to chemical stability and relatively low-phonon energy (~350 cm^{-1}), and are touted as efficient hosts for RED-CNPs. The RE and Na^+ cations have similar size and thus prevent the formation of stresses and defects in the crystal lattice [32]. $Na(RE)F_4$ crystalizes in two polymorphic forms: cubic (α-) and hexagonal (β-) phases. The crystal phase of RED-CNPs affects the luminescence efficiency. The α-phase crystal has high-symmetry since the cation sites are randomly distributed among rare-earth (RE^{3+}) and Na^+. The corresponding β-phase has two defined cation sites, which render a less symmetric electron cloud [33]. The distortion on the electron cloud results in a strong electron–lattice coupling, which is related to the B parameter. Thus, lower rate of multiphonon relaxation is expected for the β-phase than for the α-one.

6.2.3 Surface Quenchers

A critical issue for RED-CNPs in bioapplications is the thermal quenching of luminescence caused by OH oscillators present in aqueous environment and organic molecules at the particle' surface, which are necessary for keeping the colloidal stability. In general, the discussion on luminescence efficiency of conventional RE-based materials is limited to the phonon energy of the ionic lattices. The formation of covalent bonds is indeed limited by the shielding of valence electrons in RE ions, which hampers the interaction with surrounding atoms. Therefore, the ionic character dominates the metal–ligand interactions involving 4f-orbitals [34]. Differently than bulk materials, the increased surface-to-volume ratio of nanoobjects expose a larger number of RE dopants to molecular quenchers [35, 36]. Investigations on the influence of covalent systems to the multiphonon relaxation found that the vibrational energy of OH and other oscillators (N–H, C–H and C=O) are inversely proportional to the ΔE between energy levels involved in a radiative transition [37]. In this sense, the "phonon" energy in a covalent system is represented by molecular vibrations. The influence is decided by the value $\partial\mu/\partial r$, which corresponds to the dynamic movement alongwith a certain distance (r) of electric dipole moments (μ) [38]. The most efficient quencher is the OH oscillator. The energy of water molecules vibrations, with strong absorption at around 3400 cm^{-1}, are huge when compared with the phonon energy of inorganic crystalline hosts (<1000 cm^{-1}). This means that, for a non-radiative transition across two levels with ΔE to occur, less phonons are required in a covalent system when compared to an ionic one.

From a practical perspective, as the material's size is reduced, from bulk to nanoscale dimensions, the influence of surrounding phonons is extended beyond the host lattice. The highest emission intensity of RED-CNPs is observed in hydrophobic organic solvents, (*e.g.*, cyclohexane), which can be explained by their lower vibrational frequencies. The electron delocalization along the hydrocarbon chain reduces the ionic character of metal interaction with the ligand field, which means a reduced B value. The transfer of RED-CNPs into aqueous environment is accompanied by strong luminescence quenching, which is associated with the stronger ionicity, *i.e.*, a clear dipole moment because of the localization of the charges. Therefore, much lower B value is expected for a hydrophobic solvent with respect to an aqueous one.

However, the role of phonons for RED-CNPs is not always unfavorable for the luminescence efficiency. Optical processes assisted by phonons can emerge for RED-CNPs, since optical properties are not only dependent on "ceramic host", as observed in bulk materials. An interesting phenomenon, observed at the nanoscale, is the thermal enhancement of brightness. Recent works propose that phonons from the vibrating environment at nanomaterials' surface can assist anti-Stokes emissions in visible [36, 39], and NIR [40] ranges. These heat-favorable phonons sustain the emission intensity with rising temperatures. The same thermal factor may increase the luminescence quenching (W_{MPR}). Therefore, developing nanometric bioprobes is a challenging task, given the complexity of its environment and understanding whether heating is favorable or not for luminescence efficiency. Zhou et al. [36] reported an investigation on RED-CNPs' surface engineering for surface-phonon-enhanced UCL. The authors proposed that the oxygen of surrounding molecules chelates to the RED-CNPs' surface dopants. The phonon energy is trapped in such [RE···O] coordination structure at vibration frequencies (510–560 cm^{-1}) higher than the phonon energy of $NaYF_4$ host, as evidenced by infrared and Raman spectra. The "phonon confinement" effect reduces the energy mismatch between levels of sensitizer and activator and facilitates the energy transfer at elevated temperatures. The UCL enhancement of around 2,000-fold was demonstrated for ultrasmall (<10 nm) RED-CNPs based on $NaYF_4$:Yb^{3+}:Tm^{3+} at 453 K. The mechanisms involved in such an unexpected phenomenon remains unclear [40]. Further investigations are indispensable for understanding the viability of RED-CNPs as bioimaging probes.

6.2.4 Dopant Elements

The RE-dopant selection determines the spectral output of RED-CNPs. The spectral profile is not affected by variations in host lattice, size, and morphology of RED-CNPs, because of the weak interactions of electrons with the ligand field [41]. The energy level diagrams shown in Fig. 6.1a are the well-known basis for discussion on which $4f^{N-1}$ unfilled electronic states may lead to the intended optical transitions. Within the RE elements (Fig. 6.1b), the non-optically active ions are Sc^{3+}, Y^{3+}, La^{3+} and Lu^{3+} for not having unpaired 4f electrons, thus usually being components of host materials. Therefore, the lanthanide ions from Ce^{3+} to Yb^{3+} are the

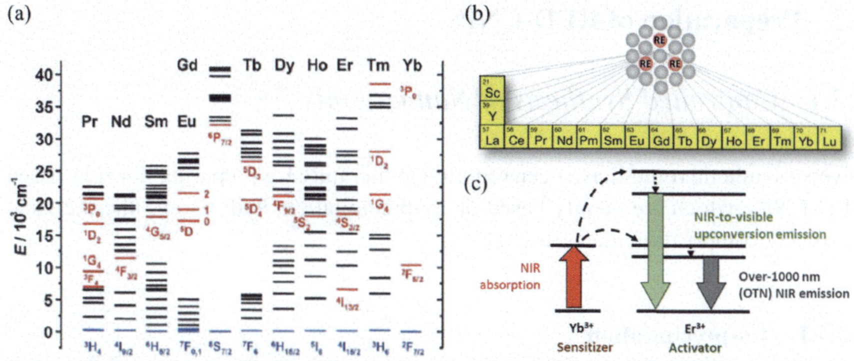

Fig. 6.1 **a** Partial energy level diagrams of optically active RE aquo ions, displaying the main luminescent levels (red) and the fundamental ones (blue). Adapted from [1] with permission from The Royal Society of Chemistry. **b** Schematic representation of a RED-CNP and dopant candidates. **c** Simplified energy level scheme showing an energy transfer from sensitizer (Yb^{3+}) to activator (Er^{3+}) and mechanisms of UCL and OTN-NIR luminescence under NIR excitation

feasible candidates to be doped as luminescent centers, offering electronic transitions at UV, (*e.g.*, Gd^{3+}), visible (*e.g.*, Eu^{3+}, Tb^{3+}, Sm^{3+}, Dy^{3+}), and NIR (*e.g.*, Nd^{3+}, Er^{3+},Yb^{3+}) wavelength ranges [4]. Er^{3+}, Tm^{3+} and Ho^{3+} are the most commonly used activator ions for UCL because of their ladder-like energy levels, which are appropriate for sequential absorption of photons. Yb^{3+} is the most common sensitizer, as it has a single excited level, which is easily populated by lasers operating at 980 nm and matches the intermediate levels of activator ions [31]. Nd^{3+} is an alternative sensitizer in bioprobes for excitation at 800 nm, rather than at 980 nm. The water absorption is significantly lower at 800 nm compared to that at 980 nm, and thereby the overheating of biological tissues is minimized [42, 43]. The optimal activator concentration for maximal luminescence intensity is usually achieved at concentrations low enough to prevent luminescence quenching by dopant–dopant interactions. The relative emission intensities are adjusted by varying the relative dopant concentrations [44]. For instance, the optimal relative molar concentrations for maximum brightness and minimal quenching is around 20% for Yb^{3+} and 2% for Er^{3+} [32]. Such pair of dopants is particularly interesting for designing a multimodal bioimaging system, for visible UCL and OTN-NIR luminescence.

6.3 Preparation of RED-CNPs

6.3.1 Controlled Synthesis of Nanocrystals

Diverse synthetic routes have been reported in literature for preparing fluoride-based RED-CNPs, which are mainly based on co-precipitation, hydro(solvo)thermal, and thermal decomposition methods [31].

6.3.1.1 Co-precipitation

The first reports on synthesis of a fluoride-based RED-CNPs ($NaYF_4$) with reduced size (<50 nm) used co-precipitation routes [14–16]. In this method, the RE ions are co-precipitated in aqueous solutions in presence of a precipitator agent (*e.g.*, NaF) and a size-controlled capping agent (*e.g.*, ethylenediaminetetraacetic acid, EDTA), under a moderate heating (typically 70–90 °C). The controlled synthesis of nanofluorides by co-precipitation requires control over the pH of reaction solution, order of reactant addition, and the relative concentration Na:RE.

The preparation of $NaYF_4:Yb^{3+}:Er^{3+}$ via co-precipitation of RE ions with NaF in presence of the coordinating ligand EDTA was preliminarily reported by Yi et al. [14]. The synthesis resulted in crystalline polymorphs with large particle size and weak UCL intensity. The size was adjusted in the range of 37–166 nm by varying the ratio between RE precursor and coordinating ligand. The thermal annealing of RED-CNPs powder (400–600 °C for 5 h) was necessary to observe some UCL signal, which was directly related to the formation of a hexagonal crystalline phase. Thermal treatment at 400–600 °C induced partial crystal phase transformation from cubic to hexagonal, resulting in a polymorphic form. Further temperature increase to 700 °C led to a pure cubic phase, indicating that the partial hexagonal phase was converted back into cubic phase.

Heer et al. [15] were able to prepare smaller $NaYF_4$ (<30 nm) RED-CNPs by a co-precipitation method using amines as solvents. The obtained fluoride-based RED-CNPs displayed UCL in transparent colloidal solutions with several tens of times higher efficiency than a phosphate-based one. The pH-dependent effect of EDTA on the particle size of $NaYF_4:Yb^{3+}:Tm^{3+}$ nanoparticles were investigated by Wei et al. [45]. The capping agent coordinates to RE ions, thereby binding to the growing nanocrystal surface. The pH value of reactional environment modulates the chelating ability of the capping agent. The strength of such interaction between surface ligand and RE is relevant parameter for modulating the particle size. In acidic medium, at which most of EDTA molecules were partially deprotonated, particles of 42 and 33 nm were generated for the molar ratios EDTA:RE of 1:1 and 2:1, respectively. The neutral and basic conditions favored the presence of fully deprotonated EDTA species, which is desirable for improving its chelating ability and binding strength. The particle size was reduced to 28 nm and 20 nm, as the pH increased to 6.8 and

10.0, respectively. However, the colloidal dispersion was not stable at such basic pH value.

In general, the co-precipitation methods require a post-synthesis annealing to remove water and improve the crystallinity. Despite the removal of OH oscillators (quenchers) being advantageous to enhance the luminescence intensity, the thermal annealing enlarges the crystallite size [46]. Therefore, RED-CNPs obtained by this method presents broad particle size distribution and irregular shape, which limits their biomedical use.

6.3.1.2 Hydro(solvo)thermal

The hydro(solvo)thermal methods are performed in autoclave in order to apply high-pressure to a mixture of RE, Na, and F precursors with coordinating ligands or surfactants. The precipitated products are more crystalline when prepared in autoclave. In the case of hydrothermal reaction, water-dispersible particles are obtained. The solubility of the products can be adjusted to polar or non-polar solvents by varying the ligands. Different sizes and morphologies are obtained by varying the capping agent and relative concentrations of reactants [47]. The drawbacks are related to long reaction times and lack of control over the particle size and shape.

Zheng et al. [48] reported the size- and morphology-controlled synthesis of hexagonal $NaYF_4:Yb^{3+}:Er^{3+}$ RED-CNPs, either hydrothermally or solvothermally, by using different ligands and surfactants. The use of cetyltrimethylammonium bromide (CTAB) favored the formation of nanorods, while EDTA rendered spherical shape. The control over particle size was possible by adjusting the EDTA:RE ratio, which determines the release of RE cations from the coordination structure EDTA-RE and allows separation of nucleation and growth steps. The particles obtained with aid of CTAB were not suitable for bioapplications in terms of size and morphology (nanorods of ~70 nm × 150–175 nm). Particles of 50 nm prepared in ethanol stabilized by EDTA showed bright UCL and potential for biological labeling use.

The preparation of $NaYF_4:Nd^{3+}:Yb^{3+}:Er^{3+}$ RED-CNPs via hydrothermal route using oleic acid as a size- and shape-modulator was investigated by Shang et al. [49]. First, the authors demonstrated the effect of reaction time on gradual phase transfer from cubic (3 h) to hexagonal (6 h and 12 h) phase. The pure hexagonal-phase was achieved by extending the reaction to 24 h. The preferential attachment of oleic acid to some crystal planes favors the anisotropic crystal growth. The particle morphology evolved with ratio of OA:RE, from rods (20:1) to microprisms (30:1 and 40:1). Furthermore, the increase of the molar relation between F and RE precursors (F:RE) rendered diverse sizes and morphologies, such as microspheres (1:1), hexagonal microprisms (2:1 and 3:1), nanorods (5:1). The synthetic parameters (OA:RE and F:RE) and corresponding $NaYF_4$ morphologies are summarized in Fig. 6.2.

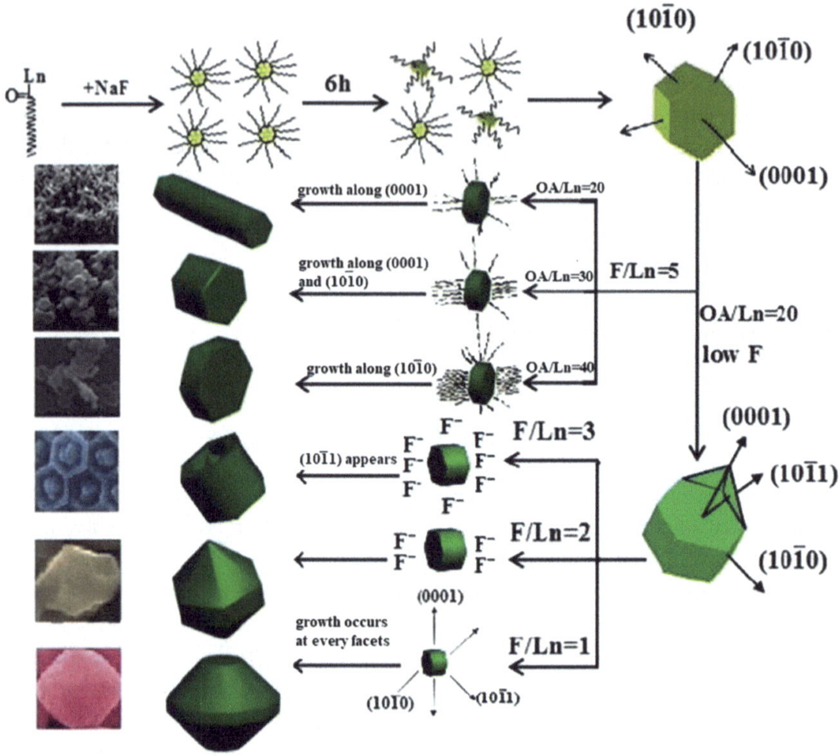

Fig. 6.2 Possible morphologies of $NaYF_4$ RED-CNPs and formation mechanisms according to OA:RE and F:RE molar ratio conditions. Reprinted from [49] with permission from MDPI

6.3.1.3 Thermal Decomposition

The well-established protocols for preparing RED-CNPs with high-quality are based on thermal decomposition method. The method consists of the thermolysis of RE precursors at high temperatures (usually between 280 and 320 °C) in high boiling point solvents. This method enables preparing nanocrystals with homogeneous size distribution, morphology, crystal phase, and stoichiometric composition [31]. The synthesis of fluoride-based RED-CNPs was first introduced for non-doped LaF_3 nanocrystals [50]. Dopants were then embedded into the host lattice to yield luminescence [13]. The preparation of $NaYF_4$ nanocrystals by thermal decomposition were first elaborated by reacting Na and RE trifluoroacetate precursors in a mixture of a coordinating ligand (oleic acid) and a non-coordinating solvent (1-octadecene) [51]. The protocol was then improved by hot-injecting the trifluoroacetate precursors at a defined rate. Such control over the reaction kinetics rendered $NaYF_4$ nanocrystals with superior size and shape homogeneity [52]. Furthermore, improvements in nucleation and growth stages were achieved by hot-injecting precursors at defined rates and temperatures [53]. Such method refinement has opened up to a range of possible morphologies, such as nanorods, nanospheres, nanoprisms, nanoplates, and nanodisks, which may undergo different pathways in a living organism [54].

The choice of capping agents is limited by the high reaction temperatures required for thermolysis. The size and morphology are tailored by varying reaction temperature and time, and the ratio between oleic acid and RE [55, 56]. The synthesis of pure crystalline β-phase of $NaYF_4$ requires high synthetic temperatures to provide the activation energy for α-to-β phase transition. Extending the reaction also leads to β-phase via Ostwald-ripening processes. Along with β-phase, these strategies also favor broadened particle size distribution with increased average size [49]. Therefore, the sub-10 nm ultrasmall $NaYF_4$ RED-CNPs, desirable for bioimaging, are generally in α-phase, which are less bright compared to β-phase. Such small particles also suffer from a significant surface quenching effect. The combined use of oleylamine and oleic acid is a strategy to reduce the particle size [57–60]. Another approach for size control is varying the ratio of precursors (*e.g.*,, RE/Na/F). The saturation with Na and F precursors favors the nucleation process, which provides ultrasmall crystals with narrow size distribution [61, 62]. Dopant selection is another approach to favor the β-phase formation. Doping $NaYF_4$ with a RE larger than Y^{3+} (1.159 Å) favors a less symmetric hexagonal structure. Gd^{3+} (1.193 Å) as a dopant in $NaYF_4$ facilitates the obtention of crystals at β-phase and reduced size. The possibility of a transition $\alpha \rightarrow \beta$ at lower temperature and shorter time enables the preparation of smaller nanocrystals at β-phase [56, 33].

Multiple steps of thermal decomposition are usually performed to grow epitaxially fluoride-based layers. The optically inert shell overcomes the quenching of RE exposed at the surface [63, 64]. The core–shell design can be further elaborated to precisely modulate energy transfer pathways among RE ions doped throughout multiple layers [65–68]. Such design, that combines specific dopants in different shells, enables better control over dopant interactions, reduces cross-relaxation processes, and improves the efficiency of an intended luminescence mode.

Skripka et al. developed RED-CNPs in a core/shell/shell design for dual-mode excitation (upon 806 or 980 nm laser irradiation) and emission (UCL and OTN-NIR) [69]. The $NaGdF_4$ nanocrystals were synthesized via successive thermal decomposition steps from multilayers with different dopant compositions (Fig. 6.3a). The RE precursor solution in oleic acid and 1-octadecene (Solution B) was injected to an initial reaction vessel containing only the solvents at high temperature (Solution A). The temperature was kept at 315 °C for 1 h to allow the growth of monodispersed core with 17 nm, as shown in the transmission electron microscopy (TEM) images (Fig. 6.3b). The shells #1 and #2 were consecutively prepared in a similar way. The core was doped with Er^{3+}, Ho^{3+}, and Yb^{3+}, and the intermediate layer with Nd^{3+} and Yb^{3+}. The external layer was a passivating $NaGdF_4$ shell. Upon 806 nm irradiation, Nd^{3+} absorbs the energy and transfers it to Yb^{3+}, which act as a bridge to sensitize the activator ions (Er^{3+} and Ho^{3+}). The bioimaging under 806 nm, allowed by core/shell/shell structure, eliminates the overheating caused by the conventional 980 nm excitation route for both emission modes, UCL and OTN-NIR. In all such cases of multilayer design, additional surface modification is required to modify the hydrophobic surface [70].

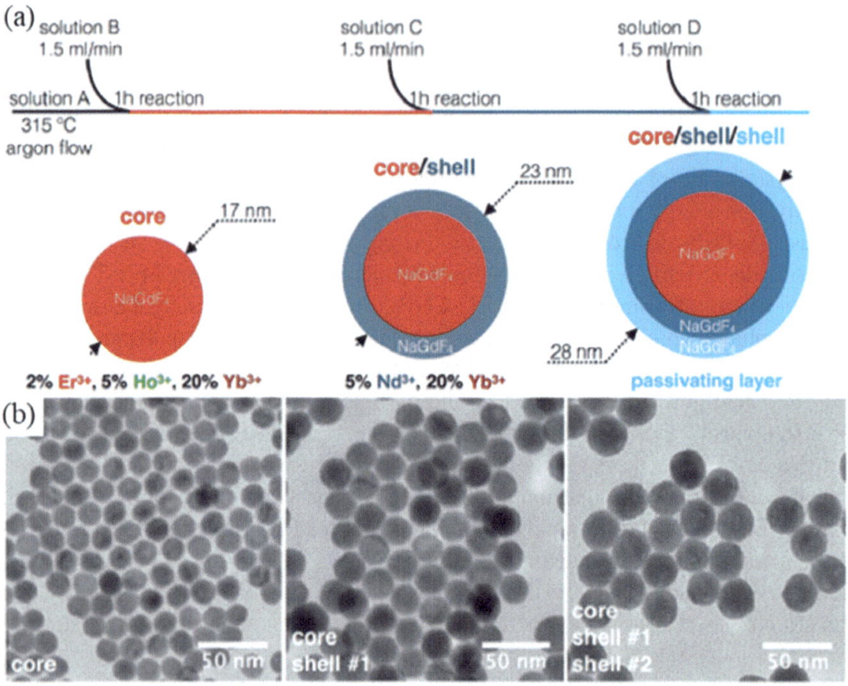

Fig. 6.3 **a** Preparation pathway of core/shell/shell designed RED-CNPs via thermal decomposition method. **b** TEM images of core (17 nm dia.), core/shell (3 nm shell #1 thickness), and core/shell/shell (2.5 nm shell #2 thickness) structures. Reproduced from [69] with permission from The Royal Society of Chemistry

6.3.2 Surface Modification of Hydrophobic Nanocrystals

The synthetic strategies for the preparation of high-quality fluoride nanocrystals are based on oil-phase methods, thus yielding hydrophobic particles. The post-synthesis surface modification is then necessary to introduce the RED-CNPs in complex physiological environments [71]. In pure water, the colloidal suspension is maintained by simply repulsion among common charges at the surface of RED-CNPs. Once in the physiological fluid, ions and local charges cancel the charges at particle' surface and destabilize the electrostatic repulsion. The particles tend to form aggregates with several hundred nanometers. Moreover, the exogenous RED-CNPs are readily attracted by endogenous proteins and biomolecules to form large aggregates [72]. The particle size dictates the metabolism pathway. The body clearance of ultrasmall (<10 nm) particles occurs efficiently through the kidneys followed by urinary excretion. As the particle size increases (>100 nm), a more complex mechanism takes place involving the liver and spleen. The particles are trapped by the macrophages (phagocyte) of reticuloendothelial system (RES) [73, 74].

Taking advantage of higher tumor vasculature and lower lymphatic drainage, in comparison with a normal tissue, is a simple approach for tumor targeting. This phenomenon, known as enhanced permeability and retention effect (EPR), can be employed as a passive targeting of nanosystems towards tumor cells [75, 76]. The EPR has greater effect on large particles, which escape from fast renal metabolism, and are retained in tumor tissues to be slowly excreted.

Diverse cell uptake pathways are possible and are still not well-understood. Two main pathways are involved in cellular internalization: (i) the physical adhesion, by simple electrostatic interaction at positively- or negatively-charged binding sites in cell membrane domains; (ii) endocytosis via specific receptors [77]. Of course, variations on the biological response also depend on the cell and organ types. Therefore, the surface modification of hydrophobic nanocrystals is more complex than simply dispersing particles in pure water. The physico-chemical surface properties will determine several pharmacological factors, including bioavailability, cell uptake, and excretion pathway. Zhang et al. [77] demonstrated the cell uptake efficiency for $NaYF_4:Yb^{3+}:Er^{3+}$ with different size, shape, and surface charge, the latter was found to be the most influential factor for smaller particles. As the particle size increases, the morphology becomes more important.

The aqueous mono-dispersion of hydrophobic RED-CNPs is mostly achieved by coating with a biocompatible and hydrophilic polymer. The polymeric coating provides steric repulsion and is also a strategy to improve biocompatibility [78]. Polyethylene glycol (PEG) and its derivatives are the classical choices for nanoparticle coating. Depending on the intended application, a chemically reactive surface may be designed for further functionalization through terminal bio reactive groups available in the polymeric shell. The positively charged surface of RED-CNPs can be coated by electrostatic adsorption by using a negatively-charged polymer. For instance, carboxylic acids present in PEG with a block copolymer poly(acrylic acid) (PEG-*b*-PAAc) can be attached to the surface of RED-CNPs [79]. The as-synthesized RED-CNPs are usually oleate-capped, in the case of synthesis by thermal decomposition method. The oleate ligands can be replaced by an amphiphilic biocompatible polymer to render hydrophilicity. In this regard, Kamimura et al. proposed a multistep ligand exchange protocol for coating RED-CNPs with PEG-*b*-PAAc. The reaction pathway is illustrated in Fig. 6.4a [80]. In the first step, BF_4^- anions displace the weakly coordinated carboxylic groups of oleic acid. The RED-CNPs migrates from a non-polar solvent (cyclohexane) to a polar solvent (*N,N*-dimethylformamide, DMF). The PEG-*b*-PAAc is then added. The multiple binding COOH groups of PAAc block attach firmly to RED-CNP' surface. The PEG block allows the dispersion in water by the steric repulsion. Wortmann et al. in a similar preparation, demonstrated by dynamic light scattering (DLS) measurements that colloidal stability was maintained in all steps (Fig. 6.4b) [81].

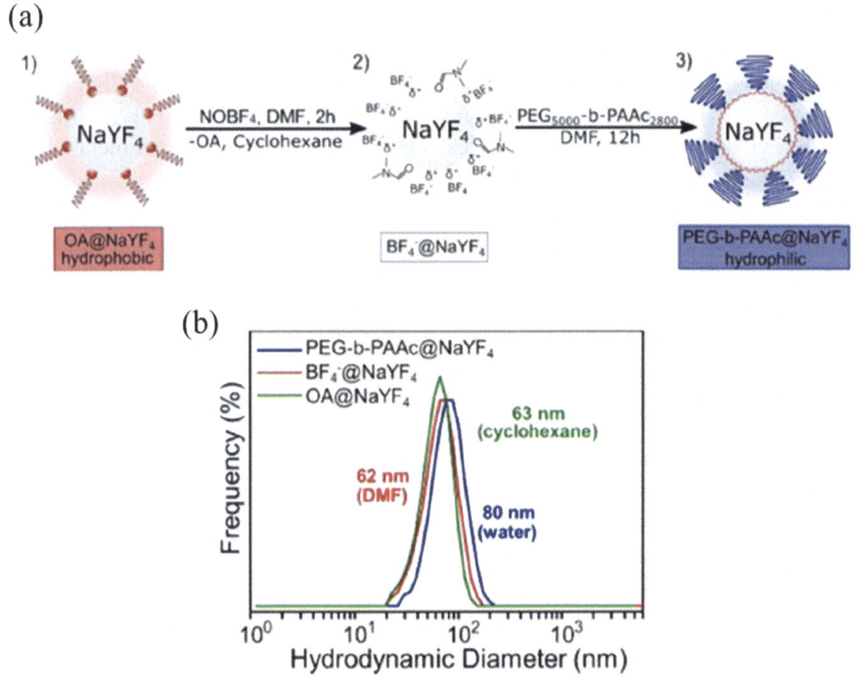

Fig. 6.4 **a** Schematic illustration of $NaYF_4$ RED-CNPs in a surface modification process via multi-step ligand exchange. First, the oleic acid is removed by $NOBF_4$, and RED-CNPs are transferred into a polar solvent (DMF). Then, PEG-b-PAAc is coordinated to RED-CNPs surface and renders hydrophilicity to the system. **b** Size distribution of RED-CNPs at different steps showing colloidal stability. Reprinted from [81] with permission from Elsevier

6.3.3 *Surface Functionalization*

The surface of RED-CNPs can be designed for a more selective cell-targeting approach, by anchoring bioactive molecules [82]. Different cell populations can be actively targeted by attaching the specific biomolecule for molecular recognition, such as antibodies [83, 84], peptides [85], folic acid [86, 87], and aptamers [88, 89]. The strategies to conjugate these cell-targeting agents are mainly based on adsorption, covalent conjugation, and biotin-avidin interaction. Besides its simplicity, the adsorption does not provide sufficient stability and reproducibility. On the other hand, the covalent strategies require prior activation of nanoparticle surface to undergo a selective reaction with the biomolecule. The previous functionalization of the biomolecule is required or not, depending on the coupling chemistry approach. The covalent methods are mainly based on carbodiimide chemistry, maleimide chemistry, or "click chemistry" [90].

The surface designed for therapeutic purposes usually contains a nanocomponent to serve as a carrier for drug delivery. In core–shell nanosystems, the luminescent core is coated with a carrier shell. The carriers are in general based on polymers [91, 92] or mesoporous silica [21, 93], with enough loading capacity. Usually, these coatings also render the particles dispersible in aqueous environments. Molecules are retained in carrier-shell by electrostatic interactions, H-bonds, or reversible chemical reactions. The encapsulated drugs can be passively released by mere desorption. Stimuli-responsive drug delivery systems provide better control on drug release, which is triggered by changes in pH [94], temperature [95], light irradiation [96], complementary molecules or functional groups [97] or magnetic field [98, 99].

6.4 Biomedical Applications of RED-CNPs

6.4.1 Bioimaging

The main contributions of RED-CNPs for bioimaging are related to deeper penetration of NIR light into biological samples. NIR-to-NIR luminescence is advantageous in comparison with UCL because of deeper tissue propagation of emission, *i.e.*, reduced optical loss of signals to be detected. In the NIR wavelength range, the optimal contrast from the probe in living specimens is obtained. The transparency of biological tissues is maximized in NIR because of reduced absorption of light by water and light attenuation coefficients of human skin, blood, and fatty tissue. Spectral regions, known as biological transparency windows, are classified according to the endogenous components and associated optical loss. The first biological window (NIR-I, 700–900 nm) offers less transparency with respect to OTN wavelengths, in the region where are located the second (NIR-II, 1000–1350 nm) and third (NIR-III 1500–1800 nm) biological windows [100].

The current imaging probes are excited in the NIR-I, usually via sensitizers Yb^{3+} and Nd^{3+}, under 980 nm or 800 nm, respectively. The excitation in the NIR-II region may cause overheating of tissues, because of larger absorption coefficient of water [101]. For this reason, few examples in literature have explored the NIR-II for exciting a RED-CNPs in bioimaging [102, 103]. As for the green or red Er^{3+}-based UCL, when excited under 1550 nm (NIR-II), the process involves absorption of three or four photons, and is thereby less efficient than conventional 980 nm (NIR-I)-excited UCL process involving two photons. On the other hand, the NIR-II excitation light reduces the losses related to scattering, and thus improves the spatial resolution in deep tissue imaging [104]. The use of NIR-II-excited RED-CNPs emitting down-shifted NIR-to-NIR luminescence in bioimaging is also limited considering the Stokes shifted emission and the limit of wavelength detection by InGaAs CCD cameras (800–2200 nm) [23]. In view of these facts, the most suitable wavelength range available for a sensitive bioimaging is the short-wavelength infrared region.

Naczynski et al. [105] proposed RED-CNPs for bioimaging in the OTN-NIR with tunable emission over the wavelength range 1,000–1,700 nm. The core–shell design consisted of a luminescent core of $NaYF_4$ co-doped with Yb^{3+}:RE (RE = Er^{3+},

Ho^{3+}, Tm^{3+}, Pr^{3+}) surrounded by undoped shell of $NaYF_4$ for surface passivation. The emission spectral profile was adjusted by changing the RE activator dopant. The OTN-NIR emissions occurred after energy transfer from sensitizer (Yb^{3+}) to the RE ions via resonant levels. The optical efficiency was compared among RE activator dopants. The most efficient system, co-doped with Yb^{3+}:Er^{3+}, was chosen to further compare the efficiency of bioimaging in terms of spectral range, *i.e.,* NIR-to-visible and NIR-to-NIR. The optical efficiency of Er^{3+} for transmitting OTN-NIR emission (at 1525 nm) through blood and pigmented tumor tissue was significantly greater than the UCL at 550 nm. The UCL bioimaging was detected under high and low power densities. The bioimaging at lower densities are hindered by lack of bright signals. The high-power density results in light scattering. Such effect is clearly observed by the lower resolution of visible emission (high power) in comparison with the OTN-based image.

The luminescence of RED-CNPs exploited in the time domain not only eliminates the autofluorescence effect, but also enables efficient multiplexed bioimaging. The multiple signal detection is frequently hampered in the steady-state measurement mode by overlapping of emission lines originated from different probes. The use of time-dependent luminescence can overcome this issue, since it does not operate in spectral domain. This approach is applicable for RED-CNPs emitting in both modes, visible UCL [106] and downshifted ONT-NIR. The luminescence lifetime of RED-CNPs in the NIR-II region was used by Ortgies et al. [107] to design an accurate probe for in vivo multiplexed bioimaging. The probes are based on core–shell $NaYF_4$:Yb^{3+}:Nd^{3+}/CaF_2 nanoparticles. The presence of Nd^{3+} and Yb^{3+} enables emission at 980 nm under 808 nm excitation. The Yb^{3+} luminescence lifetime was tuned (0.1–1.5 ms) by varying the Nd^{3+} doping concentration (Fig. 6.5a,b). The time-gated imaging technique was used for detecting different lifetime-based signals from the obtained RED-CNPs (Fig. 6.5c). The lifetime-based signals emitted from different Nd^{3+}-doping samples were isolated from a single image (Fig. 6.5d,e).

The RED-CNPs are excellent luminescent probes for single-particle imaging at subcellular level. Tracking a single particle enables detecting with high-resolution for efficient and early diagnosis. The photostability of RED-CNPs enables the long-term tracking of a single particle for hours without blinking and bleaching [108], and at low pumping power [109]. The single-particle spectroscopy integrates optical microscopy (confocal/wide-field fluorescence) with microscopic techniques, such as in situ atomic-force microscopy (AFM), or ex-situ transmission and scanning electron microscopy (TEM/SEM) [110]. The single-particle approach has been proposed to estimate the real influence of particle size, morphology, physicochemical surface properties, stoichiometric composition, geometric orientation, and the local environment in biomedical applications. The conventional bioimaging leads to an average understanding of a group of non-homogeneous particles since the well-controlled synthesis of RED-CNPs is a relevant issue. Gargas et al. [35] used single-particle characterization to investigate the UCL properties of RED-CNPs at different sizes. Zhang et al. [77] employed single-particle microscopy to investigate cellular uptake for different size, shape, and surface charge of $NaYF_4$:Yb^{3+}:Er^{3+}.

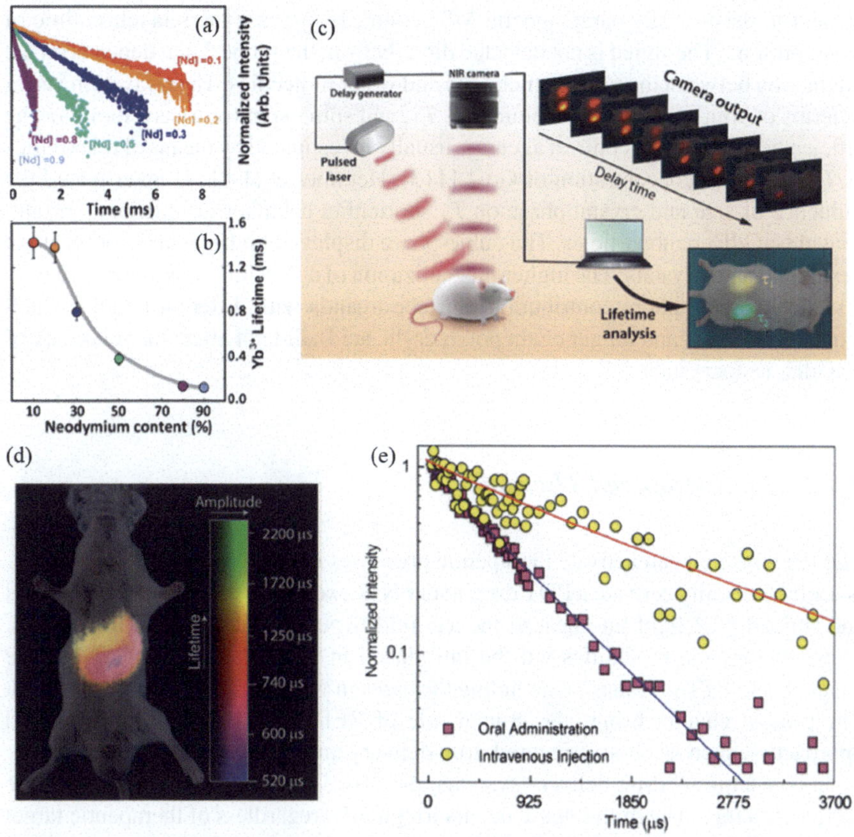

Fig. 6.5 Multiplexed in vivo imaging encoded by OTN-NIR luminescence lifetime of RED-CNPs. **a** Decay curves and **b** lifetimes upon Nd^{3+} excitation at 808 and consecutive Yb^{3+}emission at 980 nm for $NaYF_4$:Yb^{3+}:Nd^{3+}/CaF_2 core–shell NPs at different Nd^{3+} dopant concentrations. **c** Time-gated imaging set up synchronizing an excitation pulsed laser and an InGaAs camera for in vivo NIR imaging. The inset mouse image reveals regions with distinct lifetimes (τ_1 and τ_2), to which samples of RED-CNPs with different contents of Nd^{3+} were subcutaneously injected. **d** OTN-NIR lifetime-based image of a mouse after oral and intravenous injection of RED-CNPs with Nd^{3+} contents of 30% and 20%, respectively, and **e** corresponding luminescence decay curves. Adapted from [107] with permission from American Chemical Society

The RED-CNPs-based probes are versatile materials for designing multi-modal imaging systems. The rational selection of RE dopants enables modulation over optical and magnetic properties. The combination of both optical and magnetic resonance imaging modalities using a single RE-based probe is widely reported in literature for NaY(Gd)F_4:RE [84, 111]. The presence of paramagnetic Gd^{3+} ions enable the use of RED-CNPs as MRI contrast with high efficiency [112]. The complementary information obtained from magnetic signals is relevant for a more accurate diagnosis. The contrast in MRI is obtained from differences in relaxation times of water protons present in tissues. The paramagnetic MRI contrasts increase the relaxation time of water protons. The signal is not detected directly from the probe, but it depends on the interaction between the paramagnetic ion and water molecules. The relaxation varies in terms of spin–lattice (or longitudinal) T_1, and spin–spin (or transverse) T_2. The efficiency of such MRI contrast agents is usually investigated by means of relaxivities ($1/T$) per molar concentration of Gd^{3+} [113]. Hemmer et al. [114] investigated the influence of size and crystal phase on T_1 shortening behavior of ultrasmall citrate-coated $NaGdF_4$ nanoparticles. The cubic-phase displayed higher relaxivity (r_1) than hexagonal nanocrystals. The higher magnetization of α-$NaGdF_4$ was associated with a stronger inner sphere contribution. Surface ligands with different lengths, small citrate molecules, and longer chain poly(acrylic acid), did not affect the relaxivity of $NaGdF_4$ nanocrystals.

6.4.2 Light-Induced Therapies

The RE emission can activate therapeutic processes in different ways. Visible UCL is useful to locally activate a PDT drug under NIR excitation. Moreover, RED-CNPs can convert NIR light into heat to induce mild hyperthermia and kill cancer cells. These therapeutic modalities will be introduced in the next chapters (PDT in the Chap. 9, and PTT in Chap. 7), including the cases in which RED-CNPs are involved. The present chapter brings the crucial role of RED-CNPs to other light-induced approaches, such as photo-triggered drug delivery and in situ photopolymerization.

In conventional drug delivery systems, the active molecules are adsorbed in the carrier structure. The drug release occurs irregularly, regardless of therapeutic target and pharmacokinetics. A possible solution for uncontrolled drug release consists of employing a stimuli-responsive carrier in order to perform the drug release on-demand. The NIR light can be used as an external stimulus to trigger the drug release. This is possible if the nanocomposite used as a drug carrier offers light-dependent properties. Under NIR irradiation, the luminescence of RED-CNPs induces changes in carrier's physical or chemical structure, and drug is thereby released [25, 115]. The mechanisms mainly involve the photolysis [93], photocleavage [116], and photoisomerization [117] of a carrier structure.

The RED-CNPs, when added to composite resins used to fabricate dental filling materials, can improve the depth of cure under NIR irradiation [118]. The UCL activates the initiator to generate free radicals and trigger the polymerization of monomers. The coherent NIR source enables precise control over polymerization for advanced fabrication processes. In this regard, RED-CNPs-assisted photopolymerization shows potential for simultaneous NIR-imaging and real-time 3D printing [119, 120]. The in situ formation of hydrogels as tissue scaffolds is a promising approach for better adhesion of the biomaterial. Moreover, PDT photosensitizers can be incorporated in the scaffolds for photoinactivation of bacteria via $^1O_2$ generation [121, 122].

Xiao et al. [122] reported the use of UCL of $NaYF_4:Er^{3+}:Yb^{3+}$ RED-CNPs to assist in situ photopolymerization of poly(ethylene glycol) diacrylate (PEGDA) hydrogels. Moreover, the possibility to deliver visible light to deep tissue layers via UCL was proposed for local activation of a photosensitizer agent (zinc(II) phthalocyanine, ZnPc) used in PDT. The Fig. 6.6a illustrates both NIR-activated processes. The green UCL (between 520 and 554 nm) overlaps with the absorption band of the photoinitiator Eosin Y in the range 460–540 nm. The TEM and SEM images (Fig. 6.6b) reveals polymerization of PEGDA over 980 nm irradiation time. The red UCL overlaps well with the absorption of ZnPc (650–680 nm), as shown in Fig. 6.6c. Such red emission was quenched in the presence of ZnPc, suggesting the potential for generation of singlet oxygen ($^1O_2$), the main cytotoxic agent in PDT.

Kamimura et al. [123] took advantage of UCL to form a polymer shell on the RED-CNPs surface. The non-polar polystyrene (PSt) coating aimed at reducing the surface quenching caused by polar solvents, such as water. For preparing the polymeric shell, the RED-CNPs were mixed with a monomer (styrene) and photopolymerization initiators (Eosin Y and triethanolamine). The photopolymerization under NIR irradiation was based on the green UCL of $NaYF_4:Yb^{3+}:Ho^{3+}:Er^{3+}$ RED-CNPs, which

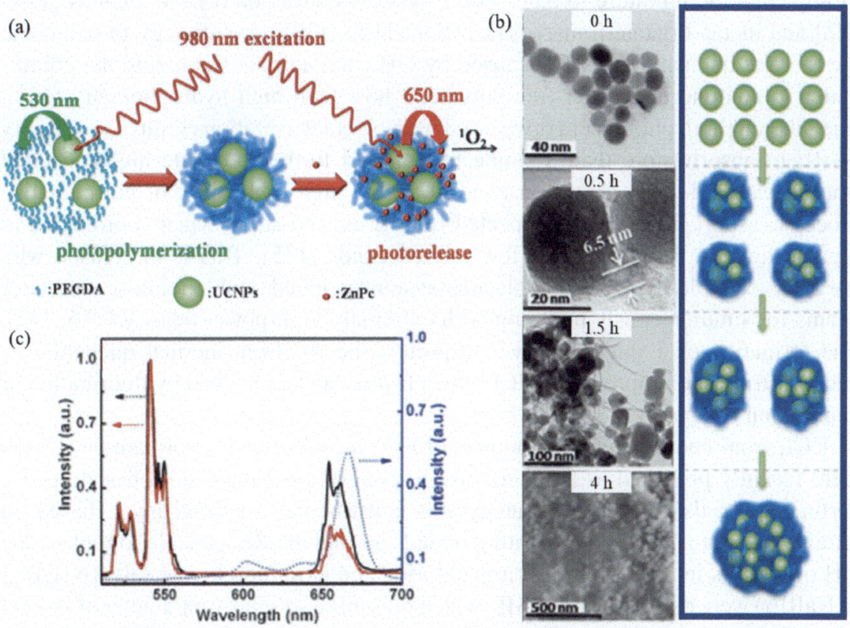

Fig. 6.6 **a** Schematic illustration hybrid microspheres of UCNPs–PEGDA emitting UCL for NIR-activation of photopolymerization of PEGDA hydrogels and photo-induced release of $^1O_2$ from a loaded photosensitizer (ZnPc). **b** TEM and SEM images of UCNPs–PEGDA hybrids prepared under 980 nm irradiation for 0–4 h. **c** Upconversion emission spectra of UCNPs–PEGDA hybrid (black line) and UCNPs–PEGDA loaded with ZnPc (red line), and absorption spectrum of ZnPc (blue dotted line). Adapted from [122] with permission from Royal Society of Chemistry

overlaps the absorption band of Eosin Y. The energy transfer from RED-CNPs to Eosin Y was evidenced by decreased green emission intensity. Additionally, the emissions of Er^{3+} and Ho^{3+} in the OTN-NIR region (at 1150 and 1550 nm, respectively) can be used for deep tissue bioimaging and nanothermometry.

6.5 Emerging Molecular-Based Systems

The miniaturization of RE-based luminophores to the molecular level offers advantages over conventional solid-state compounds in terms of optical transparency. The structural versatility of such molecular systems is desirable for designing hybrid materials [124]. Thus, flexible and transparent luminescent biomaterials can be developed as an alternative to conventional solid-based ones. On the other hand, the influence of surrounding phonons in a such molecule-based system is more pronounced with respect to RED-CNPs. Therefore, the major challenge for the implementation of molecule systems as bioprobes relies on minimizing the non-radiative losses. In this sense, the molecular-based systems for UCL is an emerging research topic with few examples in the literature to date. These systems consist on organic ligands [125–128] and metal–organic frameworks (MOF) [129, 130]. One strategy to overcome the thermal quenching of UCL caused by OH oscillators is stabilizing the coordination compound in a water-free polymeric host with high hydrophobicity [131]. The inherent high phonon energy caused by molecular vibrations limits the capacity of RE to absorb more than a single photon and further populate higher excited state levels [132–134]. As a consequence of the high probability of non-radiative processes, short lifetimes are expected for RE excited-states when coordinated to organic ligands, reaching up to few microseconds [135]. This is the reason why the early examples of UCL in molecular systems applied double femtosecond laser beams for simultaneously pumping with extremely high-power density [136, 137]. The formation of a shielding layer to protect the RE from thermal quenching is another strategy to improve the UCL. Such layers can be obtained by fluorination or deuteration of ligands [132].

UCL from coordination compounds of Er^{3+} in *N,N*-dimethylformamide (DMF) were recently proposed as an alternative to conventional ionic solid-based upconverting materials [138]. The strategy was constructing a special layer, based on tetrafluoroborate ions as coordinating ligands around the RE, capable to replace the OH oscillators in the system. Erbium chloride and tetrafluoroborate salts (NH_4BF_4 or $NaBF_4$) were dissolved into DMF, which presents a solubility parameter of around 12. If the value was close to 8, as the case of cyclohexane, the system would be too hydrophobic, and would not be able to dissolve the ions. The Cl^- was replaced by BF_4^- ligands for Er^{3+} in the systems and removed as a precipitate of NH_4Cl or NaCl, as evidenced by X-ray diffraction (XRD) analysis. UCL was observed in transparent liquid materials. The rational design was guided by a systematic spectroscopic investigation for each type of complex prepared, in order to unravel the species grafted to RE. The emission properties of the complexes indicated that the non-linear UCL was achieved because BF_4^- ligands were able to fulfil the first coordination sphere and hamper the luminescence quenching caused by OH oscillators. The chloride ligand replacement by tetrafluoroborate induced a change in the local polarization, decreasing the covalency character of the coordination bonding. The larger BF_4^- ion may also contribute to shield RE ions from OH coordination.

6.6 Summary

The advances in RED-CNPs research have a great impact on the biomedical field for several modalities of diagnosis and therapies. Their excellent optical properties make RED-CNPs a first-line candidate for novel biomedical techniques. The well-established synthesis methods permit the rational control over luminescent properties, size and shape, surface modification. From a contrary perspective, single-particle spectroscopy emerges in scientific investigations to overcome particle heterogeneity issues. The strategies to increase the luminescence quantum yield rely on maximizing absorption and minimizing the non-radiative energy losses. The miniaturization of luminescent materials in the last decades makes imperative the extension of the phonon concept to molecular systems, and raises the intriguing possibility of heat-favorable phonons at the surface of RED-CNPs, which not only prevent the thermal quenching of luminescence, but even enhance the brightness.

Beyond the optical properties, the rational design of biofunctional RED-CNPs involves size, shape, and physico-chemical surface properties, which are pertinent in different aspects of particles' interactions with biological components, such as surface adsorption of ions and proteins, cellular uptake, biodistribution, and clearance mechanisms. The main challenges for a real application of these probes are related to developing adequate and safe pharmaceutical formulations for efficient administration. Moreover, the establishment of light dosimetry is important to ensure the safety of light-induced therapies. Therefore, the aspects of cell and nanoobjects interactions must be investigated at the interface between synthesis of nanomaterials and cell biology. Such interactions can be modulated by adjusting the surface chemistry, size, and morphology of RED-CNPs. The implementation of RED-CNPs in biomedical applications require the efforts of multidisciplinary research teams.

References

1. J.C.G. Bünzli, C. Piguet, Taking advantage of luminescent lanthanide ions. Chem. Soc. Rev. **34**, 1048 (2005). https://doi.org/10.1039/b406082m
2. J.C.G. Bünzli, S.V. Eliseeva, Intriguing aspects of lanthanide luminescence. Chem. Sci. **4**, 1939 (2013). https://doi.org/10.1039/c3sc22126a
3. T.L. Doane, C. Burda, The unique role of nanoparticles in nanomedicine: imaging, drug delivery and therapy. Chem. Soc. Rev. **41**, 2885 (2012). https://doi.org/10.1039/c2cs15260f

4. Y. Liu, D. Tu, H. Zhu et al., Lanthanide-doped luminescent nano-bioprobes: from fundamentals to biodetection. Nanoscale **5**, 1369 (2013). https://doi.org/10.1039/c2nr33239f
5. F. Auzel, Upconversion and anti-stokes processes with f and d ions in solids. Chem. Rev. **104**, 139 (2004). https://doi.org/10.1021/cr020357g
6. G.S. He, J.D. Bhawalkar, C.F. Zhao et al., Upconversion dye-doped polymer fiber laser. Appl. Phys. Lett. **68**, 3549 (1996). https://doi.org/10.1063/1.116633
7. G. Zhou, X. Wang, D. Wang et al., Upconversion fluorescence and optical power limiting effects based on the two-and three-photon absorption process of a new organic dye BPAS. Appl. Opt. **41**, 1120 (2002). https://doi.org/10.1364/ao.41.001120
8. G. Zhuo, X. Wang, D. Wang et al., Two-photon induced optical-power limiting and upconverted superradiance properties of a new organic dye HEASPI. Opt. Laser Technol. **33**, 209 (2001). https://doi.org/10.1016/s0030-3992(01)00025-1
9. L.A. Padilha, J. Fu, D.J. Hagan et al., Two-photon absorption in CdTe quantum dots. Opt. Express **13**, 6460 (2005). https://doi.org/10.1364/opex.13.006460
10. J. Ke, X. Li, Q. Zhao et al., Upconversion carbon quantum dots as visible light responsive component for efficient enhancement of photocatalytic performance. J. Colloid Interface Sci. **496**, 425 (2017). https://doi.org/10.1016/j.jcis.2017.01.121
11. H.J.M.A.A. Zijlmans, J. Bonnet, J. Burton et al., Detection of cell and tissue surface antigens using up-converting phosphors: a new reporter technology. Anal. Biochem. **267**, 30 (1999). https://doi.org/10.1006/abio.1998.2965
12. J. Hampl, M. Hall, N.A. Mufti et al., Upconverting phosphor reporters in immunochromatographic assays. Anal. Biochem. **288**, 176 (2001). https://doi.org/10.1006/abio.2000.4902
13. J.W. Stouwdam, F.C. van Veggel, Near-infrared emission of redispersible Er^{3+}, Nd^{3+}, and Ho^{3+} doped LaF_3 nanoparticles. Nano Lett. **2**, 733 (2002). https://doi.org/10.1021/nl025562q
14. G. Yi, H. Lu, S. Zhao et al., Synthesis, characterization, and biological application of size-controlled nanocrystalline $NaYF_4$:Yb, Er infrared-to-visible up-conversion phosphors. Nano Lett. **4**, 2191 (2004). https://doi.org/10.1021/nl048680h
15. S. Heer, K. Kömpe, H.-U. Güdel et al., Highly efficient multicolour upconversion emission in transparent colloids of lanthanide-doped $NaYF_4$ nanocrystals. Adv. Mater. **16**, 2102 (2004). https://doi.org/10.1002/adma.200400772
16. H. Lu, G. Yi, S. Zhao et al., Synthesis and characterization of multi-functional nanoparticles possessing magnetic, up-conversion fluorescence and bio-affinity properties. J. Mater. Chem. **14**, 1336 (2004). https://doi.org/10.1039/b315103d
17. J.V. Frangioni, *In vivo* near-infrared fluorescence imaging. Curr. Opin. Chem. Biol. **7**, 626 (2003). https://doi.org/10.1016/j.cbpa.2003.08.007
18. S. Wu, N. Duan, Z. Shi et al., Simultaneous aptasensor for multiplex pathogenic bacteria detection based on multicolor upconversion nanoparticles labels. Anal. Chem. **86**, 3100 (2014). https://doi.org/10.1021/ac404205c
19. D. Ni, W. Bu, S. Zhang et al., Single Ho^{3+}-doped upconversion nanoparticles for high-performance T_2-weighted brain tumor diagnosis and MR/UCL/CT multimodal imaging. Adv. Funct. Mater. **24**, 6613 (2014). https://doi.org/10.1002/adfm.201401609
20. L. Cheng, K. Yang, Y. Li et al., Facile preparation of multifunctional upconversion nanoprobes for multimodal imaging and dual-targeted photothermal therapy. Angew. Chem. Int. Ed. **50**, 7385 (2011). https://doi.org/10.1002/anie.201101447
21. P. Zhang, W. Steelant, M. Kumar et al., Versatile photosensitizers for photodynamic therapy at infrared excitation. J. Am. Chem. Soc. **129**, 4526 (2007). https://doi.org/10.1021/ja0700707
22. Q. Zhan, S. He, J. Qian et al., Optimization of optical excitation of upconversion nanoparticles for rapid microscopy and deeper tissue imaging with higher quantum yield. Theranostics **3**, 306 (2013). https://doi.org/10.7150/thno.6007
23. D. Jaque, C. Richard, B. Viana et al., Inorganic nanoparticles for optical bioimaging. Adv. Opt. Photonics **8**, 1 (2016). https://doi.org/10.1364/aop.8.000001
24. J. Chen, T. Fan, Z. Xie, et al., Advances in nanomaterials for photodynamic therapy applications: status and challenges. Biomaterials 119827 (2020). https://doi.org/10.1016/j.biomateri als.2020.119827

25. B. Yang, H. Chen, Z. Zheng et al., Application of upconversion rare earth fluorescent nanoparticles in biomedical drug delivery system. J. Lumin. **223**, 117226 (2020). https://doi.org/10.1016/j.jlumin.2020.117226
26. G. Blasse, B.C. Grabmaier, A general introduction to luminescent materials, in *Luminescent Materials*. Heidelberg, Springer (1994). https://doi.org/10.1007/978-3-642-79017-1_1
27. R.C. Powell, Physics of solid-state laser materials. New York: Springer (1998). https://doi.org/10.1007/978-1-4612-0643-9_7
28. L.A. Riseberg, H.W. Moos, Multiphonon orbit-lattice relaxation of excited states of rare-earth ions in crystals. Phys. Rev. **174**, 429 (1968). https://doi.org/10.1103/physrev.174.429
29. K. Soga, W. Wang, R.E. Riman et al., Luminescent properties of nanostructured Dy^{3+}- and Tm^{3+}-doped lanthanum chloride prepared by reactive atmosphere processing of sol-gel derived lanthanum hydroxide. J. Appl. Phys. **93**, 2946 (2003). https://doi.org/10.1063/1.1540229
30. M.J. Weber, Multiphonon relaxation of rare-earth ions in yttrium orthoaluminate. Phys. Rev. B **8**, 54 (1973). https://doi.org/10.1103/physrevb.8.54
31. F. Zhang, Photon Upconversion Nanomaterials. Springer, Berlin. (2015). https://doi.org/10.1007/978-3-662-45597-5
32. M. Haase, H. Schäfer, Upconverting nanoparticles. Angew. Chem. Int. Ed. **50**, 5808 (2011). https://doi.org/10.1002/anie.201005159
33. F. Wang, Y. Han, C.S. Lim et al., Simultaneous phase and size control of upconversion nanocrystals through lanthanide doping. Nature **463**, 1061 (2010). https://doi.org/10.1038/nature08777
34. M.L. Neidig, D.L. Clark, R.L. Martin, Covalency in f-element complexes. Coord. Chem. Rev. **257**, 394 (2013). https://doi.org/10.1016/j.ccr.2012.04.029
35. D.J. Gargas, E.M. Chan, A.D. Ostrowski et al., Engineering bright sub-10-nm upconverting nanocrystals for single-molecule imaging. Nat. Nanotechnol. **9**, 300 (2014). https://doi.org/10.1038/nnano.2014.29
36. J. Zhou, S. Wen, J. Liao et al., Activation of the surface dark-layer to enhance upconversion in a thermal field. Nat. Photon. **12**, 154 (2018). https://doi.org/10.1038/s41566-018-0108-5
37. A. Beeby, I.M. Clarkson, R.S. Dickins, et al., Non-radiative deactivation of the excited states of europium, terbium and ytterbium complexes by proximate energy-matched OH, NH and CH oscillators: an improved luminescence method for establishing solution hydration states. J. Chem. Soc., Perkin Trans. **2**, 493 (1999). https://doi.org/10.1039/a808692c
38. B. Stuart, Infrared spectroscopy: fundamentals and applications. Wiley (2004). https://doi.org/10.1002/0470011149
39. Z. Wang, J. Christiansen, D. Wezendonk et al., Thermal enhancement and quenching of upconversion emission in nanocrystals. Nanoscale **11**, 12188 (2019). https://doi.org/10.1039/c9nr02271f
40. C. Mi, J. Zhou, F. Wang et al., Thermally enhanced NIR–NIR anti-Stokes emission in rare earth doped nanocrystals. Nanoscale **11**, 12547 (2019). https://doi.org/10.1039/c9nr03041g
41. S.V. Eliseeva, J.C.G. Bünzli, Lanthanide luminescence for functional materials and biosciences. Chem. Soc. Rev. **39**, 189 (2010). https://doi.org/10.1039/b905604c
42. Q. Shao, X. Li, P. Hua et al., Enhancing the upconversion luminescence and photothermal conversion properties of ~800 nm excitable core/shell nanoparticles by dye molecule sensitization. J. Colloid Interface Sci. **486**, 121 (2017). https://doi.org/10.1016/j.jcis.2016.09.067
43. X. Xie, N. Gao, R. Deng et al., Mechanistic investigation of photon upconversion in Nd^{3+}-sensitized core–shell nanoparticles. J. Am. Chem. Soc. **135**, 12608 (2013). https://doi.org/10.1021/ja4075002
44. V. Nguyen, P. Dawson, M. Romanowski, Spectral engineering of UV luminescence of upconverting nanoparticles, in *Proceedings of SPIE 11255, Colloidal Nanoparticles for Biomedical Applications XV, SPIE BiOS*, pp. 11255. San Francisco, California, United States (2020). https://doi.org/10.1117/12.2546826

45. Y. Wei, F. Lu, X. Zhang et al., Synthesis and characterization of efficient near-infrared upconversion Yb and Tm codoped $NaYF_4$ nanocrystal reporter. J. Alloys Compd. **427**, 333 (2007). https://doi.org/10.1016/j.jallcom.2006.03.014
46. J.S. Lee, Y.J. Kim, The effects of preparation conditions on the structural and up-conversion properties of $NaYF_4$: Yb^{3+}, Er^{3+} nano powders. Opt. Mater. **33**, 1111 (2011). https://doi.org/10.1016/j.optmat.2010.10.011
47. P. Rahman, M. Green, The synthesis of rare earth fluoride based nanoparticles. Nanoscale **1**, 214 (2009). https://doi.org/10.1039/b9nr00089e
48. J.H. Zeng, J. Su, Z.H. Li et al., Synthesis and upconversion luminescence of hexagonal-phase $NaYF_4$: Yb, Er^{3+} phosphors of controlled size and morphology. Adv. Mater. **17**, 2119 (2005). https://doi.org/10.1002/adma.200402046
49. Y. Shang, S. Hao, J. Liu et al., Synthesis of upconversion β-$NaYF_4$: Nd^{3+}/Yb^{3+}/Er^{3+} particles with enhanced luminescent intensity through control of morphology and phase. Nanomaterials **5**, 218 (2015). https://doi.org/10.3390/nano5010218
50. J. Zhou, Z. Wu, Z. Zhang et al., Study on an antiwear and extreme pressure additive of surface coated LaF_3 nanoparticles in liquid paraffin. Wear **249**, 333 (2001). https://doi.org/10.1016/s0043-1648(00)00547-0
51. J.C. Boyer, F. Vetrone, L.A. Cuccia et al., Synthesis of colloidal upconverting $NaYF_4$ nanocrystals doped with Er^{3+}, Yb^{3+} and Tm^{3+}, Yb^{3+} via thermal decomposition of lanthanide trifluoroacetate precursors. J. Am. Chem. Soc. **128**, 7444 (2006). https://doi.org/10.1021/ja061848b
52. J.C. Boyer, L.A. Cuccia, J.A. Capobianco, Synthesis of colloidal upconverting $NaYF_4$: Er^{3+}/Yb^{3+} and Tm^{3+}/Yb^{3+} monodisperse nanocrystals. Nano Lett. **7**, 847 (2007). https://doi.org/10.1021/nl0619958
53. M. Wang, Y. Zhang, Q. Yao et al., Morphology control of lanthanide doped $NaGdF_4$ nanocrystals via one-step thermolysis. Chem. Mater. **31**, 5160 (2019). https://doi.org/10.1021/acs.chemmater.9b01155
54. P. Decuzzi, B. Godin, T. Tanaka et al., Size and shape effects in the biodistribution of intravascularly injected particles. J. Control. Release **141**, 320 (2010). https://doi.org/10.1016/j.jconrel.2009.10.014
55. Z. Li, Y. Zhang, An efficient and user-friendly method for the synthesis of hexagonal-phase $NaYF_4$: Yb, Er/Tm nanocrystals with controllable shape and upconversion fluorescence. Nanotechnology **19**, 345606 (2008). https://doi.org/10.1088/0957-4484/19/34/345606
56. H. Na, K. Woo, K. Lim et al., Rational morphology control of β-$NaYF_4$: Yb, Er/Tm upconversion nanophosphors using a ligand, an additive, and lanthanide doping. Nanoscale **5**, 4242 (2013). https://doi.org/10.1039/c3nr00080j
57. Q. Liu, Y. Sun, T. Yang et al., Sub-10 nm hexagonal lanthanide-doped $NaLuF_4$ upconversion nanocrystals for sensitive bioimaging in vivo. J. Am. Chem. Soc. **133**, 17122 (2011). https://doi.org/10.1021/ja207078s
58. T. Cheng, R. Marin, A. Skripka et al., Small and bright lithium-based upconverting nanoparticles. J. Am. Chem. Soc. **140**, 12890 (2018). https://doi.org/10.1021/jacs.8b07086
59. G. Chen, T.Y. Ohulchanskyy, R. Kumar et al., Ultrasmall monodisperse $NaYF_4$: Yb^{3+}/Tm^{3+} nanocrystals with enhanced near-infrared to near-infrared upconversion photoluminescence. ACS Nano **4**, 3163 (2010). https://doi.org/10.1021/nn100457j
60. B. Chen, W. Kong, N. Wang et al., Oleylamine-mediated synthesis of small $NaYbF_4$ nanoparticles with tunable size. Chem. Mater. **31**, 4779 (2019). https://doi.org/10.1021/acs.chemmater.9b01050
61. H. Li, L. Xu, G. Chen, Controlled synthesis of monodisperse hexagonal $NaYF_4$: Yb/Er nanocrystals with ultrasmall size and enhanced upconversion luminescence. Molecules **22**, 2113 (2017). https://doi.org/10.3390/molecules22122113
62. T. Rinkel, J. Nordmann, A.N. Raj et al., Ostwald-ripening and particle size focussing of sub-10 nm $NaYF_4$ upconversion nanocrystals. Nanoscale **6**, 14523 (2014). https://doi.org/10.1039/c4nr03833a

63. H.X. Mai, Y.W. Zhang, L.D. Sun et al., Highly efficient multicolor up-conversion emissions and their mechanisms of monodisperse $NaYF_4$: Yb, Er core and core/shell-structured nanocrystals. J. Phys. Chem. C **111**, 13721 (2007). https://doi.org/10.1021/jp073920d
64. F. Vetrone, R. Naccache, V. Mahalingam et al., The active-core/active-shell approach: a strategy to enhance the upconversion luminescence in lanthanide-doped nanoparticles. Adv. Funct. Mater. **19**, 2924 (2009). https://doi.org/10.1002/adfm.200990081
65. Q. Su, S. Han, X. Xie et al., The effect of surface coating on energy migration-mediated upconversion. J. Am. Chem. Soc. **134**, 20849 (2012). https://doi.org/10.1021/ja3111048
66. J. Shen, G. Chen, A. Vu et al., Engineering the upconversion nanoparticle excitation wavelength: cascade sensitization of tri-doped upconversion colloidal nanoparticles at 800 nm. Adv. Opt. Mater. **1**, 644 (2013). https://doi.org/10.1002/adom.201300160
67. F. Wang, R. Deng, J. Wang et al., Tuning upconversion through energy migration in core–shell nanoparticles. Nat. Mater. **10**, 968 (2011). https://doi.org/10.1038/nmat3149
68. Y.E.S. Correales, C. Hazra, S. Ullah et al., Precisely tailored shell thickness and Ln^{3+} content to produce multicolor emission from Nd^{3+}-sensitized Gd^{3+}-based core/shell/shell UCNPs through bi-directional energy transfer. Nanoscale Adv. **1**, 1936 (2019). https://doi.org/10.1039/c9na00006b
69. A. Skripka, R. Marin, A. Benayas et al., Covering the optical spectrum through collective rare-earth doping of $NaGdF_4$ nanoparticles: 806 and 980 nm excitation routes. Phys. Chem. Chem. Phys. **19**, 11825 (2017). https://doi.org/10.1039/c7cp01167a
70. S. Han, R. Deng, X. Xie et al., Enhancing luminescence in lanthanide-doped upconversion nanoparticles. Angew. Chem. Int. Ed. **53**, 11702 (2014). https://doi.org/10.1002/anie.201403408
71. A.E. Nel, L. Mädler, D. Velegol et al., Understanding biophysicochemical interactions at the nano–bio interface. Nat. Mater. **8**, 543 (2009). https://doi.org/10.1038/nmat2442
72. J. Zhang, F. Liu, T. Li et al., Surface charge effect on the cellular interaction and cytotoxicity of $NaYF_4$:Yb^{3+}, E^{r3}+@Si_O2 nanoparticles. RSC Adv. **5**, 7773 (2015). https://doi.org/10.1039/c4ra11374h
73. H.S. Choi, W. Liu, P. Misra et al., Renal clearance of nanoparticles. Nat. Biotechnol. **25**, 1165 (2007). https://doi.org/10.1038/nbt1340
74. M. Longmire, P.L. Choyke, H. Kobayashi, Clearance properties of nano-sized particles and molecules as imaging agents: Considerations and caveats. Nanomedicine **3**, 703 (2008). https://doi.org/10.2217/17435889.3.5.703
75. A.K. Iyer, G. Khaled, J. Fang et al., Exploiting the enhanced permeability and retention effect for tumor targeting. Drug Discov. Today **11**, 812 (2006). https://doi.org/10.1016/j.drudis.2006.07.005
76. D. Kalyane, N. Raval, R. Maheshwari et al., Employment of enhanced permeability and retention effect (EPR): Nanoparticle-based precision tools for targeting of therapeutic and diagnostic agent in cancer. Mater. Sci. Eng. C **98**, 1252 (2019). https://doi.org/10.1016/j.msec.2019.01.066
77. D. Zhang, L. Wei, M. Zhong et al., The morphology and surface charge-dependent cellular uptake efficiency of upconversion nanostructures revealed by single-particle optical microscopy. Chem. Sci. **9**, 5260 (2018). https://doi.org/10.1039/c8sc01828f
78. L.G. Justino, K. Nigoghossian, T.S. Capote et al., Multifunctional $EuYVO_4$ nanoparticles coated with mesoporous silica. J. Lumin. **179**, 197 (2016). https://doi.org/10.1016/j.jlumin.2016.07.011
79. M. Kamimura, D. Miyamoto, Y. Saito et al., Design of poly (ethylene glycol)/streptavidin coimmobilized upconversion nanophosphors and their application to fluorescence biolabeling. Langmuir **24**, 8864 (2008). https://doi.org/10.1021/la801056c
80. M. Kamimura, T. Matsumoto, S. Suyari et al., Ratiometric near-infrared fluorescence nanothermometry in the OTN-NIR (NIR II/III) biological window based on rare-earth doped β-$NaYF_4$ nanoparticles. J. Mater. Chem. B **5**, 1917 (2017). https://doi.org/10.1039/c7tb00070g

81. L. Wortmann, S. Suyari, T. Ube et al., Tuning the thermal sensitivity of β-$NaYF_4$: Yb^{3+}, H^{o3}+, ^{Er}3+ nanothermometers for optimal temperature sensing in OTN-NIR (NIR II/III) biological window. J. Lumin. **198**, 236 (2018). https://doi.org/10.1016/j.jlumin.2018.01.049
82. X.L. Hu, N. Kwon, K.C. Yan et al., Bio-conjugated advanced materials for targeted disease theranostics. Adv. Funct. Mater. **30**, 1907906 (2020). https://doi.org/10.1002/adfm.201907906
83. C.M. Liu, L.Y. Zhang, L. Li et al., Specific detection of latent human blood fingerprints using antibody modified $NaYF_4$: Yb, Er, Gd fluorescent upconversion nanorods. Dyes Pigm. **149**, 822 (2018). https://doi.org/10.1016/j.dyepig.2017.11.050
84. R. Kumar, M. Nyk, T.Y. Ohulchanskyy et al., Combined optical and MR bioimaging using rare earth ion doped $NaYF_4$ nanocrystals. Adv. Funct. Mater. **19**, 853 (2009). https://doi.org/10.1002/adfm.200800765
85. L. Xiong, Z. Chen, Q. Tian et al., High contrast upconversion luminescence targeted imaging in vivo using peptide-labeled nanophosphors. Anal. Chem. **81**, 8687 (2009). https://doi.org/10.1021/ac901960d
86. L. Zeng, L. Luo, Y. Pan et al., In vivo targeted magnetic resonance imaging and visualized photodynamic therapy in deep-tissue cancers using folic acid-functionalized superparamagnetic-upconversion nanocomposites. Nanoscale **7**, 8946 (2015). https://doi.org/10.1039/c5nr01932j
87. Q. Chen, X. Wang, F. Chen et al., Functionalization of upconverted luminescent $NaYF_4$: Yb/Er nanocrystals by folic acid–chitosan conjugates for targeted lung cancer cell imaging. J. Mater. Chem. **21**, 7661 (2011). https://doi.org/10.1039/c0jm04468g
88. Q. Yuan, Y. Wu, J. Wang et al., Targeted bioimaging and photodynamic therapy nanoplatform using an aptamer-guided G-quadruplex DNA carrier and near-infrared light. Angew. Chem. **52**, 13965 (2013). https://doi.org/10.1002/anie.201305707
89. M.M. Rubner, D.E. Achatz, H.S. Mader et al., DNA "nanolamps": "Clicked" DNA conjugates with photon upconverting nanoparticles as highly emissive biomaterial. ChemPlusChem **77**, 129 (2012). https://doi.org/10.1002/cplu.201100055
90. A.C. Marques, P.J. Costa, S. Velho et al., Functionalizing nanoparticles with cancer-targeting antibodies: A comparison of strategies. J. Control. Release **320**, 180 (2020). https://doi.org/10.1016/j.jconrel.2020.01.035
91. M. Kamimura, A. Omoto, H.C. Chiu, et al., Enhanced red upconversion emission of $NaYF_4$: Yb^{3+}, Er^{3+}, Mn^{2+} nanoparticles for near-infrared-induced photodynamic therapy and fluorescence imaging Chem. Lett. **46**, 1076 (2017). 10.1246/cl.170322
92. Y.C. Tsai, P. Vijayaraghavan, W.H. Chiang et al., Targeted delivery of functionalized upconversion nanoparticles for externally triggered photothermal/photodynamic therapies of brain glioblastoma. Theranostics **8**, 1435 (2018). https://doi.org/10.7150/thno.22482
93. T. Zhao, L. Chen, Q. Li et al., Near-infrared light triggered drug release from mesoporous silica nanoparticles. J. Mater. Chem. B **6**, 7112 (2018). https://doi.org/10.1039/c8tb01548a
94. W. Cheng, J. Nie, L. Xu et al., A pH-sensitive delivery vehicle based on folic acid-conjugated polydopaminemodified mesoporous silica nanoparticles for targeted cancer therapy. ACS Appl. Mater. Interfaces **9**, 18462 (2017). https://doi.org/10.1021/acsami.7b02457
95. J. Dong, J.I. Zink, Light or heat? The origin of cargo release from nanoimpeller particles containing upconversion nanocrystals under IR irradiation. Small **11**, 4165 (2015). https://doi.org/10.1002/smll.201500607
96. E. Aznar, L. Mondragón, J.V. Ros-Lis et al., Finely tuned temperature-controlled cargo release using paraffin-capped mesoporous silica nanoparticles. Angew. Chem. Int. Ed. Engl. **50**, 11172 (2011). https://doi.org/10.1002/anie.201102756
97. J. Lee, S. Han, J. Lee et al., Stimuli-responsive α-helical peptide gatekeepers for mesoporous silica nanocarriers. New J. Chem **41**, 6969 (2017). https://doi.org/10.1039/C7nj00124j
98. E. Ruiz-Hernández, A. Baeza, M. Vallet-Regí, Smart drug delivery through DNA/magnetic nanoparticle gates. ACS Nano **5**, 1259 (2011). https://doi.org/10.1021/nn1029229
99. M. Arruebo, R. Fernández-Pacheco, M.R. Ibarra et al., Magnetic nanoparticles for drug delivery. Nano Today **2**, 22 (2007). https://doi.org/10.1016/s1748-0132(07)70084-1

100. E. Hemmer, A. Benayas, F. Légaré et al., Exploiting the biological windows: Current perspectives on fluorescent bioprobes emitting above 1000 nm. Nanoscale Horiz. **1**, 168 (2016). https://doi.org/10.1039/c5nh00073d
101. S.L. Jacques, Optical properties of biological tissues: A review. Phys. Med. Biol. **58**, R37 (2013). https://doi.org/10.1088/0031-9155/58/11/r37
102. M. Yamanaka, H. Niioka, T. Furukawa et al., Excitation of erbium-doped nanoparticles in 1550-nm wavelength region for deep tissue imaging with reduced degradation of spatial resolution. J. Biomed. Opt. **24**, 070501 (2019). https://doi.org/10.1117/1.jbo.24.7.070501
103. H. Zhang, Y. Fan, P. Pei et al., Tm^{3+}-sensitized NIR-II fluorescent nanocrystals for *in vivo* information storage and decoding. Angew. Chem. **131**, 10259 (2019). https://doi.org/10.1002/ange.201903536
104. Y.T. Lim, S. Kim, A. Nakayama et al., Selection of quantum dot wavelengths for biomedical assays and imaging. Mol. Imaging **2**, 50 (2003). https://doi.org/10.1162/15353500200302163
105. D.J. Naczynski, M.C. Tan, M. Zevon et al., Rare-earth-doped biological composites as in vivo shortwave infrared reporters. Nat. Commun. **4**, 1 (2013). https://doi.org/10.1038/ncomms3199
106. H. Dong, L.D. Sun, W. Feng et al., Versatile spectral and lifetime multiplexing nanoplatform with excitation orthogonalized upconversion luminescence. ACS Nano **11**, 3289 (2017). https://doi.org/10.1021/acsnano.7b00559
107. D.H. Ortgies, M. Tan, E.C. Ximendes et al., Lifetime-encoded infrared-emitting nanoparticles for in vivo multiplexed imaging. ACS Nano **12**, 4362 (2018). https://doi.org/10.1021/acsnano.7b09189
108. S. Wu, G. Han, D.J. Milliron et al., Non-blinking and photostable upconverted luminescence from single lanthanide-doped nanocrystals. Proc. Natl. Acad. Sci. **106**, 10917 (2009). https://doi.org/10.1073/pnas.0904792106
109. Q. Liu, Y. Zhang, C.S. Peng et al., Single upconversion nanoparticle imaging at sub-10 W cm^{-2} irradiance. Nat. Photon. **12**, 548 (2018). https://doi.org/10.1038/s41566-018-0217-1
110. J. Zhou, A.I. Chizhik, S. Chu et al., Single-particle spectroscopy for functional nanomaterials. Nature **579**, 41 (2020). https://doi.org/10.1038/s41586-020-2048-8
111. N.J. Johnson, W. Oakden, G.J. Stanisz et al., Size-tunable, ultrasmall $NaGdF_4$ nanoparticles: insights into their T_1 MRI contrast enhancement. Chem. Mater. **23**, 3714 (2011). https://doi.org/10.1002/chin.201144006
112. Y. Chen, Y. Fu, X. Li et al., Peptide-functionalized $NaGdF_4$ nanoparticles for tumor-targeted magnetic resonance imaging and effective therapy. RSC Adv. **9**, 17093 (2019). https://doi.org/10.1039/c9ra02135c
113. A.J.L. Villaraza, A. Bumb, M.W. Brechbiel et al., Macromolecules, dendrimers, and nanomaterials in magnetic resonance imaging: The interplay between size, function, and pharmacokinetics. Chem. Rev. **110**, 2921 (2010). https://doi.org/10.1021/cr900232t
114. N. Liu, R. Marin, Y. Mazouzi et al., Cubic versus hexagonal–effect of host crystallinity on the T_1 shortening behaviour of $NaGdF_4$nanoparticles. Nanoscale **11**, 6794 (2019). https://doi.org/10.1039/c9nr00241c
115. G. Lee, Y.I. Park, Lanthanide-doped upconversion nanocarriers for drug and gene delivery. Nanomaterials **8**, 511 (2018). https://doi.org/10.3390/nano8070511
116. G. Jalani, R. Naccache, D.H. Rosenzweig et al., Photocleavable hydrogel-coated upconverting nanoparticles: a multifunctional theranostic platform for NIR imaging and on-demand macromolecular delivery. J. Am. Chem. Soc. **138**, 1078 (2016). https://doi.org/10.1021/jacs.5b12357
117. C. Yao, P. Wang, X. Li et al., Near-infrared-triggered azobenzene-liposome/upconversion nanoparticle hybrid vesicles for remotely controlled drug delivery to overcome cancer multidrug resistance. Adv. Mater. **28**, 9341 (2016). https://doi.org/10.1002/adma.201503799
118. M. Uo, E. Kudo, A. Okada et al., Preparation and properties of dental composite resin cured under near infrared irradiation. J. Photopolym. Sci. Technol. **22**, 551 (2009). https://doi.org/10.2494/photopolymer.22.551
119. V.V. Rocheva, A.V. Koroleva, A.G. Savelyev et al., High-resolution 3D photopolymerization assisted by upconversion nanoparticles for rapid prototyping applications. Sci. Rep. **8**, 1 (2018). https://doi.org/10.1038/s41598-018-21793-0

120. K. Wang, J. Peña, J. Xing, Upconversion nanoparticles assisted photopolymerization. Photochem. Photobiol. (2020). https://doi.org/10.1111/php.13249
121. K. Nigoghossian, S. Saska, L. Christovam et al., Upconversion 3D printed composite with multifunctional applications for tissue engineering and photodynamic therapy. J. Braz. Chem. Soc. **31**, 638 (2020). https://doi.org/10.21577/0103-5053.20190228
122. Q. Xiao, Y. Ji, Z. Xiao et al., Novel multifunctional $NaYF_4$: Er^{3+}, Yb^{3+}/PEGDA hybrid microspheres: NIR-light-activated photopolymerization and drug delivery. Chem. Commun. **49**, 1527 (2013). https://doi.org/10.1039/c2cc37620b
123. M. Kamimura, Y. Yano, S. Kuraoka et al., Near-infrared to visible upconversion emission induced photopolymerization: Polystyrene shell coated $NaYF_4$ nanoparticles for fluorescence bioimaging and nanothermometry. J. Photopolym. Sci. Technol. **30**, 265 (2017). https://doi.org/10.2494/photopolymer.30.265
124. D.W. Bruce, D. O'Hare, R.I. Walton, Eds., Molecular materials, Wiley (2011). https://doi.org/10.1002/9780470686058.
125. L. Aboshyan-Sorgho, C. Besnard, P. Pattison et al., Near-infrared→ visible light upconversion in a molecular trinuclear d–f–d complex. Angew. Chem. Int. Ed. **50**, 4108 (2011). https://doi.org/10.1002/anie.201100095
126. A. Nonat, C.F. Chan, T. Liu et al., Room temperature molecular up conversion in solution. Nat. Commun. **7**, 1 (2016). https://doi.org/10.1038/ncomms11978
127. A. Nonat, S. Bahamyirou, A. Lecointre, et al., Molecular upconversion in water in heteropolynuclear supramolecular Tb/Yb assemblies. J. Am. Chem. Soc. **141**, 1568 (2019). https://doi.org/10.1021/jacs.8b10932
128. B. Golesorkhi, H. Nozary, A. Fürstenberg, et al., Erbium complexes as pioneers for implementing linear light-upconversion in molecules. Mater. Horiz. **7**, 1279 (2020). https://doi.org/10.1039/C9MH01899A
129. X. Zhang, B. Li, H. Ma et al., Metal–organic frameworks modulated by doping Er^{3+} for upconversion luminescence. ACS Appl. Mater. Interfaces **8**, 17389 (2016). https://doi.org/10.1021/acsami.6b03841
130. M. Li, S. Gul, D. Tian et al., Erbium(III)-based metal–organic frameworks with tunable upconversion emissions. Dalton Trans. **47**, 12868 (2018). https://doi.org/10.1039/c8dt02329h
131. D. Manzani, K. Nigoghossian, M.F. Iastrensk, et al., Luminescent silicone materials containing Eu3+-complexes for photonic applications. J. Mater. Chem. C**6**, 8258 (2018). https://doi.org/10.1039/c8tc01798k
132. Y. Suffren, B. Golesorkhi, D. Zare et al., Taming lanthanide-centered upconversion at the molecular level. Inorg. Chem. **55**, 9964 (2016). https://doi.org/10.1021/acs.inorgchem.6b00700
133. L.J. Charbonnière, Bringing upconversion down to the molecular scale. Dalton Trans. **47**, 8566 (2018). https://doi.org/10.1039/c7dt04737a
134. A.M. Nonat, L.J. Charbonnière, Upconversion of light with molecular and supramolecular lanthanide complexes. Coord. Chem. Rev. **409**, 213192 (2020). https://doi.org/10.1016/j.ccr.2020.213192
135. A. Reinhard, H.U. Güdel, High-resolution optical spectroscopy of $Na_3[Ln(dpa)_3]\cdot 13H_2O$ with $Ln = Er^{3+}$, Tm^{3+}, Yb^{3+}. Inorg. Chem. **41**, 1048 (2002). https://doi.org/10.1021/ic0108484
136. X. Xiao, J.P. Haushalter, G.W. Faris, Upconversion from aqueous phase lanthanide chelates. Opt. Lett. **30**, 1674 (2005). https://doi.org/10.1364/ol.30.001674
137. O.A. Blackburn, M. Tropiano, T.J. Sorensen et al., Luminescence and upconversion from thulium(III) species in solution. Phys. Chem. Chem. Phys. **14**, 13378 (2012). https://doi.org/10.1039/c2cp42228j
138. K. Nigoghossian, T. Miyashita, A. Omura et al., Infrared to visible upconversion luminescence of trivalent erbium tetrafluoroborate complexes. Opt. Mater. Express **10**, 1749 (2020). https://doi.org/10.1364/ome.10.001749

Part III
Application of NIR Light for Biomedical Photonics

Chapter 7
Nanothermometry for Deep Tissues by Using Near-Infrared Fluorophores

Masakazu Umezawa and Karina Nigoghossian

Abstract Temperature is a fundamental quantity that can regulate various biological phenomena and thus important in clinical biology as well as sport and health sciences. This chapter reviews how the body temperature is regulated in the organisms (mainly in mammals including humans), and how the body temperature of not only the surface but also the deep tissues. As an emerging technique, the luminescence nanothermometry that works in the over-thousand-nanometer (OTN) near-infrared (NIR) wavelength range, which allows us to look the biological tissues transparently, is being developed to visualize and reveal the mechanisms of dynamic time-dependent changes in body temperature distribution in deep tissues. The data and knowledge collected with the new techniques will provide insights into body temperature control in biology and its management in biomedical and engineering fields.

Keywords Core temperature · Temperature distribution · Contactless luminescence thermometry · Self-referencing sensors · Ratiometric thermometry · Fluorescence lifetime

7.1 Body Temperature of Deep Tissues in Biology

7.1.1 Body Temperature

Temperature is a fundamental physical quantity that governs every chemical reaction in biological tissues. Body temperature is normally controlled within very small range as well as 35–37 °C on the surface and 36.5–37.5 °C in deep tissues (the core)

M. Umezawa (✉) · K. Nigoghossian
Department of Materials Science and Technology, Faculty of Industrial Science and Technology, Tokyo University of Science, 6-3-1 Niijuku, Katsushika, Tokyo 125-8585, Japan
e-mail: masa-ume@rs.tus.ac.jp

K. Nigoghossian
ICGM, Univ. Montpellier, CNRS, ENSCM, UM, Bât 17, cc1701, Place Eugène Bataillon, 34095 Montpellier, France

K. Soga et al. (eds.), *Transparency in Biology*,
https://doi.org/10.1007/978-981-15-9627-8_7

in humans. Temperature control is an important part of homeostatic mechanism that keeps the organism at optimum operating temperature because the temperature influences the rate of chemical and enzymatic reactions. If the body temperature was too low, the rate of chemical reactions would become too low to maintain living activities. If the temperature was too high, proteins would be denatured and life functions could not be maintained. Of course, thermophilic bacteria also called hyperthermophiles such as *Thermus aquaticus* (*Taq*) [1], which live in high-temperature environments such as thermal springs and hydrothermal vents, possess thermostable proteins that work without denaturation at high temperatures. *Taq* polymerase, originally isolated in 1976 [2], and newly found enzyme [3] have been used for the polymerase chain reaction, a fast in vitro method to make millions to billions of copies of specific regions of deoxynucleic acid (DNA) [4, 5] that requires an enzymatic amplification reaction at high temperature (>90 °C). Further, in vivo biological phenomena with signal transductions are essentially regulated by molecular association and dissociation, such as heat shock proteins in Raf signaling [6, 7] and in MEK5-ERK5 pathway for cell proliferation [8], and other proteins in Rel/NF-kB/IkB signaling [9], Nrf2-ARE signaling [10], and heme-related signal transduction [11]. Physically the molecular association/dissociation likely to be also affected by the temperature.

Biologically, cellular functions such as cell division is affected by temperature [12]. It is very popular that temperature regulates the timing of multiple developmental processes and have dramatic effects on plant growth on architecture [13]. Incubation temperature of fertilized eggs influences the sex determination over genotypic factor in certain species (e.g., *Amphibolurus muricatus*) in reptile [14]. In another case, ovulation-dependent temperature difference within the female genital tract may be related to sperm guidance towards the egg [15]. Because biological reactions responsible for cellular functions occur at particular locations within an organism, temperature distributions inside a body and a cell and their relation with functions of cellular components have been much focused [16].

7.1.2 Mechanisms of Body Temperature Control (Mainly in Humans)

Human body temperature varies by the circadian rhythm [17–19], diets, lifestyles, hormones [19], aging, physical activities [18], and surrounding environments. The core temperature is regulated more constantly than the surface temperature, which is more easily influenced by the surroundings [20]. The body temperature is controlled by the balance between internal heat production (thermogenesis) and controlled heat dissipation. The thermogenesis occurs along with the production of adenosine triphosphate (ATP), which is so-called "molecular unit of currency" of intracellular energy transfer, at mitochondria, which are present at about 300–400 per cell (~2000 per hepatocyte) on average. Thus, body heat is generated mostly in deep organs, especially the liver, brain, and in contraction of heart and skeletal muscles, which are metabolically active and need much energy. The heat generated is distributed to the whole body by circulating blood (convection). Following, the heat dissipation occurs by the convection as well as conduction, radiation, and evaporation (Fig. 7.1).

The core temperature of humans is regulated and stabilized by the hypothalamus, which is a region of the brain and is responsible for the regulation of most of metabolic processes and activities of the autonomic nervous system including the sympathetic system [20]. The hypothalamus senses the core temperature and work as like a thermostat to maintain it near a desired range [21, 22]. While the heat production along with the contraction of skeletal muscle is stimulated by motor neurons, the other heat source, the metabolism in the internal organs, is majorly dominated by the autonomic nerve system. When the blood temperature rises above desired range, hypothalamus

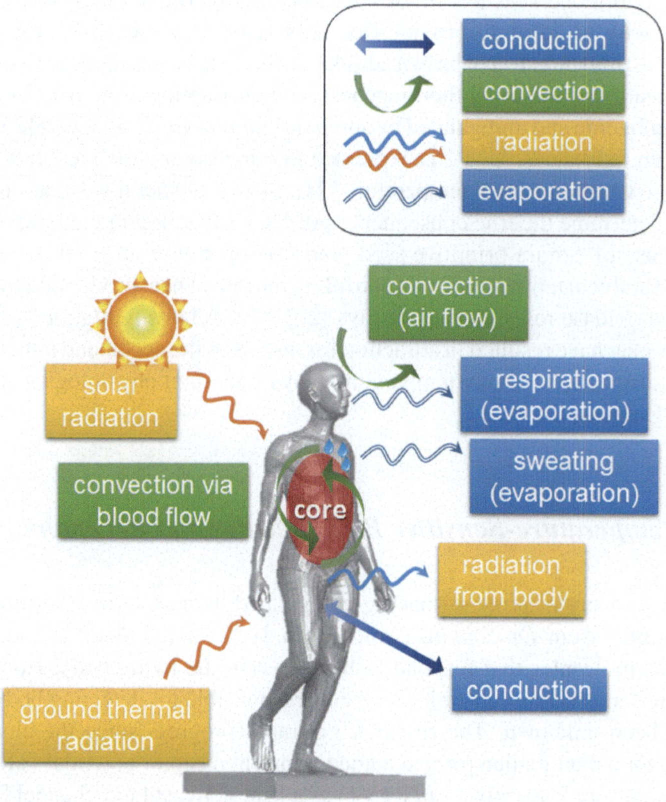

Fig. 7.1 Schematic image of heat transfer between the body and the environment. The body temperature is controlled by the balance between internal heat production (thermogenesis) and controlled heat dissipation. Thermal energy is provided by contraction of skeletal muscle, metabolism in internal organs, and radiation and conduction from the environment. The heat generated is distributed to the whole body by circulating blood (convection). Following, the heat dissipation occurs by the convection as well as conduction, radiation, and evaporation

stimulates sweating and vasodilation to cool the temperature by enhancing heat loss via evaporation, convection, and conduction from the body.

In cold conditions, heat is lost from the skin via the conduction and convection mainly through the hands and feet, while sweat production is decreased to suppress the evaporation. Therefore the hypothalamus stimulates heat production and retention to maintain the body temperature. By activating the sympathetic nervous system, the arterioles that receives noradrenalin by the adrenergic α_1-receptor can constrict to reduce the carrying amount of blood to superficial capillaries under the skin and to suppress the heat loss from the surface. Skeletal muscles can also receive information from the hypothalamus via motor neurons to cause shivering, which increases thermogenesis by exothermic reaction of the ATP production. In addition, noradrenalin also reaches adrenergic β_3-receptor on adipose tissues to enhance the activity of intracellular hormone-sensitive lipase for metabolizing triglycerides, which are fatty acid esters with glycerol, to provide free fatty acids. Released free fatty acids are provided into mitochondria of brown adipose tissue. The brown adipose tissue mainly produces heat (non-shivering thermogenesis) by uncoupling protein 1 (UCP1), also called thermogenin and upregulated by adrenergic signal via β_3-receptor [23], instead of ATP from fatty acids. UCP1 is expressed in other organs such as brain and may contribute to regulating local temperature [24], though further investigations are still needed to determine the true consequences of UCP1 function in the brain [25].

In the case of certain primitive seed plants, temperature of the flowers remains higher than ambient temperature by controlling the rate of heat production by involvement of the oxidase respiratory pathways [26]. This relatively high temperature of the flowers may have resulted in attraction for insects as the nectar and pollen rewards are today [26]. Thus the body temperature plays important physiological and social roles in organisms.

7.1.3 Temperature-Sensitive Biomolecules and Receptors

Since the discovery of proteins that are upregulated by heat stress, thus named heat shock proteins, from *Drosophila* (fruit fly) in 1974 [27], it has been known that endogenous molecules that respond to heat exist in the living body. However, the endogenous temperature sensor biomolecule responsible for detecting body temperature had been unknown. The research got rapid progress since the cloning was succeeded for a mammalian protein named transient receptor potential (TRP) cation channel subfamily *V* member 1 (TRPV1) as a heat-activated ion channel [28]. TRP channels are a group of ion channels that respond to various mechanical and physicochemical stimuli such as pain, pressure, osmolarity, vision, tastes, chemicals (acids, lipids, etc.), and temperature [29, 30]. They are expressed in various cells and tissues including sensory neurons, skin, and brain [30, 31]. The name TRP, originally found as a photoreceptor in *Drosophila*, is derived from the transient response to light in the *trp*-mutant fruit flies, which lack a functional copy of trp gene, whereas wild-type strains demonstrate a sustained activity in their photoreceptor cell response [32].

There are several different temperature-sensitive TRPs, whose temperature thresholds for activation are widely distributed in the physiological temperature range [30].

Especially TRPV channels are famous for their responsiveness to capsaicin, the main pungent ingredient of 'hot' chilli peppers [28]. TRPV1, which responds to high temperature (>42 °C or lower threshold modified by in vivo factors such as inflammatory condition [33]), likely to work as a thermosensor in the hypothalamus and be involved in body temperature control, as shown by increase in body temperature with TRPV1 blockade in rats and other mammals [34]. TRPV1 knockout mice lack the secretion of vasopressin secretion, which promotes water retention in the body, in response to hyperthermia [35]. Another TRP channel, TRPV4, that responds to lower (warm) temperature (>27–35 °C) is expressed in the hypothalamus where central thermoregulatory centers reside and speculated to be a warmth-sensitive temperature sensor that responds to changes in body temperature [30, 36]. However, TRPV4 knockout mice show normal body temperature control in response to rapid ambient temperature change [37], suggesting TRPV4 is dispensable in the central temperature sensing [30].

7.1.4 Significance of Local Body Temperature in Clinical Biology

7.1.4.1 Fever

Fever, also referred to as pyrexia, is defined as having a body temperature above the normal range due to elevated set point of the temperature. The mechanism of fever is different from hyperthermia in hot and humid environments, called heat stroke, which is caused by insufficient or impaired heat loss. It can be induced by infections of viruses, bacteria, and parasites and by non-infectious causes such as vasculitis, deep vein thrombosis, and cancer. The growth and replication of some microorganisms and viruses are affected by temperature above 37 °C, while some host response mechanisms including immunological processes are enhanced at higher body temperature [38]; therefore, treatment to reduce the fever is generally not needed [39]. However, treatment of associated pain, discomfort, delirium, and lethargy by antipyretics is helpful to improve the person's well-being [38].

The central set point of body temperature is elevated by the effect of prostaglandin E_2 (PGE_2), a physiologically active molecule synthesized from an unsaturated fatty acid, arachidonic acid. Arachidonic acid is rich in the phospholipids composed of cell membrane and is released by an enzyme phospholipase A_2 (PLA_2), which is activated by intracellular phosphorylation via inflammatory stimuli and specifically recognizes the *sn*-2 acyl bond of phospholipids to hydrolyze it [40, 41]. Then arachidonic acid is converted to PGE_2 by cyclooxygenases (COX) and PGE synthase. COX-2 expression and PGE_2 production is induced in the cerebral endothelial cells

[42, 43] and astrocytes [44] in response to circulating inflammatory cytokines such as interleukin (IL)-1β, IL-6, and tumor necrosis factor (TNF), which are increased in infectious and some other disease conditions. Notably, produced PGE_2 acts at its receptor EP3 [45], which is expressed on GABAergic neurons inhibiting thermogenic responses in the hypothalamus [46], to suppress the function of the inhibitory neurons and causes fever responses [47], mainly heat generation via the shivering and the contraction of vessels to suppress the heat dissipation (Fig. 7.2).

7.1.4.2 Medical Use of Hypothermia for Newborn Brain Injury

Induction of hypothermia is used for purposes of treatment of brain injury, especially ischemic newborn brain injury, also called neonatal hypoxic encephalopathy. The main purpose of cooling is to reduce the metabolism that requires oxygen. Especially, the brain is the organ most susceptible to ischemia and hypoxia; therefore, brain cooling results in a reduction in oxygen requirements and protects the organ from injury induced by hypoxia [48]. Even though the hypothermia therapy does not appear to be useful for the injuries in adults [49, 50], it is especially effective for newborn brain injury to reduce brain damage and future disability [51]. The baby is cooled down to around 33 °C for several days in this hypothermia therapy for the brain injury.

7.1.4.3 Potential Use of Body Cooling in Sport Science

Local and whole body cold therapy is also used in a training regimen usually following a period of intense exercise. Previous reports suggested that cold water immersion may reduce muscle damage and discomfort [52]. However, the method is so far controversial with a risk-adverse side effects such as undesired hypothermia and shock. This situation is also pointed out in the field of rehabilitation [53]. Other studies suggested that cold water immersion may slow muscle growth and interfere long-term training gains [54]. At this stage, we have to say that there is remarkably little evidence investigated on the effect of the cooling (cool water immersion and ice-pack contact) after exercises, including potentials of suppression of excess inflammation and tissue damage, though the cold application may induce immediate vasoconstriction with reflexive vasodilation. The visualization of the effects of cooling on a time-dependent change in the temperature of deep tissues (i.e., muscles and brains with blood vessels) may help to overcome the limitations of the cooling therapies.

7.1.4.4 Photothermal Therapy (PTT) for Cancer

Monitoring the temperature of tissues is also important for hyperthermia therapy, including an emerging technique of photothermal therapy, of diseased tissues. The

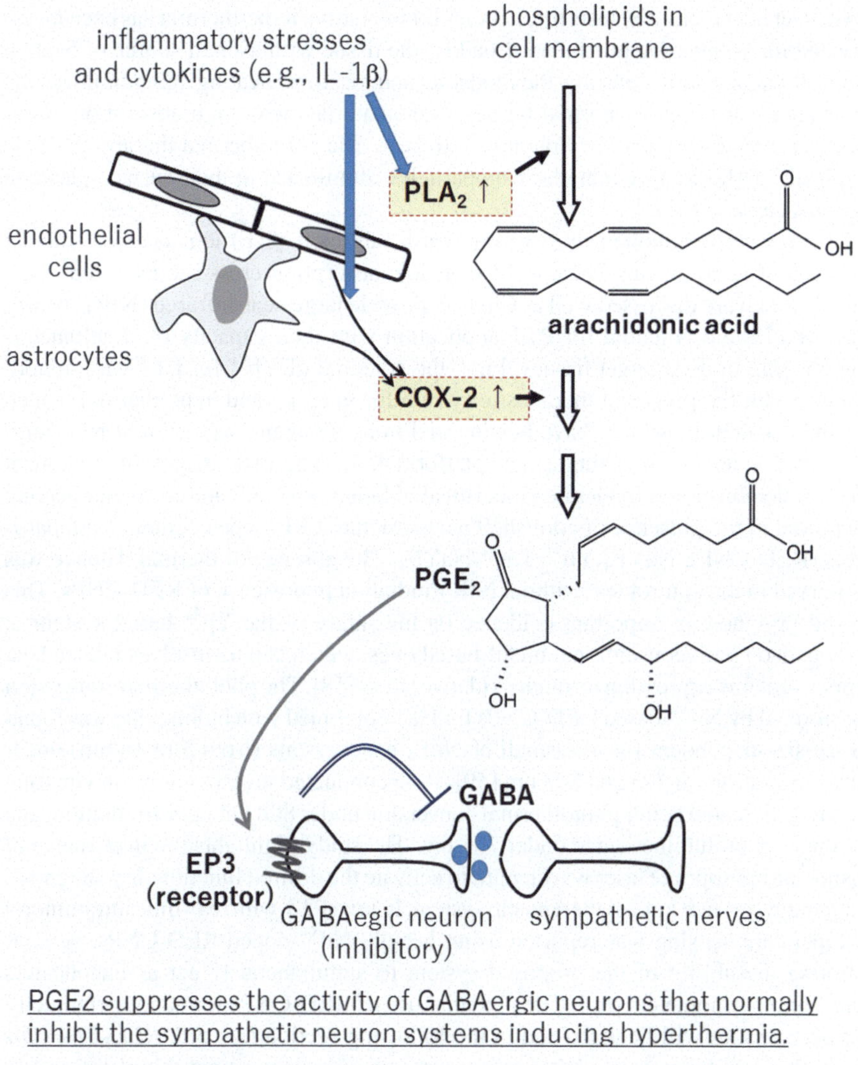

Fig. 7.2 Molecular mechanism of fever is regulated in the hypothalamus during inflammation and infection. Prostaglandin E_2 (PGE_2) production likely to be increased via upregulation of phospholipase A_2 (PLA_2) and cyclooxygenase-2 (COX-2) by inflammatory cytokines. In the hypothalamus, it suppresses the activity of GABAergic neurons that regulates the sympathetic nervous systems, and results in fever

hyperthermia therapy is based on heating of a target tissue above its normal level up to a cytotoxic range. Conventional methods for thermal ablation of cancer use irradiations at ultrasound [55] and microwave [56] frequencies. Because such external irradiation source causes overheating of surrounding normal tissues, alternative method to deliver heat locally by inserting needles for interstitial hyperthermia has been investigated for an invasive procedure. Loading the tissue with a small source of heat, at subcellular size scale, ensures the moderate and localized heating and minimizes the damage to surrounding normal tissues. Nanomaterials with high absorption cross-section can convert absorbed photons into heat. The photothermal therapy (PTT) is typically performed by metallic nanoparticles illuminated at their surface plasmon resonance band [57].

Interestingly, although most of rare-earth rare-earth (RE) ions are not efficient sources of local heating because of their low absorption cross-sections, Yb^{3+} and Nd^{3+}, which are discussed well *in* Chap. 6, provide large near-infrared (NIR) absorption and have a potential for PTT application with their capacity of simultaneous bioimaging in deep target tissues. First, the potential of Yb^{3+} for PTT was demonstrated with the proposed theranostic system for imaging and hyperthermia, which is simpler than those that have heating and imaging agents as different nanocomposites assembled in a single nanoplatform [58]. Thermal images of biological tissues demonstrated moderate (subcritical ablation, <42 °C) and localized heating achieved in the presence of core/shell nanostructured RE-doped ceramic nanoparticles (RED-CNP), $NaYF_4$:Yb^{3+}:Tm^{3+}/$NaYF_4$. The absence of thermal damage was observed in the control test without NIR irradiation in presence of RED-CNPs. This result provided an important evidence on the safety of the Yb^{3+}-based system in comparison with conventional metal-based ones, which can themselves induce heat shock proteins expression through oxidative stress [58]. The photothermal conversion performed by Nd^{3+}-based RED-CNPs for PTT combined with bioimaging was found to be size-dependent for ultrasmall $NaNdF_4$ nanocrystals (from 5 to 16 nm) under laser excitations at 793 and 808 nm [59]. Both conducted investigations, in vitro and in vivo, indicated better photothermal conversion under 808 nm laser irradiation, and stronger photoluminescence under 793 nm. The study highlights the importance of using the appropriate laser wavelength to activate the desired functionality, diagnosis or therapy, in such kind of theranostic system. In vivo PTT with real-time intratumoral temperature sensing was reported using heavily Nd^{3+}-doped RED-CNPs, demonstrating the ability of the proposed system to simultaneously act as nanoheaters and nanothermometers [60]. The potential use of RED-CNPs for nanothermometry of deep tissues will be discussed in the following section (*see 3.3. RED-CNPs for over-thousamd nanometer NIR luminescent nanothermometry*) of this chapter.

Another NIR phosphor, single-walled carbon nanotubes (SWCNTs, see Chap. 5 in detail), also works as a fluorescent agent for in vivo deep imaging and a heater that can be operated by photoexcitation for tumor elimination [61]. The capacity of SWCNTs for photothermal effects and simultaneously deep imaging has been investigated after their intravenous [61, 62] and intratumor injection [63] towards imaging-guided therapy. Combination use of such materials with thermometers for deep tissues will enhance the efficiency and safety control of PTT.

7.2 Present Techniques of Body Temperature Measurement

Traditional techniques of body temperature measurement have been contact thermometers, and infrared (IR) thermography. These techniques are convenient and help monitor the temperatures very easily at every place including homes and hospitals.

Contact thermometers include mercury thermometer, thermocouple, and thermistor. Recently, body medical thermometers are composed of the thermistor, a resistor whose electric resistance largely changes with temperature, instead of mercury thermometer. Surgical insertion of a contact thermometer has been required for direct measurement of the core body temperature. Therefore, the rectal or vaginal temperature, additionally in the mouth (oral) and the ear (tympanic), measured with thermometers is generally considered to give the most accurate assessment of the core body temperature [17, 18, 64]. Thermochromism, a property of substances to change their color by temperature change, is an important phenomena to know that the optical properties such as light absorption can change due to a change in temperature [65], though it is not often used for body temperature measurement. Some kinds of dyes, polymers, and inorganic materials have thermochromism as liquid crystals which are used in temperature indicators such as 'thermometer sticker'. The limitation of contact thermometry methods is that they are invasive and unsuitable for remote temperature measurement.

IR thermography is based on the detection of blackbody radiation in IR, emitted by all objects above the absolute zero (0 K), for which the wavelength of its spectra is inversely related to intensity that depends only on the object's temperature. This type of thermometer essentially consists of a lens to focus the IR thermal radiation on to a detector, which converts the detected signal to temperature to display after being compensated for ambient temperature. This method is also convenient with the advantage of enabling remote temperature sensing, as used for health screening in public spaces such as international airports. In the medical field, the possibility of remote temperature sensing has been investigated on, for example, temporal artery in hospitalized adults [66] and ear of children [67], although the accuracy is considered to be still poor. This techniques also has a limitation that it visualizes only the surface temperature because mid-IR is largely absorbed by water contained in biological tissues; however, remote sensing techniques will help collect a large amount of data over time.

7.3 Contactless Luminescence Thermometry (LNTh)

7.3.1 Luminescence Thermometry

Luminescence thermometry has been studied for the development of contactless and remote temperature measurement, widely reported in the fields of materials science and engineering [68]. It is well known that luminescence intensity of fluorescent molecules and materials tends to change with the temperature, because of thermal influence on non-emissive relaxations and energy transfers of excited photons. This review focused on the thermometry that works in the temperature range of physiological conditions.

7.3.1.1 LNTh Demands in Biomedical Field

Especially, luminescent nanothermometry (LNTh) is an emerging research field dealing with the nano-sized (sub-micrometric scaled) materials or polymers that exhibit temperature-dependent luminescent properties. The new technique can be used to measure the temperature of an object which is smaller than a micrometer. Temperature sensors with high performance are highly demanded in the nanomedicine field [69]. Monitoring temperature variations is crucial for scientific investigations in biomedicine since cell events can be indicated by temperature gradients, such as cell division, gene expression, enzymatic reactions, metabolism, and pathological states. The action of an exogenous agent, such as a drug, may also be accompanied by temperature variations [70]. The cancer cells can be distinguished from normal ones by an increased temperature, which is related to increased rate of metabolic activity [71]. Moreover, measuring temperature with high accuracy is highly demanded to control hyperthermia procedures [72]. Thus, the nanothermometry field has the biomedicine as one of its main focus [73]. The required improvements in sensor performance are mainly related to the spatial resolution at the submicron level, and the ability to remotely detect temperature, without physical contact between probe and specimen. Such contactless and high-resolution methods can be achieved by using probes that offer thermal-dependent luminescence as a detectable signal. The development of remote temperature sensors with high spatial and temporal resolutions are of great importance to early diagnosis, maximize the therapeutic benefits, and reduce side effects. Hybrid systems combining heater and thermometer nanocomposites have been proposed for real-time monitoring the therapeutic parameters of hyperthermia [74, 75].

7.3.1.2 LNTh Visualizing Intracellular Temperature Gradient

Fluorescence probes which show temperature-dependent emission properties have been designed as fluorescent nanothermometers [76–80]. As shown in an example of the mechanism of temperature-dependent change in visible luminescence of Mn^{2+}-doped semiconductor nanocrystals (quantum dots) [81] (Fig. 7.3), the non-emissive relaxation and energy transfer dependent on temperature provide significant changes in the luminescence. These effects can be exploited for measuring temperature. The physical principle of temperature-dependent change in luminescence of organic

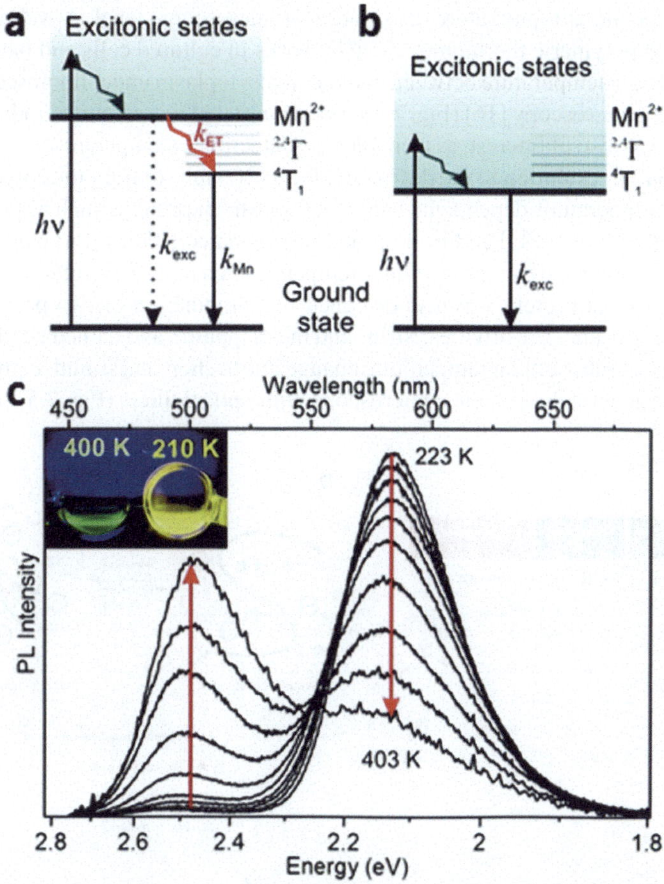

Fig. 7.3 Photoluminescence of colloidal Mn^{2+}-doped semiconductor nanocrystals and its temperature-dependent change. (**a**, **b**) Schematic representation of electronic structures related to photoluminescence in the nanocrystals. **a** When Mn^{2+} states reside within the semiconductor gap, efficient energy transfer (ET) quenches excitonic emission and sensitizes $Mn^{2+}\,{}^4T_1 \rightarrow {}^6A_1$ luminescence. **b** In narrower-gap semiconductors, all Mn^{2+} excited states are located outside of the gap, and the nanocrystals show excitonic luminescence. **c** Variable-temperature photoluminescence spectra of colloidal $Zn_{1-x}Mn_xSe$/ZnCdSe nanocrystals collected in 20-K intervals and normalized to total integrated intensity, showing intensity transfer between Mn^{2+} and excitonic bands as a function of temperature. Inset: Color photograph of the photoluminescence of the nanocrystals at 210 and 400 K (ultraviolet excitation). Adapted with permission from [81]. Copyright (2010) American Chemical Society

dyes, quantum dots, rare-earth (RE) (lantanide)-based phosphors, and polymer-based systems has been well-reviewed with their thermometric properties in previous publications [78].

Organic materials for LNTh, including polymer-based systems, organelle-targeting dyes, and genetically-encoded fluorescent proteins, have shown impressive data on intracellular temperature distribution at microscopic level. A well-designed fluorescence polymeric thermometer (FPT) works in cultured cells and have shown the difference in temperature between nucleus and cytoplasm under fluorescence lifetime imaging microscopy [16] (Fig. 7.4). The potential of temperature gradient in the cytoplasm was also of interest, as it enables the detection of temperature distributions at a high spatial resolution to the diffraction limit (200 nm) of light microscope [82]. The use of temperature-dependent change in fluorescent proteins such as green fluorescent protein (GFP) [83] and its modified form (named tsGFP) [84] has been also investigated for visualizing intracellular temperature change. Self-referencing ratiometric fluorescent protein was also designed by combination of two proteins with different temperature sensitivities, Sirius and mT-Sapphire, and named gTEMP [85]. It visualized an intracellular temperature change due to chemical stimuli to mitochondria under a general fluorescence microscope with optical filters (Fig. 7.5). The most

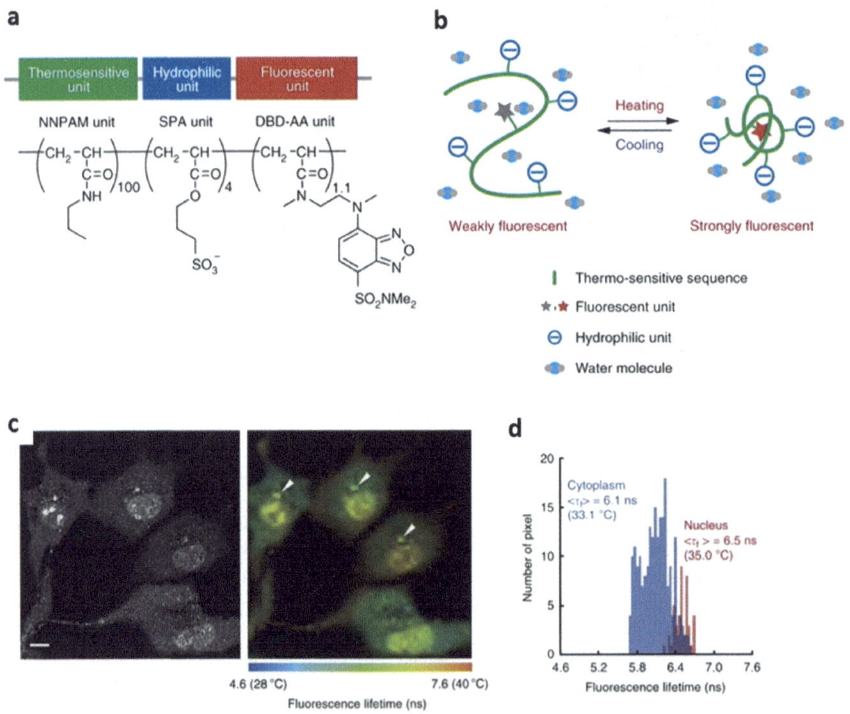

Fig. 7.4 Intracellular temperature mapping with a fluorescent polymeric thermometer (FPT) and fluorescence lifetime imaging microscopy. **a** Chemical structure of the FPT composed of NNPAM (poly-*N*–*n*-propylacrylamide), SPA (3-sulfopropyl acrylate), and DBD-AA (*N*-{2-[(7-*N*,*N*-dimethylaminosulfonyl)-2,1,3-benzoxadiazol-4-yl](methyl)amino}ethyl-*N*-methylacrylamide) units. **b** Functional diagram of the FPT in an aqueous medium. **c** Confocal fluorescence image (left) and fluorescence lifetime image (right) of FPT. **d** Histograms of the fluorescence lifetime in the nucleus and in the cytoplasm in a representative cell (the leftmost cell in **c**) showing higher temperature in the nucleus. Adapted from Reference [16] with permission from Springer Nature

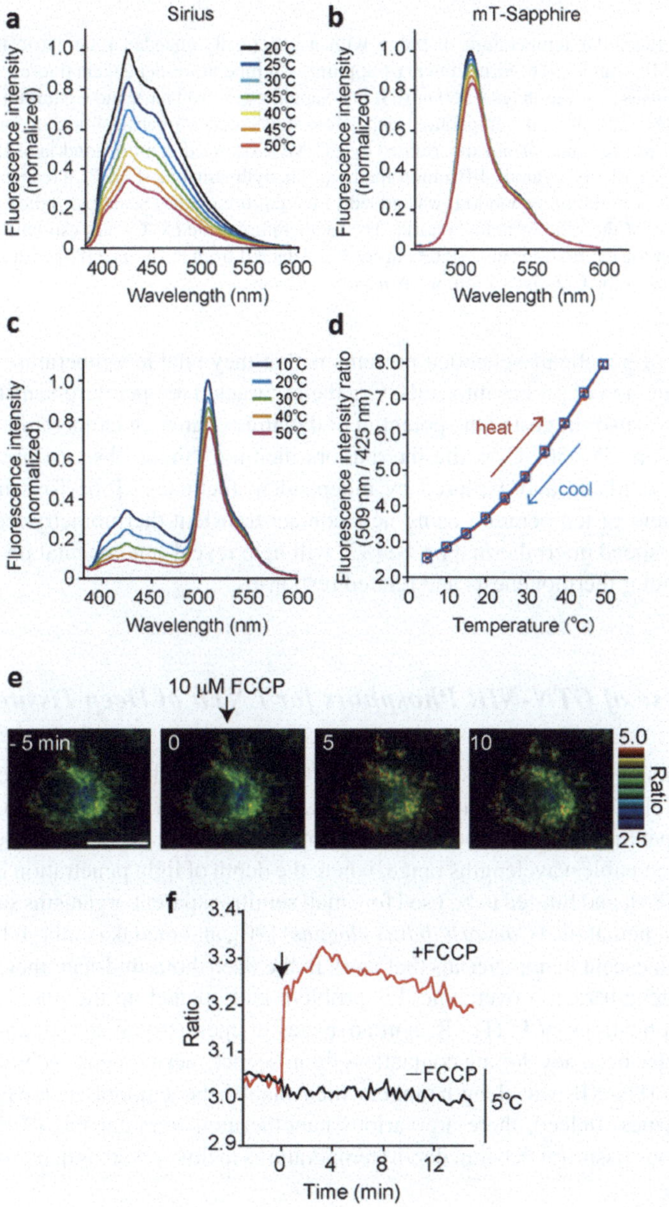
a
Sirius
Fluorescence intensity (normalized)
20°C
25°C
30°C
35°C
40°C
45°C
50°C
Wavelength (nm)
b
mT-Sapphire
Fluorescence intensity (normalized)
Wavelength (nm)
c
Fluorescence intensity (normalized)
10°C
20°C
30°C
40°C
50°C
Wavelength (nm)
d
Fluorescence intensity ratio (509 nm / 425 nm)
heat
cool
Temperature (°C)
e
10 μM FCCP
- 5 min
0
5
10
5.0
Ratio
2.5
f
Ratio
+FCCP
−FCCP
5°C
Time (min)

◀**Fig. 7.5** Intracellular temperature mapping with a genetically-encoded ratiometric fluorescent protein gTEMP, composed of Sirius and mT-Sapphire. **a** Temperature-dependent fluorescence spectrum of (**a**) Sirius [excitation (ex): 360 nm], **b** mT-Sapphire (ex: 400 nm), and **c** equimolar mixture of the two (ex: 360 nm). **d** Temperature-dependent fluorescence intensity ratio of the mixture. **e** Monitoring temperature change determined by gTEMP expressed in mitochondria of HeLa cells. At time 0 min, carbonyl cyanide 4-(trifluoromethoxy)phenylhydrazone (FCCP), a reagent inducing heat production in the mitochondria, was added to the culture media. Scale bar indicates 20 μm. **f** Time course of the ratio in mitochondria. The temperature scale (5 °C) was estimated from the slope value of the calibration curve (0.031 ratio/°C). Adapted from Reference [85] with permission with the terms of the Creative Commons Attribution License

advantage of genetically-encoded proteins is that they enable temperature imaging without injection of probes into cells. Organelle-targeted temperature-sensitive fluorescence dye also suggested the potential of dynamic change in temperature of mitochondria from 38–50 °C, while the authors mention "these observations need to be further validated and explored by independent methods." [86] To validate the microgradient of temperature using new contact transient thermometry techniques like a high-speed microthermal probe [87] will help reveal fundamental phenomena of intracellular thermogenesis and related responses.

7.3.2 *Use of OTN-NIR Phosphors for LNTh of Deep Tissues*

LNTh utilizing lights in the visible wavelength has been used in in vitro experiments. However, it is not suitable for in vivo deep tissues because of the problem in terms of light transparency. Most luminescent thermometers are using excitation and emission lights in the visible wavelengths range, where the depth of light penetration is limited to <1 mm [88], and limited to be used for small semitransparent organisms such as fly larvae [89], nematode (*Caenorhabditis elegans*) [90], and medaka embryo [85]. The use of luminescent nanomaterials that work in the over-thousamd-nanometer (OTN) NIR wavelength region overcame this problem and opened up the possibilities of in vivo applications of LNTh. Recent advances in chemistry, material, and optical sciences have been developing contactless fluorescence thermometry techniques that utilize the OTN-NIR wavelength range, which enables the temperature measurement at deep tissues. Indeed, there are various nanothermometers capable of recording subcutaneous tissue (at ≈1-mm depth) temperatures in this wavelength region shown as follows.

7.3.3 *Rare-Earth-Doped Ceramic Nanoparticles (RED-CNPs) for OTN-NIR LNTh*

7.3.3.1 Mechanism and Development of RED-CNPs for OTN-NIR LNTh

The thermal-dependence of luminescent properties of RE-based fluorescent materials can be further exploited for sensing temperature at the nanoscale level simultaneously with bioimaging [91]. Temperature variations can affect different RE luminescence parameters, such as: (i) population of coupled levels following Boltzmann equilibrium; (ii) energy transfer probabilities related to spectral overlap between donor and acceptor transitions; (iii) spectral shift, bandshape, or bandwidth of a given transition; (iv) integrated emission intensity of a single transition or of a pair of transitions; (v) lifetime measurements. The several excited levels of RE ions enable the development of temperature probes based on multiple temperature-dependent emission lines. The temperature measurement from the ratio of two emission lines is more reliable than using a single luminescence signal. The self-referencing methods, including ratiometric type using two emission bands and lifetime-based type [92] (Fig. 7.6), are not affected by variations in probe concentration, excitation power, detection system, among other factors [78].

Prior to down-conversion OTN-NIR luminescence, upconversion luminescence (UCL) of RE-doped ceramic nanoparticles (RED-CNPs), $NaYF_4$: co-doped with Yb^{3+} and Er^{3+} ($NaYF_4$:Yb^{3+}:Er^{3+}), was first reported to be used for thermal reading in living cells [93] The population of Er^{3+} excited levels $^2H_{11/2}$ and $^4S_{3/2}$, from which the green UCL lines are originated, are in thermal equilibrium according to Boltzmann statistics. Therefore, the luminescence intensity ratio (LIR) between Er^{3+} green UCL lines have a strong temperature dependence. The HeLa cancer cells incubated with polyethyleneimine (PEI)-capped RED-CNPs were subjected to a pump-probe experiment. A 980 nm diode laser (pump beam) was first used for heating. Subsequently, a 488 nm Ar^+ ion laser beam scanned the sample (heating-free probe beam, since it is not absorbed by water). LIR was monitored by a fiber-coupled spectrometer. Thermal images were obtained by using a confocal fluorescence microscopy. Temperature of cells was measured by monitoring the LIR in the vicinity of physiological temperature range, from 25 °C to its thermally induced death at 45 °C. However, nanothermometry based on visible emission via UCL is limited for in vivo thermal imaging by the low propagation of visible signals.

7.3.3.2 Ratiometric Nanothermometry Using Yb^{3+} as a Photosensitizer

The RED-CNPs-based system, β-$NaYF_4$ co-doped with Yb^{3+}, Ho^{3+}, and Er^{3+} [94], is the simplest fluorescent ratiometric nanothermometer (Fig. 7.6a) that works in the OTN-NIR wavelength range. It demonstrates temperature-dependent luminescence intensity ratio (LIR) between Ho^{3+} ($^5I_6 \rightarrow {}^5I_8$) and Er^{3+} ($^4I_{13/2} \rightarrow {}^4I_{15/2}$) emissions

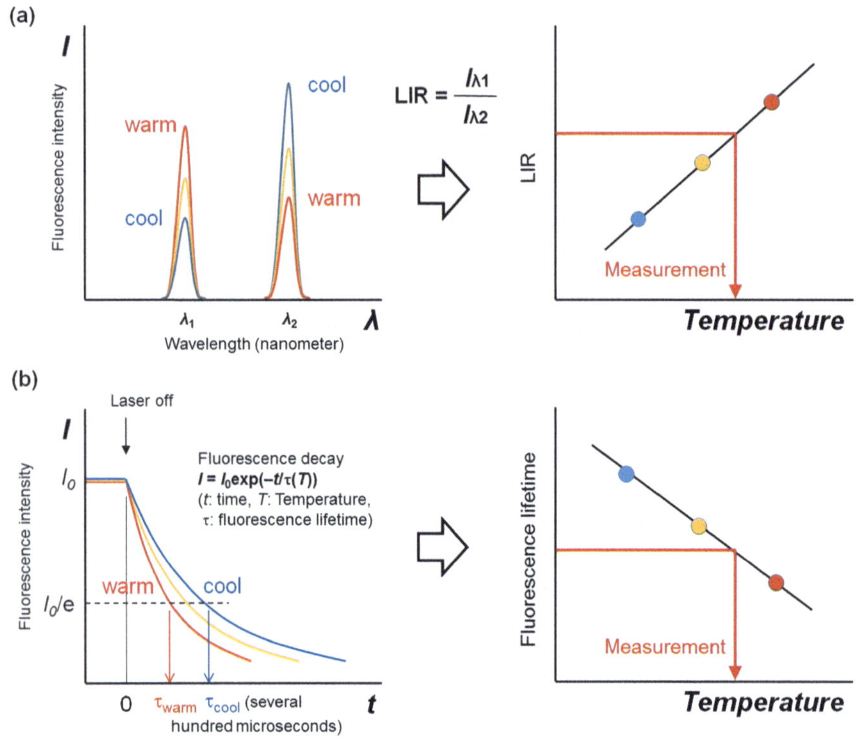

Fig. 7.6 Two principles of self-referencing luminescence nanothermometry. **a** A typical example of temperature-dependent change in fluorescence spectra of OTN-NIR phosphor, that allows self-referencing ratiometric nanothermometry using two wavelength bands (λ_1 and λ_2). **b** Relation between the fluorescence decay and temperature that allows fluorescence lifetime-based thermometry. Self-referencing can be done by using the decay images taken at different delay times (*t*) after cutting off the excitation laser

at 1150 and 1550 nm, respectively, under 980 nm excitation for Yb^{3+} ($^2F_{7/2} \rightarrow {}^2F_{5/2}$) [94] (Fig. 7.7a, b). The performance of such nanothermometer was further optimized by investigating the factors affecting the thermal sensitivity (*S*), e.g., dopant concentration, particle size, and RE interactions with surrounding medium [95]. The *S* parameter is defined as the rate of change of LIR with temperature, and indicates the ratiometric thermometer performance. Smaller nanocrystals were more sensitive to temperature variations, because of increased surface to volume ratio, and therefore more susceptible to surface quenching. As expected, emission intensity in water was lower compared to cyclohexane, and thus higher *S* values were observed in aqueous environments. The dopant concentration was found to be a tuning parameter for LIR, so as for *S*. The deep tissue probing ability of these RED-CNPs was confirmed in a later study, by thermal imaging experiments performed at the peritoneal cavity [96] (Fig. 7.7c–e), which is at deeper layer than the subcutaneous tissue level.

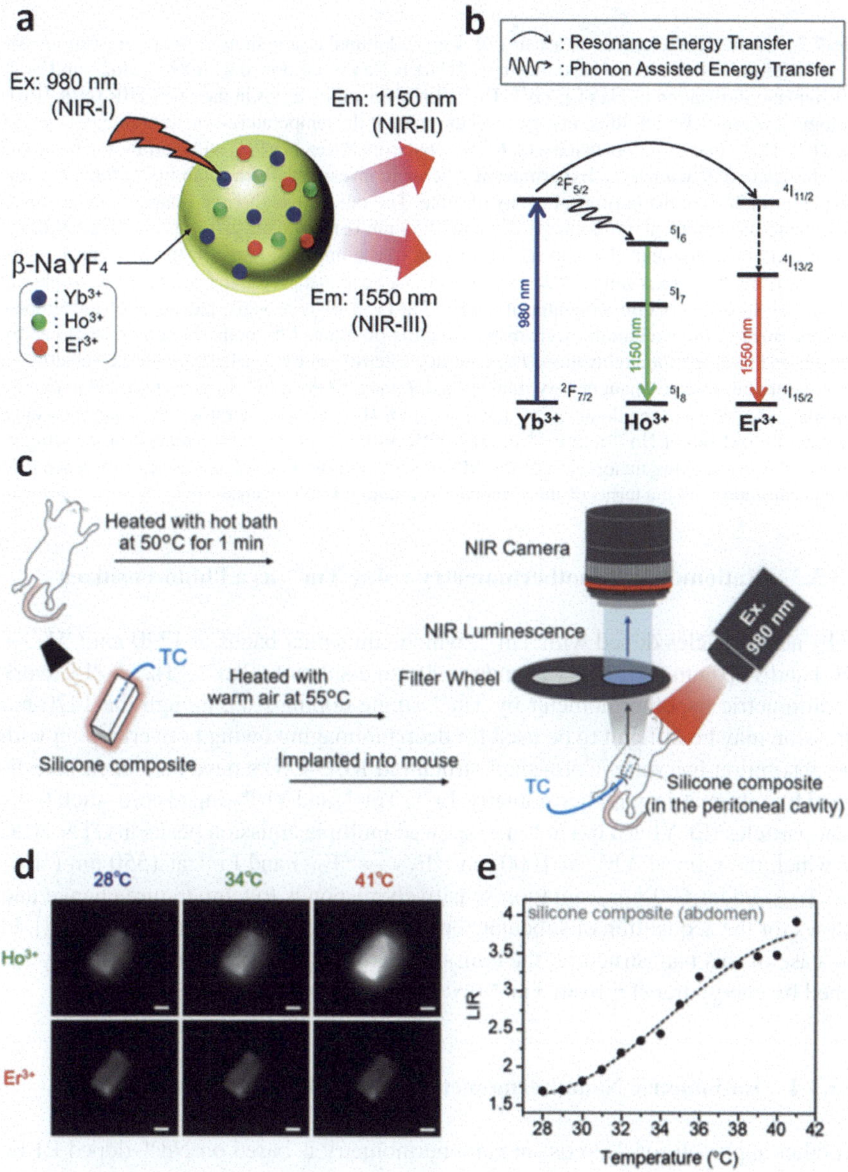
a
Ex: 980 nm
(NIR-I)
Em: 1150 nm
(NIR-II)
Em: 1550 nm
(NIR-III)
β-NaYF4
: Yb3+
: Ho3+
: Er3+
b
: Resonance Energy Transfer
: Phonon Assisted Energy Transfer
980 nm
1150 nm
1550 nm
Yb3+
Ho3+
Er3+
c
Heated with hot bath
at 50°C for 1 min
TC
Heated with
warm air at 55°C
Silicone composite
Implanted into mouse
NIR Camera
NIR Luminescence
Filter Wheel
Ex.
980 nm
TC
Silicone composite
(in the peritoneal cavity)
d
28°C
34°C
41°C
Ho3+
Er3+
e
silicone composite (abdomen)
LIR
Temperature (°C)

◀**Fig. 7.7** Contactless temperature sensing of deep abdominal region in mice by using a ratiometric OTN-NIR luminescence nanothermometry (LNTh), β-$NaYF_4$ co-doped with Yb^{3+}, Ho^{3+}, and Er^{3+}. **a** Schematic illustration of $NaYF_4$:Yb^{3+}, Ho^{3+}, Er^{3+} that emits lights in the OTN-NIR (NIR II/III) biological window. **b** Simplified energy level diagram of the temperature-sensitive energy levels of the Yb^{3+}, Ho^{3+}, and Er^{3+} dopant ions. (**a, b**) Adapted from Reference [94] with permission from The Royal Society of Chemistry. **c** Experimental scheme to investigate the availability of the nanothermometer to LNTh in the peritoneal cavity of mice. The ratiometric thermal images were acquired using band-pass filters at 1150 nm (Ho^{3+}) and 1550 nm (Er^{3+}) under 980-nm laser excitation for temperature measurement. **d** Fluorescence images of the implanted area in the peritoneal cavity acquired using band-pass filters at 28, 34, and 41 °C. Scale bars represent 5 mm. In hydrophobic matrix-like silicone, the emission intensity of Ho^{3+} at 1150 nm is strongly dependent on the temperature because of the excitation mechanism, which is dominated by both phonon-assisted energy transfer and multi-phonon relaxation. The emission intensity of Er^{3+} at 1550 nm is less sensitive to temperature because resonant energy transfer is dominant in the Yb^{3+}–Er^{3+} system due to energy gap matching between the donor (Yb^{3+}) and acceptor (Er^{3+}) [94]. Therefore, their luminescence intensity ratio (LIR) of Ho^{3+} and Er^{3+} in OTN-NIR provides a good representation of the temperature of the surrounding nanoregion of the NPs as shown in (**e**). **c–e** Adapted from Reference [96] with permission with the terms of the Creative Commons CC BY license

7.3.3.3 Ratiometric Nanothermometry Using Tm^{3+} as a Photosensitizer

LaF_3 nanoparticles doped with Tm^{3+}, which emits dual bands at 1230 nm ($^3H_5 \rightarrow {}^3H_6$) and 1470 nm ($^3H_4 \rightarrow {}^3F_4$) under 690-nm excitation (Tm^{3+}, $^3H_6 \rightarrow {}^3F_3$, work as ratiometric nanothermometer by Tm^{3+} single doping [97], though the 1470-nm emission may be difficult to be used for deep bioimaging owing to overlapping with the absorption by water. Core–shell structured RED-CNPs have also been investigated for ratiometric nanothermometry. Er^{3+}, Tm^{3+}, and Yb^{3+}-doped core–shell LaF_3 nanoparticles (Er-Yb@Yb-Tm LaF_3) showed multiple emission peaks in OTN-NIR, of which the ratio of Yb^{3+} at 1000 nm ($^2F_{5/2} \rightarrow {}^2F_{7/2}$) and Er^{3+} at 1550 nm ($^4I_{13/2} \rightarrow {}^4I_{15/2}$) under 690-nm excitation sensitively responds to temperature change and allows for the acquisition of subcutaneous thermal videos of a living animal [98]. In the case of this nanostructure, the emission wavelength and thermal sensitivity are tuned by energy transfer from Tm^{3+} (excited by 690 nm) to Yb^{3+} and Er^{3+}.

7.3.3.4 Ratiometric Nanothermometry Using Nd^{3+} as a Photosensitizer

Another approach for deep tissue nanothermometry is based on Nd^{3+}-doped RED-CNPs. The advantage of using Nd^{3+} is the heating-free excitation route at around 800 nm ($^4I_{9/2} \rightarrow {}^4F_{5/2}$) to obtain emissions at 880–900 nm ($^4F_{3/2} \rightarrow {}^4I_{9/2}$) and in the OTN-NIR range at 1050–1070 and 1320–1350 nm ($^4F_{3/2} \rightarrow {}^4I_{11/2}$ and $^4I_{13/2}$, respectively), of which the temperature-dependence were investigated for $LiLuF_4$:Nd^{3+}/$LiLuF_4$ core–shell RED-CNPs [99]. The $LiLuF_4$ host offered low phonon energy and high crystal field strength, and thus allowed distinguishing the Stark components within these bands, which are thermosensitive. The study showed that the most intense emission at 1050 nm was more reliable for temperature measurement in terms of repeatability and thermal uncertainty. On the other hand, lower scattering of biological tissues was observed for the emission at around 1320 nm, thus being the best choice for in vivo thermal imaging [99]. The temperature dependence of particles co-doped with other REs has been also investigated. In the case of co-doping with Yb^{3+}, core/shell-structured Yb^{3+}/Nd^{3+}-doped LaF_3 nanoparticle showed emission at 1350 nm (Nd^{3+}, ${}^4F_{3/2} \rightarrow {}^4I_{13/2}$) and 1000 nm ($Yb^{3+}$, ${}^2F_{5/2} \rightarrow {}^2F_{7/2}$), whose intensity ratio is sensitive to temperature change and successfully used to measure subcutaneous thermal transients in mice [100]. A designed core–shell nanoparticle of $NaYF_4$:Yb, Er@$NaGdF_4$:Nd provides a ratiometric nanothermometer that works 1060 nm (Nd^{3+}, ${}^4F_{3/2} \rightarrow {}^4I_{11/2}$) and 980 nm ($Yb^{3+}$, ${}^2F_{5/2} \rightarrow {}^2F_{7/2}$) in NIR and simultaneously in visible range (520 and 540 nm) via upconversion [101]. Multilayered SrF_3 nanoparticle with Tm^{3+}, Nd^{3+}, Yb^{3+}, and Er^{3+} also emits 1060 nm of Nd^{3+} and 980 nm of Yb^{3+} whose intensity ratio is correlated to temperature with multi-color upconversion emission in ultraviolet and visible wavelength ranges [102]. Multi-layered $NaGdF_4$ nanoparticle with Nd^{3+}, Yb^{3+}, Er^{3+} and Ho^{3+} shows OTN-NIR emission bands centered at 1180 and 1550 nm (Ho^{3+}, ${}^5I_6 \rightarrow {}^5I_8$ and Er^{3+}, ${}^4I_{13/2} \rightarrow {}^4I_{15/2}$, respectively) that increases in the intensity with the temperature increase and at 1340 nm (Nd^{3+}, ${}^4F_{3/2} \rightarrow {}^4I_{13/2}$) whose intensity drops with the temperature increase while dispersed in aqueous environment in the range of 20–50 °C [103].

The advantage of ratiometric thermometry (Fig. 7.6a) is that it can be done by using conventional imaging systems with light source for excitation, detector, and optical filters (Fig. 7.7c). However, care should be taken for the influence of slight shifts of light absorption by biological tissues (containing water, lipids, proteins, etc.) by change in body temperature, especially for 1550-nm emission affected by the shift of water absorption peak [96]. This problem can be solved in the lifetime-based thermometry described next.

7.3.3.5 Lifetime-Based Temperature Imaging for Deep Tissues

A fluorescence lifetime-based temperature imaging (Fig. 7.6b) in OTN-NIR was recently reported for an efficient in vivo thermal imaging [104]. This approach is based on time-resolved OTN-NIR luminescence via heating-free Nd^{3+}-excitation pathway. The probe used was $NaYF_4$: co-doped with Yb^{3+} and Nd^{3+} RED-CNPs with a major emission band at around 1000 nm, attributed to Yb^{3+} transition ${}^2F_{5/2} \rightarrow {}^2F_{7/2}$, and a weaker Nd^{3+} emission peak at 1064 nm (Fig. 7.8a). The temperature imaging was based on Nd^{3+}-sensitized Yb^{3+} emission, of which luminescence mechanisms involve resonant and phonon-assisted energy transfer, multi-phonon relaxation, and back transfers (Fig. 7.8b). The fluorescence decay images were collected at different temperatures (Fig. 7.8c), and average lifetimes showed a linear decrease with temperature (Fig. 7.8d). The satisfactory S value of 0.0096 °C^{-1} was estimated from the thermal coefficient. The time-gated imaging (TGI) synchronized with a pulsed-laser for excitation enabled acquiring series of lifetime decay curves at a pixel-level, which were converted into lifetime-based images. The lifetime analysis

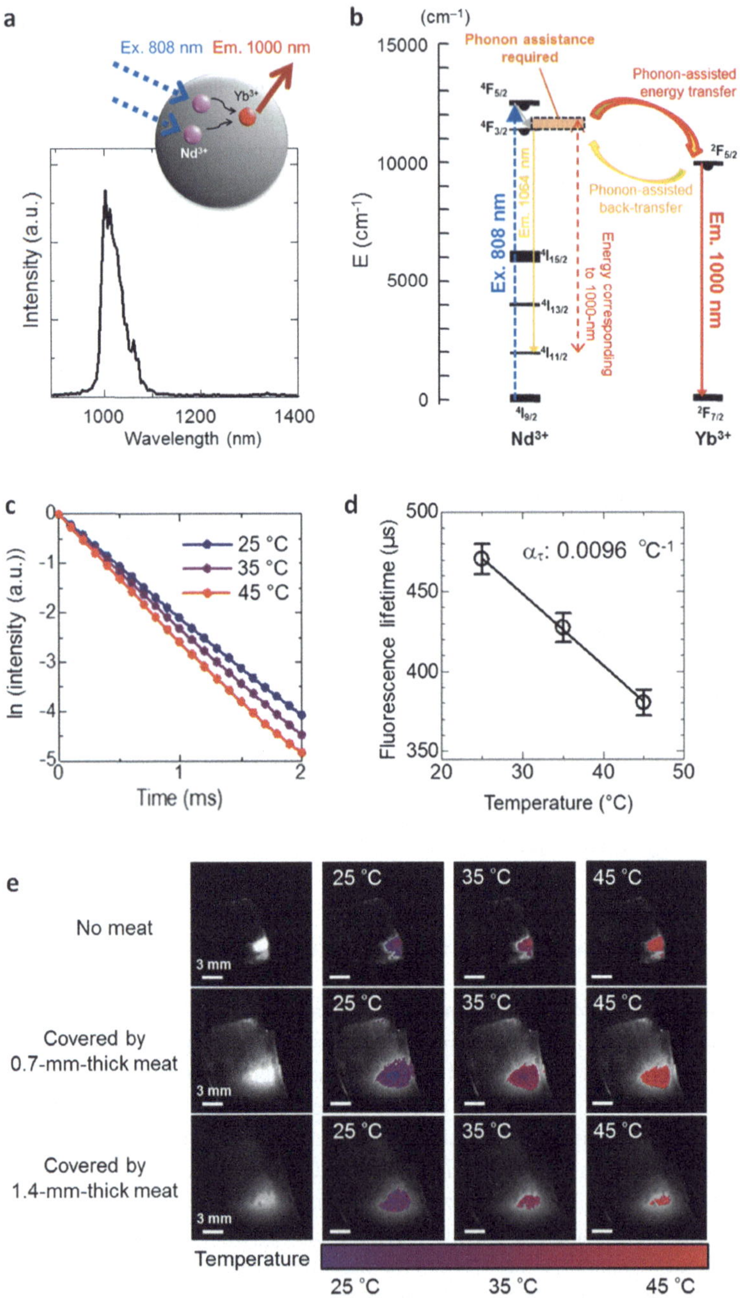
a
Ex. 808 nm Em. 1000 nm
Yb3+
Nd3+
Intensity (a.u.)
1000 1200 1400
Wavelength (nm)
b
(cm−1)
Phonon assistance required
Phonon-assisted energy transfer
Phonon-assisted back-transfer
E (cm−1)
Ex. 808 nm
Em. 1064 nm
Energy corresponding to 1000-nm
Em. 1000 nm
Nd3+
Yb3+
c
25 °C
35 °C
45 °C
ln (intensity (a.u.))
Time (ms)
d
ατ: 0.0096 °C-1
Fluorescence lifetime (μs)
Temperature (°C)
e
No meat
Covered by 0.7-mm-thick meat
Covered by 1.4-mm-thick meat
3 mm
25 °C
35 °C
45 °C
Temperature
25 °C 35 °C 45 °C

◂**Fig. 7.8** NIR Fluorescence lifetime-based temperature imaging. **a** Illustration of NIR luminescence of Nd^{3+}/Yb^{3+}-doped RED-CNPs emitting at 1000 nm under 808-nm laser excitation, and the respective emission spectrum. **b** Corresponding energy level diagram and luminescence mechanisms. **c** Fluorescence decay curves are measured in optical cuvette at different temperatures. **d** Thermal calibration curve for fluorescence lifetime detected by NIR camera. **e** Thermal images obtained by calculation from the calibration curve of the thermally sensitive phosphor ($NaYF_4$: Nd^{3+}, Yb^{3+}) while heating the biological tissue phantom from 25 °C to 45 °C with and without meat. Adapted from Reference [104] with permission with the terms of the Creative Commons CC BY license

using the TGI technique was realized because REs show relatively long lifetimes (several hundred microseconds), whereas most of organic fluorescent dyes show lifetime of several nanoseconds. The deepest observation of thermal images was achieved at 1.4 mm for ex vivo tissue (meat) [104] (Fig. 7.8e).

7.3.4 Quantum Dots for OTN-NIR LNTh

NIR-emitting quantum dots (QDs) have been also reported as luminescence nanothermometer for deep tissues. Besides photoluminescence, the exciton in QDs has a chance of relaxation by nonradiative transitions such as resonant energy transfer and polaronic quenching [105], both of which are likely to be temperature-dependent phenomena. In fact, the emission of PbS/CdS/ZnS core/shell/shell QD at 1270 nm (excitation: 808 nm) decreased as the temperature increased [106]. Because the temperature-sensitive radiative emission, that provides its thermal sensitivity, is accompanied by the nonratiative transitions, the QD converts the absorbed optical energy into phononic energy. The energy confers the QD heating capabilities applicable to PTT. The QD can simultaneously act as a photothermal agent (heater) and high-resolution fluorescent thermal sensors, making it possible to monitor and control the temperature of targets (e.g., tumors) in subcutaneous tissue in mice [106]. The temperature-dependent optical properties of Ag_2S nanocrystals have been also studied. The broad emission band of Ag_2S centered at 1200 nm (excitation: 800 nm) is dependent on temperature as it decreases as temperature increases. Thermal relaxation of subcutaneous melanoma [107] and brain [108] was visualized in live mice. However, the emission intensity is affected by not only temperature but also other factors such as attenuation of excitation light; therefore, ratiometric self-referencing nanothermometers have also been designed for QDs. A conjugated hybrid nanostructure of CdS/PbS/ZnS QD and a RED-CNP, Nd^{3+}-doped $NaGdF_4$, with a biocompatible polymer, poly(lactic-co-glycolic acid) (PLGA) was reported. This hybrid nanostructure emits both temperature-independent 1060-nm (Nd^{3+}) and highly temperature-dependent ≈1270-nm (QD) NIR luminescence, that allows for ratiometric temperature sensing, under 808-nm excitation [109]. Regarding Ag_2S

nanothermometer, core/shell nanostructured Ag/Ag_2S crystal shows temperature-induced spectral shifting and luminescence quenching of their response with temperature [110]. In particular, its emission peak is observed at 1235 nm at 15 °C and shifts to 1311 nm at 50 °C. The ratio between the emitted intensities 1311 and 1235 nm is available for LNTh of deep tissues.

7.4 Summary

This chapter reviewed the meanings, the regulation mechanisms, and current and emerging techniques for the measurement of body temperature. Body temperature is sensed by receptors like TRPV1 in the hypothalamus to be maintained homeostatically by regulating heat production and dissipation. The conventional methods for measuring body temperature has a limitation in that the spatial distribution of the temperature of deep tissues cannot be monitored. Luminescence thermometry developed recently have realized to visualize intracellular temperature distribution. Further, thermometry using OTN-NIR fluorescence, which allows the biological tissues to be seen transparently, has been developed to clarify the temporal change of the temperature distribution in deep tissues. Future research is expected to reveal the unknown functions of thermoreceptors and thermogenin molecules in biological organs, and to provide novel perspectives to the dynamic changes in body temperature distribution and their management in the biomedical and engineering fields.

References

1. T.D. Brock, H. Freeze, Thermus aquaticus gen. g. and sp. n., a nonsporulating extreme thermophile. J. Bact. **98**, 289 (1969)
2. A. Chien, D.B. Edgar, J.M. Trela, Deoxyribonucleic acid polymerase from the extreme thermophile thermus aquaticus. J. Bact. **127**, 1550 (1976)
3. K.S. Lundberg, D.D. Shoemaker, M.W. Adams et al., High-fidelity amplification using a thermostable DNA polymerase isolated from pyrococcus furiosus. Gene **108**, 1 (1991). https://doi.org/10.1016/0378-1119(91)90480-y
4. R.K. Saiki, S. Scharf, F. Faloona et al., Enzymatic amplification of beta-globin genomic sequences and restriction site analysis for diagnosis of sickle cell anemia. Science **230**, 1350 (1985). https://doi.org/10.1126/science.2999980
5. R.K. Saiki, D.H. Gelfand, S. Stoffel et al., Primer-directed enzymatic amplification of DNA with a thermostable DNA polymerase. Science **239**, 487 (1988). https://doi.org/10.1126/science.239.4839.487
6. T.W. Schulte, M.V. Blagosklonny, C. Ingui et al., Disruption of the Raf-1-Hsp90 molecular complex results in destabilization of Raf-1 and loss of Raf-1-Ras association. J. Biol. Chem. **270**, 24585 (1995). https://doi.org/10.1074/jbc.270.41.24585
7. D.S. Cissel, M.A. Beaven, Disruption of Raf-1/heat shock protein 90 complex and Raf signaling by Dexamethasone in mast cells. J. Biol. Chem. **275**, 7066 (2000). https://doi.org/10.1074/jbc.275.10.7066
8. N. Gomez, T. Erazo, J.M. Lizcano, ERK5 and cell proliferation: nuclear localization is what matters. Front. Cell Dev. Biol. **4**, 105 (2016). https://doi.org/10.3389/fcell.2016.00105

9. I.M. Verma, J.K. Stevenson, E.M. Schwarz et al., Rel/NF-kappa B/I Kappa B family: intimate tales of association and dissociation. Genes Dev. **9**, 2723 (1995). https://doi.org/10.1101/gad.9.22.2723
10. V. Krajka-Kuźniak, J. Paluszczak, W. Baer-Dubowska, The Nrf2-ARE signaling pathway: an update on its regulation and possible role in cancer prevention and treatment. Pharmacol. Rep. **69**, 393 (2017). https://doi.org/10.1016/j.pharep.2016.12.011
11. T. Shimizu, A. Lengalova, V. Martínek et al., Heme: emergent roles of heme in signal transduction, functional regulation and as catalytic centres. Chem. Soc. Rev. **48**, 5624 (2019). https://doi.org/10.1039/c9cs00268e
12. M.L. Begasse, M. Leaver, F. Vazquez, et al., Temperature Dependence of cell division timing accounts for a shift in the thermal limits of C. Elegans and C. Briggsae. Cell Rep. **10**, 647 (2015). https://doi.org/10.1016/j.celrep.2015.01.006
13. D. Patel, K.A. Franklin, Temperature-regulation of plant architecture. Plant Signal Behav. **4**, 577 (2009). https://doi.org/10.4161/psb.4.7.8849
14. D.A. Warner, R. Shine, The adaptive significance of temperature-dependent sex determination in a reptile. Nature **451**, 566 (2008). https://doi.org/10.1038/nature06519
15. A. Bahat, I. Tur-Kaspa, A. Gakamsky et al., Thermotaxis of mammalian sperm cells: a potential navigation mechanism in the female genital tract. Nat. Med. **9**, 149 (2003). https://doi.org/10.1038/nm0203-149
16. K. Okabe, N. Inada, C. Gota et al., Intracellular temperature mapping with a fluorescent polymeric thermometer and fluorescence lifetime imaging microscopy. Nat. Commun. **3**, 705 (2012). https://doi.org/10.1038/ncomms1714
17. P.A. Mackowiak, S.S. Wasserman, M.M. Levine, A Critical appraisal of 98.6 degrees F, the upper limit of the normal body temperature, and other legacies of Carl Reinhold August Wunderlich. JAMA **268**, 1578 (1992). https://doi.org/10.1001/jama.1992.03490120092034
18. G.S. Kelly, Body temperature variability (Part 1): a review of the history of body temperature and its variability due to site selection, biological rhythms, fitness, and aging. Altern. Med. Rev. **11**, 278 (2006)
19. G.S. Kelly, Body temperature variability (Part 2): masking influences of body temperature variability and a review of body temperature variability in disease. Altern. Med. Rev. **12**, 49 (2007)
20. K. Kanosue, L.I. Crawshaw, K. Nagashima et al., Concepts to utilize in describing thermoregulation and neurophysiological evidence for how the system works. Eur. J. Appl. Physiol. **109**, 5 (2010). https://doi.org/10.1007/s00421-009-1256-6
21. S. Kobayashi, Temperature-sensitive neurons in the hypothalamus: a new hypothesis that they act as thermostats, not as transducers. Prog. Neurobiol. **32**, 103 (1989). https://doi.org/10.1016/0301-0082(89)90012-9
22. J.A. Boulant, Role of the preoptic-anterior hypothalamus in thermoregulation and fever. Clin. Infect. Dis. **31**(Suppl 5), S157 (2000). https://doi.org/10.1086/317521
23. G. Mory, F. Bouillaud, M. Combes-George et al., Noradrenaline controls the concentration of the uncoupling protein in brown adipose tissue. FEBS Lett. **166**, 393 (1984). https://doi.org/10.1016/0014-5793(84)80120-9
24. Z.B. Andrews, S. Diano, T.L. Horvath, Mitochondrial uncoupling proteins in the CNS: in support of function and survival. Nat. Rev. Neurosci. **6**, 829 (2005). https://doi.org/10.1038/nrn1767
25. M.J. Gaudry, M. Jastroch, Molecular evolution of uncoupling proteins and implications for brain function. Neurosci. Lett. **696**, 140 (2019). https://doi.org/10.1016/j.neulet.2018.12.027
26. R.S. Seymour, Biophysics and physiology of temperature regulation in thermogenic flowers. Biosci. Rep. **21**, 223 (2001). https://doi.org/10.1023/A:1013608627084
27. A. Tissières, H.K. Mitchell, U.M. Tracy, Protein synthesis in salivary glands of drosophila melanogaster: relation to chromosome puffs. J. Mol. Biol. **84**, 389 (1974). https://doi.org/10.1016/0022-2836(74)90447-1
28. M.J. Caterina, M.A. Schumacher, M. Tominaga et al., The Capsaicin receptor: a heat-activated ion channel in the pain pathway. Nature **389**, 816 (1997). https://doi.org/10.1038/39807

29. R. Inoue, T. Hanano, J. Shi et al., Transient receptor potential protein as a novel non-voltage-gated Ca^{2+} entry channel involved in diverse pathophysiological functions. J. Pharmacol. Sci. **91**, 271 (2003). https://doi.org/10.1254/jphs.91.271
30. G.M. Story, The emerging role of TRP channels in mechanisms of temperature and pain sensation. Curr. Neuropharmacol. **4**, 183 (2006). https://doi.org/10.2174/157015906778019482
31. K. Uchida, K. Dezaki, T. Yoneshiro et al., Involvement of thermosensitive TRP channels in energy metabolism. J. Physiol. Sci. **67**, 549 (2017). https://doi.org/10.1007/s12576-017-0552-x
32. D.J. Cosens, A. Manning, Abnormal electroretinogram from a drosophila mutant. Nature **224**, 285 (1969). https://doi.org/10.1038/224285a0
33. T. Sugiura, M. Tominaga, H. Katsuya et al., Bradykinin lowers the threshold temperature for heat activation of vanilloid receptor 1. J. Neurophysiol. **88**, 544 (2002). https://doi.org/10.1152/jn.2002.88.1.544
34. N.R. Gavva, A.W. Bannon, S. Surapaneni et al., The vanilloid receptor TRPV1 Is tonically activated in vivo and involved in body temperature regulation. J. Neurosci. **27**, 3366 (2007). https://doi.org/10.1523/JNEUROSCI.4833-06.2007
35. R. Sharif-Naeini, S. Ciura, C.W. Bourque, TRPV1 gene required for thermosensory transduction and anticipatory secretion from vasopressin neurons during hyperthermia. Neuron **58**, 179 (2008). https://doi.org/10.1016/j.neuron.2008.02.013
36. A.D. Güler, H. Lee, T. Iida, Heat-evoked activation of the ion channel, TRPV4. J. Neurosci. **22**, 6408 (2002). https://doi.org/10.1523/JNEUROSCI.22-15-06408.2002
37. H. Lee, T. Iida, A. Mizuno et al., Altered thermal selection behavior in mice lacking transient receptor potential vanilloid 4. J. Neurosci. **25**, 1304 (2005). https://doi.org/10.1523/JNEUROSCI.4745.04.2005
38. A.S.M. El-Radhi, Why is the evidence not affecting the practice of fever management? Arch. Dis. Child. **93**, 918 (2008). https://doi.org/10.1136/adc.2008.139949
39. M. Richardson, E. Purssell, Who's afraid of fever? Arch. Dis. Child. **100**, 818 (2015). https://doi.org/10.1136/archdischild-2014-307483
40. E.A. Dennis, Diversity of group types, regulation, and function of phospholipase A2. J. Biol. Chem. **269**, 13057 (1994)
41. C.C. Leslie, Properties and regulation of cytosolic phospholipase A2. J. Biol. Chem. **272**, 16709 (1997). https://doi.org/10.1074/jbc.272.27.16709
42. K. Matsumura, C. Cao, M. Ozaki et al., Brain endothelial cells express cyclooxygenase-2 during lipopolysaccharide-induced fever: light and electron microscopic immunocytochemical studies. J. Neurosci. **18**, 6279 (1998). https://doi.org/10.1523/JNEUROSCI.18-16-06279.1998
43. H.E. de Vries. K.H. Hoogendoorn, J.van Dijk, et al., Eicosanoid production by rat cerebral endothelial cells: Stimulation by lipopolysaccharide, interleukin-1 and interleukin-6. J. Neuroimmunol. **59**, 1 (1995). https://doi.org/10.1016/0165-5728(95)00009-q
44. R. Hanada, A. Leibbrandt, T. Hanada et al., Central control of fever and female body temperature by RANKL/RANK. Nature **462**, 505 (2009). https://doi.org/10.1038/nature08596
45. F. Ushikubi, E. Segi, Y. Sugimoto et al., Impaired febrile response in mice lacking the prostaglandin E receptor subtype EP3. Nature **395**, 281 (1998). https://doi.org/10.1038/26233
46. K. Nakamura, K. Matsumura, T. Kaneko et al., The rostral raphe pallidus nucleus mediates pyrogenic transmission from the preoptic area. J. Neurosci. **22**, 4600 (2002). https://doi.org/10.1523/JNEUROSCI.22-11-04600.2002
47. M. Lazarus, K. Yoshida, R. Coppari et al., EP3 prostaglandin receptors in the median preoptic nucleus are critical for fever responses. Nat. Neurosci. **10**, 1131 (2007). https://doi.org/10.1038/nn1949
48. S. Conolly, J.E. Arrowsmith, A.A. Klein, Deep hypothermic circulatory arrest. Continuing Educ Anaesth Crit Care Pain **10**, 138 (2010). https://doi.org/10.1093/bjaceaccp/mkq024

49. M. Saxena, P.J.D. Andrews, A. Cheng, et al., Modest cooling therapies (35 to 37.5 °C) for traumatic brain injury. Cochrane Database Syst. Rev. **2014**, CD006811 (2014). https://doi.org/10.1002/14651858.CD006811.pub3
50. S.R. Lewis, D.J. Evans, A.R. Butler, et al., Hypothermia for Traumatic Brain Injury. Cochrane Database Syst. Rev. **2017** CD001048, (2017). https://doi.org/10.1002/14651858.CD001048.pub5
51. S.E. Jacobs, M. Berg, R. Hunt, et al. Cooling for newborns with hypoxic ischaemic encephalopathy. Cochrane Database Syst. Rev. **2013**, CD003311 (2013). https://doi.org/10.1002/14651858.CD003311.pub3
52. A. Ascensão, M. Leite, A.N. Rebelo et al., Effects of cold water immersion on the recovery of physical performance and muscle damage following a one-off soccer match. J. Sports Sci. **29**, 217 (2011). https://doi.org/10.1080/02640414.2010.526132
53. T.A.H. Järvinen, T.L.N. Järvinen, M. Kääriäinen et al., Muscle Injuries: optimising recovery. Best Pract. Res. Clin. Rheumatil. **21**, 317 (2007). https://doi.org/10.1016/j.berh.2006.12.004
54. L.A. Roberts, T. Raastad, J.F. Markworth et al., Post-exercise cold water immersion attenuates acute anabolic signalling and long-term adaptations in muscle to strength training. J. Physiol. **593**, 4285 (2015). https://doi.org/10.1113/JP270570
55. L. Guan, G. Xu, Damage effect of high-intensity focused ultrasound on breast cancer tissues and their vascularities. World J. Surg. Oncol. **14**, 153 (2016). https://doi.org/10.1186/s12957-016-0908-3
56. J.Y. Park, P. Choi, H.K. Kim et al., Increase in apoptotic effect of Panax ginseng by microwave processing in human prostate cancer cells: in vitro and in vivo studies. J. Ginseng Res. **40**, 62 (2016). https://doi.org/10.1016/j.jgr.2015.04.007
57. G. Baffou, R. Quidant, Thermo-plasmonics: using metallic nanostructures as nano-sources of heat. Laser Photonics Rev. **7**, 171 (2013). https://doi.org/10.1002/lpor.201200003
58. I. Krylov, R. Akasov, V. Rocheva et al., Local overheating of biotissue labeled with upconversion nanoparticles under Yb^{3+} resonance excitation. Front. Phys. **8**, 295 (2020). https://doi.org/10.3389/fchem.2020.00295
59. L. Ding, F. Ren, Z. Liu et al., Size-dependent photothermal conversion and photoluminescence of theranostic $NaNdF_4$ nanoparticles under excitation of different-wavelength lasers. Bioconjugate Chem. **31**, 340 (2020). https://doi.org/10.1021/acs.bioconjchem.9b00700
60. E. Carrasco, B. del Rosal, F. Sanz-Rodríguez et al., Intratumoral thermal reading during photo-thermal therapy by multifunctional fluorescent nanoparticles. Adv. Funct. Mater. **25**, 615 (2015). https://doi.org/10.1002/adfm.201403653
61. J.T. Robinson, K. Welsher, S.M. Tabakman et al., High performance in vivo Near-IR (>1 μm) imaging and photothermal cancer therapy with carbon nanotubes. Nano Res. **3**, 779 (2010). https://doi.org/10.1007/s12274-010-0045-1
62. A.L. Antaris, J.T. Robinson, O.K. Yaghi et al., Ultra-low doses of chirality sorted (6,5) carbon nanotubes for simultaneous tumor imaging and photothermal therapy. ACS Nano **7**, 3644 (2013). https://doi.org/10.1021/nn4006472
63. C. Liang, S. Diao, C. Wang et al., Tumor metastasis inhibition by imaging-guided photothermal therapy with single-walled carbon nanotubes. Adv. Mater. **26**, 5646 (2014). https://doi.org/10.1002/adma.201401825
64. M. Sund-Levander, C. Forsberg, L.K. Wahren, Normal oral, rectal, tympanic and axillary body temperature in adult men and women: a systematic literature review. Scand. J. Caring Sci. **16**, 122 (2002). https://doi.org/10.1046/j.1471-6712.2002.00069.x
65. I. Sage, Thermochromic liquid crystals. Liq. Cryst. **38**, 1551 (2011). https://doi.org/10.1080/02678292.2011.631302
66. P. Kiekkas, N. Stefanopoulos, N. Bakalis et al., Agreement of infrared temporal artery thermometry with other thermometry methods in adults: systematic review. J. Clin. Nurs. **25**, 894 (2016). https://doi.org/10.1111/jocn.13117
67. C. Zhen, Z. Xia, L. Long et al., Accuracy of infrared ear thermometry in children: a meta-analysis and systematic review. Clin. Pediatr. (Phila) **53**, 1158 (2014). https://doi.org/10.1177/0009922814536774

68. J.P. Feist, A.L. Heyes, Europium-doped yttria-stabilized zirconia for high-temperature phosphor thermometry. Proc. Inst. Mechanic. Eng. **214 Part L**, 7 (2000). https://doi.org/10.1177/146442070021400102
69. H. Zhou, M. Sharma, O. Berezin et al., Nanothermometry: from microscopy to thermal treatments. Chem. Phys. Chem. **17**, 27 (2016). https://doi.org/10.1002/cphc.201500753
70. T. Bai, N. Gu, Micro/nanoscale thermometry for cellular thermal sensing. Small **12**, 4590 (2016). https://doi.org/10.1002/smll.201600665
71. A.M. Stark, S. Way, The use of thermovision in the detection of early breast cancer. Cancer **33**, 1664 (1974). https://doi.org/10.1002/1097-0142(197406)33:6%3c1664::AID-CNCR2820330629%3e3.0.CO;2-7
72. D. Jaque, L.M. Maestro, B. del Rosal et al., Nanoparticles for photothermal therapies. Nanoscale **6**, 9494 (2014). https://doi.org/10.1039/C4NR00708E
73. D. Jaque, F. Vetrone, Luminescence nanothermometry. Nanoscale **4**, 4301 (2012). https://doi.org/10.1039/C2NR30764B
74. K. Nigoghossian, S. Ouellet, J. Plain et al., Upconversion nanoparticle-decorated gold nanoshells for near-infrared induced heating and thermometry. J. Mater. Chem. B **5**, 7109 (2017). https://doi.org/10.1039/c7tb01621b
75. Y. Zhang, S. Xu, X. Li et al., Temperature sensing, excitation power dependent fluorescence branching ratios, and photothermal conversion in $NaYF_4$:Er^{3+}/Yb^{3+} @$NaYF_4$:Tm^{3+}/Yb^{3+} core-shell particles. Opt. Mater. Exp. **8**, 368 (2018). https://doi.org/10.1364/OME.8.000368
76. B. del Rosal, E. Ximendes, U. Rocha et al., In vivo luminescence nanothermometry: from materials to applications. Adv. Opt. Mater. **5**, 1600508 (2017). https://doi.org/10.1002/adom.201600508
77. E. Hemmer, P. Acosta-Mora, J. Méndez-Ramos et al., Optical nanoprobes for biomedical applications: shining a light on upconverting and near-infrared emitting nanoparticles for imaging, thermal sensing, and photodynamic therapy. J. Mater. Chem. B **5**, 4365 (2017). https://doi.org/10.1039/C7TB00403F
78. C.D.S. Brites, P.P. Lima, N.J.O. Silva et al., Thermometry at the nanoscale. Nanoscale **4**, 4799 (2012). https://doi.org/10.1039/c2nr30663h
79. C.D.S. Brites, S. Balabhadra, L.D. Carlos, Lanthanide-based thermometers: at the cutting-edge of luminescence thermometry. Adv. Opt. Mater. **7**, 1801239 (2019). https://doi.org/10.1002/adom.201801239
80. K. Okabe, R. Sakaguchi, B, Shi, et al., Intracellular thermometry with fluorescent sensors for thermal biology. Eur. J. Physiol. **470**, 717 (2018). https://doi.org/10.1007/s00424-018-2113-4
81. V.A. Vlaskin, N. Janssen, J. van Rijssel et al., Tunable Dual Emission in Doped Semiconductor Nanocrystals. Nano Lett. **10**, 3670 (2010). https://doi.org/10.1021/nl102135k
82. N. Inada, N. Fukuda, T. Hayashi et al., Temperature imaging using a cationic linear fluorescent polymeric thermometer and fluorescence lifetime imaging microscopy. Nat. Protoc. **14**, 1293 (2019). https://doi.org/10.1038/s41596-019-0145-7
83. J.S. Donner, S.A. Thompson, M.A. Kreuzer et al., Mapping intracellular temperature using green fluorescent protein. Nano Lett. **12**, 2107 (2012). https://doi.org/10.1021/nl300389y
84. S. Kiyonaka, T. Kajimoto, R. Sakaguchim et al., Genetically encoded fluorescent thermosensors visualize subcellular thermoregulation in living cells. Nat. Methods **10**, 1232 (2013). https://doi.org/10.1038/nmeth.2690
85. M. Nakano, Y. Arai, I. Kotera et al., Genetically encoded ratiometric fluorescent thermometer with wide range and rapid response. PLoS ONE **12**, e0172344 (2017). https://doi.org/10.1371/journal.pone.0172344
86. D. Chrétien, P. Bénit, H.H. Ha et al., Mitochondria are physiologically maintained at close to 50 °C. PLoS Biol. **16**, e2003992 (2018). https://doi.org/10.1371/journal.pbio.2003992
87. M.C. Rajagopal, J.W. Brown, D. Gelda et al., Transient heat release during induced mitochondrial proton uncoupling. Commun. Biol. **2**, 279 (2019). https://doi.org/10.1038/s42003-019-0535-y
88. E. Hemmer, N. Venkatachalam, H. Hyodo et al., Upconverting and NIR emitting rare earth based nanostructures for NIR-bioimaging. Nanoscale **5**, 11339 (2013). https://doi.org/10.1039/c3nr02286b

89. S. Arai, Ferdinandus, S. Takeoka, et al., Micro-thermography in millimeter-scale animals by using orally-dosed fluorescent nanoparticle thermosensors. Analyst **140**, 7534 (2015). https://doi.org/10.1039/C5AN01287B
90. J.S. Donner, S.A. Thompson, C. Alonso-Ortega et al., Imaging of plasmonic heating in a living organism. ACS Nano **7**, 8666 (2013). https://doi.org/10.1021/nn403659n
91. L. Huo, J. Zhou, R. Wu et al., Dual-functional β-$NaYF_4$: Yb^{3+}, Er^{3+} nanoparticles for bioimaging and temperature sensing. Opt. Mater. Exp. **6**, 1056 (2016). https://doi.org/10.1364/OME.6.001056
92. K. Soga, M. Kamimura, K. Okubo et al., Near-infrared biomedical imaging for transparency. J. Imag. Soc. Jpn **58**, 602 (2019). https://doi.org/10.11370/isj.58.602
93. F. Vetrone, R. Naccache, A. Zamarrón et al., Temperature sensing using fluorescent nanothermometers. ACS Nano **4**, 3254 (2010). https://doi.org/10.1021/nn100244a
94. M. Kamimura, T. Matsumoto, S. Suyari et al., Ratiometric near-infrared fluorescence nanothermometry in the OTN-NIR (NIR II/III) biological window based on rare-earth doped β-$NaYF_4$ nanoparticles. J. Mater. Chem. B **5**, 1917 (2017). https://doi.org/10.1039/C7TB00070G
95. L. Wortmann, S. Suyari, T. Ube et al., Tuning the thermal sensitivity of β-$NaYF_4$: Yb^{3+}, H^{o3}+, ^{Er}3+ nanothermometers for optimal temperature sensing in OTN-NIR (NIR II/III) biological window. J. Lumin. **198**, 236 (2018). https://doi.org/10.1016/j.jlumin.2018.01.049
96. S. Sekiyama, M. Umezawa, S. Kuraoka et al., Temperature sensing of deep abdominal region in mice by using over-1000 nm near-infrared luminescence of rare-earth-doped $NaYF_4$ nanothermometer. Sci. Rep. **8**, 16979 (2018). https://doi.org/10.1038/s41598-018-35354-y
97. E.C. Ximendes, A.F. Pereira, U. Rocha et al., Thulium doped LaF_3 for nanothermometry operating over 1000 nm. Nanoscale **11**, 8864 (2019). https://doi.org/10.1039/c9nr00082h
98. E.C. Ximendes, U. Rocha, T.O. Sales et al., In vivo subcutaneous thermal video recording by supersensitive infrared nanothermometers. Adv. Funct. Mater. **27**, 1702249 (2017). https://doi.org/10.1002/adfm.201702249
99. A. Skripka, A. Morinvil, M. Matulionyte et al., Advancing neodymium single-band nanothermometry. Nanoscale **11**, 11322 (2019). https://doi.org/10.1039/c9nr02801c
100. E.C. Ximendes, W.Q. Santos, U. Rocha et al., Unveiling in vivo subcutaneous thermal dynamics by infrared luminescent nanothermometers. Nano Lett. **16**, 1695 (2016). https://doi.org/10.1021/acs.nanolett.5b04611
101. F. Xu, Z. Ba, Y. Zheng et al., Rare-earth-doped optical nanothermometer in visible and near-infrared regions. J. Mater. Sci. **53**, 15107 (2018). https://doi.org/10.1007/s10853-018-2702-9
102. P. Cortelletti, A. Skripka, C. Facciotti et al., Tuning the sensitivity of lanthanide-activated NIR nanothermometers in the biological windows. Nanoscale **10**, 2568 (2018). https://doi.org/10.1039/c7nr06141b
103. A. Skripka, A. Benayas, R. Marin et al., Double rare-earth nanothermometer in aqueous media: opening the third optical transparency window to temperature sensing. Nanoscale **9**, 3079 (2017). https://doi.org/10.1039/c6nr08472a
104. T. Chihara, M. Umezawa, K. Miyata et al., Biological deep temperature imaging with fluorescence lifetime of rare-earth-doped ceramics particles in the second NIR biological window. Sci. Rep. **9**, 12806 (2019). https://doi.org/10.1038/s41598-019-49291-x
105. E. Thimsen, B. Sadtler, M.Y. Berezin, Shortwave-infrared (SWIR) emitters for biological imaging: a review of challenges and opportunities. Nanophotonics **6**, 1043 (2017). https://doi.org/10.1515/nanoph-2017-0039
106. B. del Rosal, E. Carrasco, F. Ren et al., Infrared-emitting QDs for thermal therapy with real-time subcutaneous temperature feedback. Adv. Funct. Mater. **26**, 6060 (2016). https://doi.org/10.1002/adfm.201601953
107. H.D.A. Santos, E.C. Ximendes, M. del C. Iglesias-de la Cruz, et al., In vivo early tumor detection and diagnosis by infrared luminescence transient nanothermometry. Adv. Funct. Mater. **28**, 1803924 (2018). https://doi.org/10.1002/adfm.201803924
108. B. del Rosal, D. Ruiz, I. Chaves-Coira et al., In vivo contactless brain nanothermometry. Adv. Funct. Mater. **28**, 1806088 (2018). https://doi.org/10.1002/adfm.201806088

109. E.N. Cerón, D.H. Ortgies, B. del Rosal et al., Hybrid nanostructures for high-sensitivity luminescence nanothermometry in the second biological window. Adv. Mater. **27**, 4781 (2015). https://doi.org/10.1002/adma.201501014
110. D. Ruiz, B. del Rosal, M. Acebrón et al., Ag/Ag_2S nanocrystals for high sensitivity near-infrared luminescence nanothermometry. Adv. Funct. Mater. **27**, 1604629 (2017). https://doi.org/10.1002/adfm.201604629

Chapter 8
Computed Tomography

Toshihiro Sera

Abstract Computed tomography (CT) is widely used to visualize the cross-section of a sample with non-destructive techniques in clinical and industrial applications, and the attenuation coefficients are basically distributed in reconstructed CT images. Particularly, life science fields require the visualization of a living matter and soft tissue at high spatial resolution and in three (3D)/four (4D) dimensions. However, almost all tissues contains a lot of water, resulting that the attenuations are approximately same among tissues and smaller contrast differences in CT images. In this chapter, we review the conventional CT and new technologies in contrast agents and new imaging systems for visualizing inside of tissues.

Keywords X-ray · Fluorescence · Computed tomography · Reconstruction · Contrast agents

8.1 Computed Tomography (CT)

Computed tomography (CT) is a medical and industrial imaging procedure that uses computer-processed reconstruction from many projections from different angles to produce cross-sectional images of a sample with non-destructive techniques. The German physicist Wilhelm C. Röntgen, who earned Nobel Prize in Physics in 1901, discovered X-rays in 1985 and showed the potential applications in medical imaging. X-rays are high-energy electromagnetic radiation, and their wavelength is ranging from 0.1 to 10 nm, shorter than visible light. Thereafter, Godfrey N. Hounsfield constructed first X-ray CT scanner and earned Nobel Prize in Physiology and Medicine in 1979 with Allan M. Cormack, who developed the theoretical calculations for CT reconstructions. Today, there are many other types of CT, such as positron emission tomography (PET) and single-photon emission computed tomography (SPECT); however, the term "CT" is often used to refer to X-ray CT.

T. Sera (✉)
Department of Mechanical Engineering, Faculty of Engineering, Kyushu University, 744 Motooka, Nishi-ku, Fukuoka 819-0395, Japan
e-mail: sera@mech.kyushu-u.ac.jp

K. Soga et al. (eds.), *Transparency in Biology*,
https://doi.org/10.1007/978-981-15-9627-8_8

As the X-rays (emitted photons) pass through the object, some of them are absorbed and some are transmitted. The X-ray intensity reduction due to absorption and scatting by materials is called attenuation, and X-rays attenuated due to the interactions with the object do not reach the detector. Photons transmitted through the object at each angle are collected on the detector and reconstructed by a computer, creating a complete tomogram of the scanned object. Therefore, in CT, the reconstructed image generally represents the distribution of X-ray attenuation in the measured object as the gray value.

8.2 CT Imaging

8.2.1 Hardware

Basic CT system consists of an X-ray source, a rotary table, an X-ray detector, and a data processing unit for reconstruction and visualization and is divided into several generations depending on the scanner geometry (Figs. 8.1 and 8.2) [1, 2].

8.2.1.1 First Generation

First-generation CT systems are characterized by a single X-ray source (pencil beam) directing across the object and a single detector. Both, the source and the detector, translate simultaneously in a scan plane. This process is repeated for a number of angular rotations. The system is simplicity; however, a longer scanning time is required (several minutes).

8.2.1.2 Second Generation

Second-generation CT systems use the same translate/rotate scan geometry as the first generation. The difference is that a pencil beam and single detector are replaced by a fan beam and multiple detectors, respectively. As the result, a series of views can be acquired during each translation, and a scanning time is shorter (less than 90 s).

8.2.1.3 Third Generation

Third-generation CT systems use a fan beam and multiple detectors, which is normally same as the second generation. The difference is a rotate-only scan geometry, and the multiple detectors rotate simultaneously with the X-ray source. Scan time is faster than second-generation systems (approximately 5 s).

8.2.1.4 Fourth Generation

Fourth-generation CT systems also use a rotate-only scan motion. Here, the system consists of a stationary ring with multiple detectors and a single fan beam X-ray source which rotates around the scanned object (stationary/rotate). Scan time is a few seconds.

8.2.1.5 Fifth Generation

Fifth-generation CT systems, called as electron beam CT (EBCT), uses a stationary X-ray source and a stationary detector, resulting in avoiding any mechanically motion. In these systems, an electron beam is emitted and magnetically deflected to hit a semi-circular anode surrounding the sample. The magnetic deflection sweeps the electron beam over the target, thus generating an X-ray source that virtually rotates around the sample (Fig. 8.2).

8.2.1.6 Other Systems

Other CT technologies have been adapted to third- and fourth-generation scanners, such as helical image acquisition and a cone beam CT (CBCT). In helical CT, the patient is translated in patient direction continuously with rotating X-ray source and detector. From the perspective of the patient, the X-ray beam from the CT traces a helical (spiral) path. In CBCT, X-rays are divergent forming a cone-shaped beam, and generally, 2D detector is used. These technologies result in a three-dimensional (3D) data set. Additionally, as a non-clinical use, some commercial companies developed a microfocused and nano-focused X-ray CT with a cone-shaped beam of smaller focal spot, providing a high-resolution image down to sub-micrometer scale of tissues [2].

8.2.2 Reconstruction

The X-ray intensity is decreased due to the absorption and scatting by materials. This relationship is simply characterized by the following Lambert-Beer's law, which describes as a function of path length (L) and material linear attenuation coefficient (μ);

$$I = I_0 \mathrm{e}^{-\mu L} \tag{8.1}$$

where I_0 and I are the X-ray intensity at a source and detector, respectively. When μ is not constant in a sample and X-ray is not monochromatic (usually conventional X-ray source), the equation is expressed as followed;

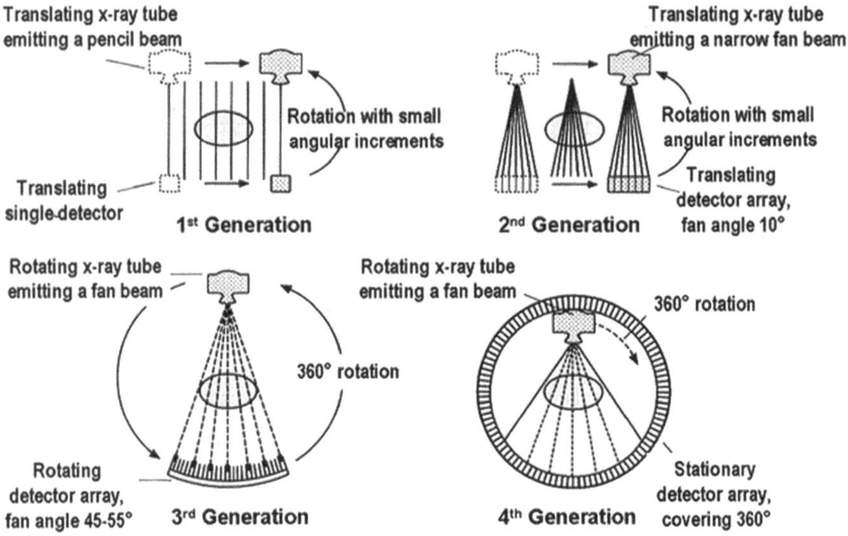

Fig. 8.1 Design and acquisition schemes of four generations of X-ray CT scanners. Reprinted from Ref. [1] with permission from Springer Nature

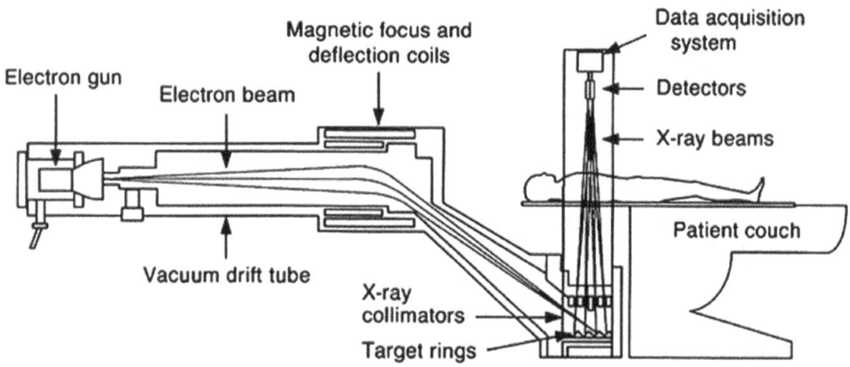

Fig. 8.2 Schematic illustration of EBCT. Reprinted from Ref. [1] with permission from Springer Nature

$$I = \int_E I_0(E)\mathrm{e}^{-\int_0^L \mu(x,E)\mathrm{d}x}\mathrm{d}E \tag{8.2}$$

where E is a X-ray energy, and $I_0(E)$ and μ (x, E) are the X-ray intensity at a source and linear attenuation coefficient at x along to path length as a function of E, respectively. These equations are one dimension as a function of x; however, μ is generally expressed in two or three dimensions (2D, 3D); therefore, a 2D or 3D image detector is used for CT imaging. As mentioned before, tomographic imaging

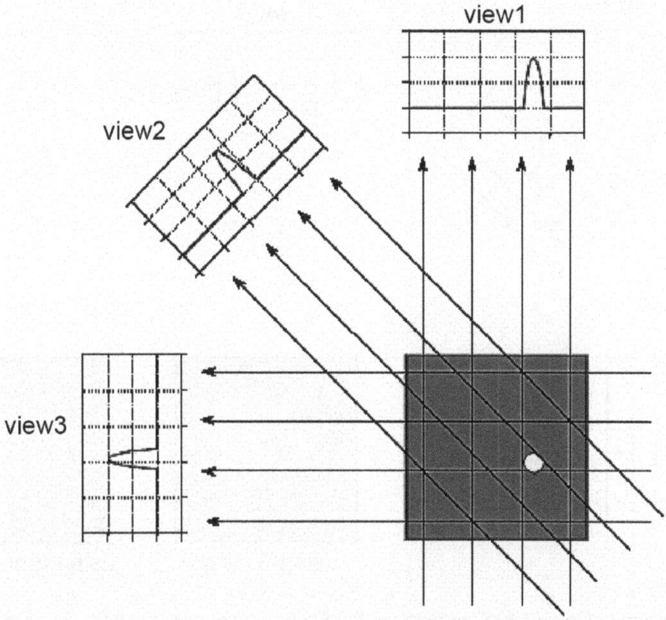

Fig. 8.3 Example of CT scan. A small pin exists in air. The views show each projected image from different angles. In this example, only three views are shown; however, a typical CT scan used hundreds of views at different orientations. Reprinted from Ref. [2] with permission

consists of directing X-rays at an object from multiple orientations. Therefore, the 2D or 3D distribution of X-ray attenuation is obtained by measuring the decrease in the intensity from different angles during sample rotation and using various CT reconstruction algorithms, such as filtered back projection (FBP) and algebraic reconstruction technique (ART). Briefly, FBP is based on a simple back projection and decreases image blurring in the CT image. Figure 8.3 shows the example of the projection in CT scan. Simple back projection smears each view along the path it was acquired to form an image, resulting in a blurred version of the correct image (Fig. 8.4). By setting all the pixel values along a ray to the same value and summing the back-projected views, the final image is obtained. To correct the blurring effect of simple back projection, filtered back projection is used (Fig. 8.5). Each view is filtered before back projection by convolution with a filter kernel to create a set of filtered views. By back projecting the filtered views, the resultant image is a better representation of the correct image, and several filters are proposed, such as Hamming filter, Shepp-Logan filter, and Ramachandran-Lakshminarayanan filter. In addition, FBP can use fast Fourier transform (FFT), leading to fast reconstruction. On the other hand, ART is based on an inverse problem and updates an assumed image by the stepwise repetition of a calculation until the image agrees with the measured image. Various methods in ART are proposed, such as iterative reconstruction (AR) [3] and simultaneous algebraic reconstruction technique (SART) [4], and the image

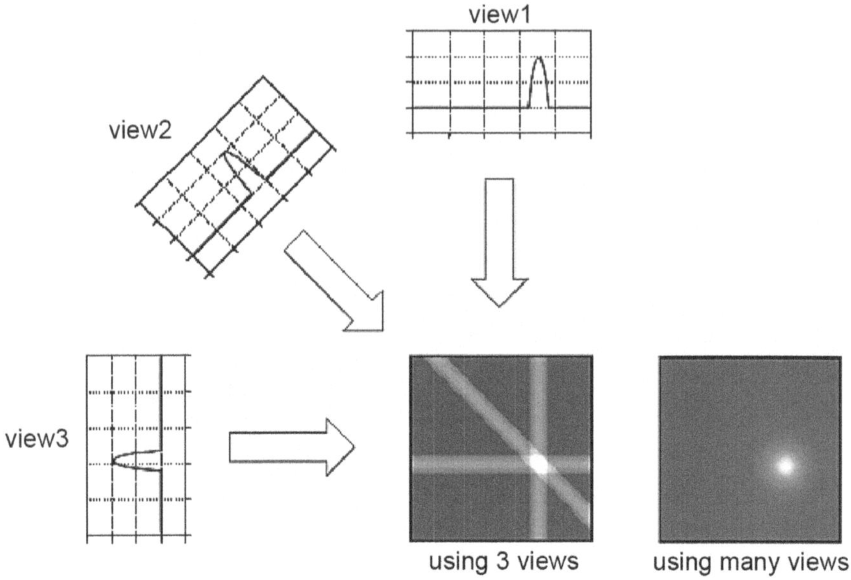

Fig. 8.4 Example of simple back projection. Simple back projection smears each view along the path, resulting in a blurred version of the correct image. Reprinted from Ref. [2] with permission

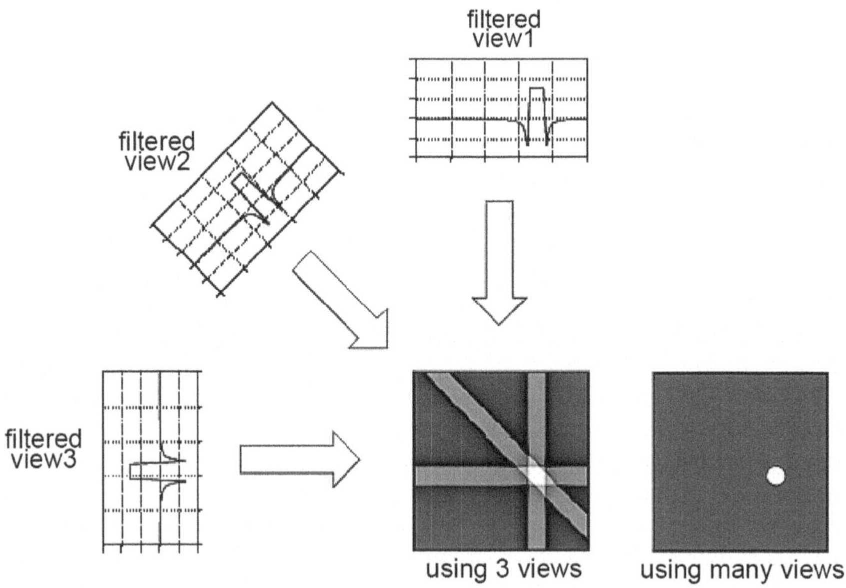

Fig. 8.5 Example of FBP. Each view is filtered before reconstruction by back projection. Reprinted from Ref. [2] with permission

quality increases through noise reduction and radiation dose decreases even though the computational cost in reconstruction increases.

8.3 Contrast Agents for CT

Materials possessing high density (ρ) or high atomic number (Z) tend to better absorb X-rays. The relationship is expressed in the following equation;

$$\mu \approx \frac{\rho Z^4}{AE^3} \tag{8.3}$$

where A is the atomic mass, E is the X-ray energy, and Z can increase the contrast greatly between different tissues and types of contrast agents.

When X-ray has energy equal or slightly greater than the binding energy of a shell electron of the atom, such as K shell, a large sudden increase in absorption coefficient is observed. This energy value is known as absorption edge (k-edge), and this value increases with atomic number of the element. Consequently, X-ray attenuating contrast media containing atoms of high atomic number (commonly iodine for blood vessels or barium for gastrointestinal tract in clinical application) is frequently used in clinical settings to obtain images of soft tissues. To generate images with the highest contrast to the surrounding tissue, the X-ray energy can be adjusted to closely match the absorption edge value (k) of the relevant imaging-agent atoms. Therefore, it is also possible to differentiate between attenuating materials by choosing the X-ray energy to the appropriate absorption edge value and lead to subtraction CT [5, 6].

Many tissues can be easily visualized by CT imaging, and the X-ray attenuation is expressed in Hounsfield units (HU) in clinical application. Water and air are defined as 0 HU and −1000 HU, respectively. The attenuation of lung tissue is closed to −750 HU due to high air content and mineralized tissues such as bone, with attenuation values of approximately 1000 HU. While different types of tissues can exhibit contrast, most soft tissues contain a lot of water. Therefore, their attenuation values fall within 25–100 HU, and it is difficult to image the tissues of interest and identify the interface between adjacent tissues and into the tissues. Hence, contrast agents are required for better visualization of the tissue by X-ray CT.

In order to achieve higher levels of X-ray attenuation than tissues, elements of higher Z are incorporated into the contrast agent molecule (Table 8.1). Iodine ($Z = 53$) has been used for CT imaging applications, in particular to visualize arterial and venous vessels (CT angiography; angio-CT). Generally, there are two groups in iodinated contrast agents, such as "ionic" and "non-ionic" iodinated agents. Although widely used in the clinic, the ionic iodinated agents possess several inherent disadvantages, such as higher osmolality, compared to non-ionic contrast agents, leading to renal toxicity and other physiological problems such as vasodilatation and pulmonary hypertension. In addition to osmolality, the viscosity of the contrast agents should

Table 8.1 Examples of X-ray contrast agents. The mass-weighted attenuation coefficient μ' is for X-rays of 80 keV energy. The mass-weighted attenuation coefficients for soft tissue and cortical bone are 0.182 and 0.223 $cm^2\ g^{-1}$, respectively

Contrast agent	μ' [cm^2/g]	Density [g/cm^3]	Application
Iohexol	1.093	0.924	Cardiac angiography, liver imaging
Iodixanol	1.156	2.295	Cardiac angiography
Diatrizoate	1.454	1.66	Cardiac angiography
Ioxaglate	1.408		Cardiac angiography
Silver iodide	2.130	5.675	Retrograde pyelography, liver imaging
Cesium chloride	1.969	3.99	Bronchography
Barium sulfate	1.616	4.5	Gastrointestinal tract
Bismuth subnitrate	1.437	4.93	Gastrointestinal tract
Bismuth sulfide	1.813	6.78	Gastrointestinal tract
Gold nanoparticle	1.72	19.32	Vascular casting, tumor
Silver nanoparticle	1.907	10.49	Vascular casting
Gadolinium oxide	2.511	7.4	Hepatobiliary
Gadolinium-DTPA[a]	0.855		Angiography
Ytterbium-DTPA[a]	0.935		Angiography
Stannic oxide	1.660	6.95	Liver imaging
Calcium tungstate	1.871	6.06	Bronchography
Thorium oxide	2.223	10.00	Cerebral arteriography, liver imaging
Tantalum oxide	2.399	8.2	Gastrointestinal tract, hypervascular tumors

Note The linear attenuation coefficient μ [cm^{-1}] is the value of the mass-weighted coefficient times the density

[a]DTPA is gadopentetic acid. Reprinted from Ref. [10] with permission from Royal Society of Chemistry

be considered. High viscosity of contrast agents may require higher volume and prolonged retention in kidneys [7].

Contrast agents increasing the residence time in blood and reducing the rate of renal clearance are proposed and called as a blood-pool agent. Liposomal contrast agent is one of the blood-pool agents, and liposomes of ~100 nm in size contain iodinated contrast agents. Nanoemulsions of lipid-soluble iodinated CT contrast agents also exhibit prolonged blood circulation times and good visualization of vasculature in rodents (Fig. 8.6) [8]. Lanthanide-based contrast agents are commonly used in magnetic resonance imaging (MRI), however, can be used as CT contrast agents due to their high Z, such as gadolinium ($Z = 64$), dysprosium ($Z = 66$), and ytterbium ($Z = 70$), and so have a potential contrast agent for multimodal MRI/CT/upconversion fluorescence imaging (Table 8.1) [9, 10]. Gold nanoparticles are also used because of their high X-ray attenuation and low toxicity, and 1.9 nm gold nanoparticles visualized the vessels less than 100 μm in diameter using a standard mammography in rodent animal experiments [11].

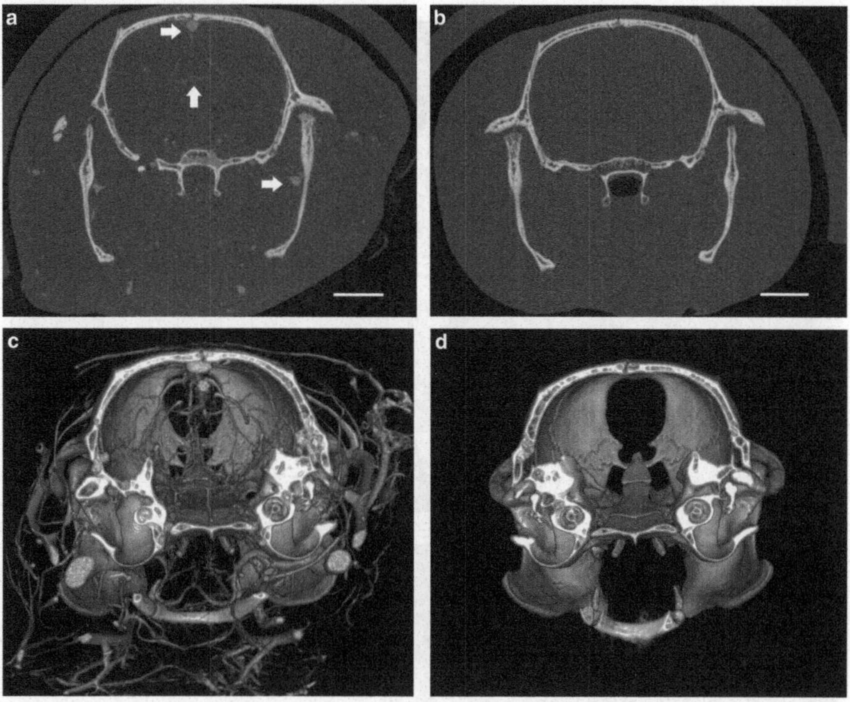

Fig. 8.6 Representative CT images (**a** and **b**) and 3D volume images (**c** and **d**) of a cerebral blood vessels of a rat. **a**, **c** the injection of nanoemulsions of lipid-soluble iodinated CT contrast agent (Fenestra™), **b** and **d** no contrast agents. Many vessels, including cerebral blood vessels and carotid arteries, are visualized clearly after injection of contrast agent (arrows in **a** and **c**). Only bones are visualized without contrast agents (**b** and **d**). Bars: 4 mm. Reprinted from Ref. [8] with permission from SPIE

On the other hand, heavy metal staining can be used such as tungsten and osmium to improve the contrast for soft tissue for micro- and nano-CT absorption imaging because of non-clinical use and low contrast due to high spatial resolution [12]. While a mouse kidney appears only with a slight contrast in the non-stained image, iodine-staining method achieves the micro-CT images with higher contrast and details (Fig. 8.7) [13].

8.4 Other CT Technology

8.4.1 Refraction Enhanced CT

When a sample is illuminated by the low divergent X-ray beam from synchrotron radiation source, such as SPring-8, and its projected image is recorded at an adequate

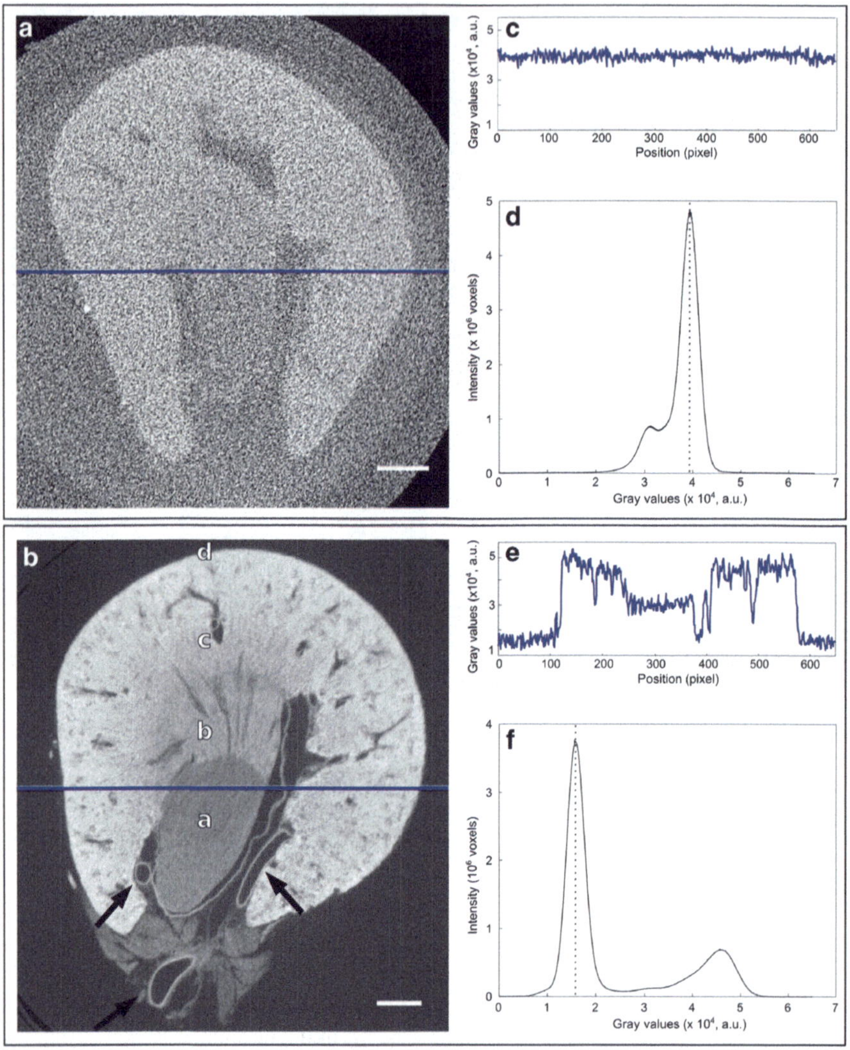

Fig. 8.7 Comparison between non-stained and iodine-stained X-ray micro-CT images of a mouse kidney. **a** a non-stained kidney, **b** an iodine-stained kidney. Kidney structures are well delineated in (**b**); the inner medulla (**a**), the inner stripe (**b**), the outer stripe of the outer medulla (**c**), the renal cortex (**d**), and some veins and arteries (arrows) are clearly identified. Profiles along the blue path in (**a** and **b**) are shown in (**c** and **e**), respectively. Distribution of voxel intensity values is shown in (**d**) for the non-stained kidney and in (**f**) for the iodine-stained kidney. Scale bars: 500 μm [13]

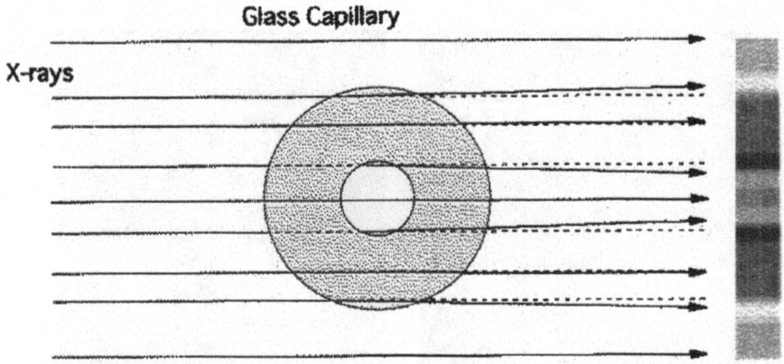

Fig. 8.8 Schematic drawing of refraction of X-rays in a glass capillary. From [15] Copyright © 1999 by John Wiley Sons, Inc. Reprinted by permission of John Wiley & Sons, Inc.

distance, an edge-enhancing effect is observed. When a ray passes into the medium with refraction index of n from air, the equation is expressed by the Snell's law of refraction,

$$\cos\theta_1 = n\cos\theta_2 \tag{8.4}$$

where θ_1 and θ_2 are incident and refraction angles, respectively. In visible lights, n of biological samples is basically ranging between 1.3 and 1.5 [14], resulting in $\theta < \theta_2$. Conversely, in X-ray, $\theta_1 > \theta_2$ because n is slightly smaller than 1. Figure 8.8 shows the principle of the edge enhancement, and the outward deflection causes a bright fringe outside the outer edge of the capillary, while a bright fringe appears at the inside of the inner edge. Large refraction is expected at an interface between materials with a large difference in refractive indices. Thus, when applied to a body of an animal, this technique is powerful to visualize boundaries between air and tissues, such as lung [15–17]. Furthermore, this refraction edge enhancement is observed in CT images in small animals, suggesting the utility of a high-resolution tomographic image with low dose (Fig. 8.9) [18–20]. This edge enhancement can be considered as an in-line holography of the X-rays; therefore, this technique based on absorption contrast can be classified into the phase-contrast imaging techniques in Sect. 8.4.1, propagation-based phase-contrast imaging [17, 21]. A monochromatic X-ray with a plane wave front can generate the projected images containing contrast from attenuation and phase shift through Fresnel diffraction through the object. The phase retrieval algorithm prior to CT reconstruction provides high contrast CT images with low radiation dose [22].

8.4.2 *Phase-Contrast CT*

X-rays can also be considered as waves, and the wave front distortion is occurred since the wave speed is changed due to the interaction inside of the medium, resulting a local

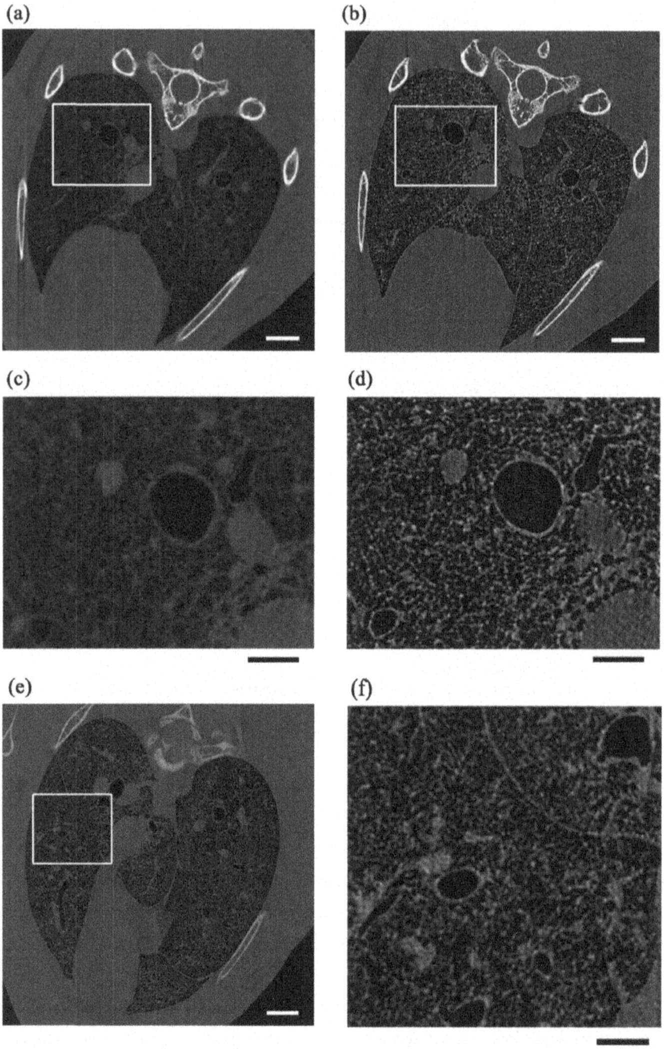

Fig. 8.9 Tomographic cross-sections of a rabbit thorax. **a** Recorded with an X-ray energy of 40 keV and a specimen-to-detector distance of 40 cm. **b** Same slice as (**a**) but recorded with $E = 40$ keV and $L = 480$ cm. **c** An enlarged view of part of (**a**). **d** An enlarged view of part of (**b**). **e** Recorded at $E = 70$ keV and $L = 260$ cm. **f** An enlarged view of part of (**e**). The bars in (**a**, **b**, and **e**) correspond to 5.0 mm, while those in (**c** and **d**) 2.5 mm, and that in (**f**) 2.0 mm. From [18] Copyright © 2005 by John Wiley Sons, Inc. Reprinted by permission of John Wiley & Sons, Inc.

phase shift in the X-ray after the object. Physically, local phase shifts of X-ray waves are determined by the local electron density in the object. When X-rays pass through the object, the internal structural information is encoded into the corresponding wave front distortions. Thus, the structural information can be speculated by detecting the wave front distortion in X-ray waves [23, 24]. At X-ray energies, every material has an energy-dependent complex refractive index as followed,

$$n = 1 - \delta + i\beta \tag{8.5}$$

where δ is the refractive index decrement, which is responsible for X-ray phase shift, $\beta = \lambda\mu/4\pi$ is the absorption index, and λ is the X-ray wavelength. X-ray imaging conventionally depends on β alone for absorption contrast, whereas phase contrast is sensitive to both δ and β. For biological tissue at diagnostic energies, δ is typically around three to four orders of magnitude larger than β [25]. As a result, phase contras can achieve significantly greater image contrast than conventional absorption contrast. Additionally, δ is inversely proportional to the square of the X-ray energy, while β decreases as the fourth power of energy [26]. Therefore, at higher X-ray energy, phase contrast is dominated more than absorption contrast, leading to the dose reduction. Several methods have been used to extract wave front distortion, such as interferometric technique and refraction-based technique, and show the potential to visualize the biological tissue without contrast agents [17, 27]. This phase-contrast technique is applied to CT imaging [28]. Particularly, interferometry is most comprehensive to detect the phase shift and so is involved in not only synchrotron radiation-based CT but also micro-CT [29] to visualize the soft tissue, such as fetus [30, 31], eye lens [32], brain [33–35], and atherosclerotic plaque [36]. Figure 8.10 shows the X-ray phase-contrast CT system based on the Talbot grating interferometer at SPring8, and the phase shift is estimated by the Moiré fringes generated by two gratings (an absorbance and phase grating) [30]. Indeed,

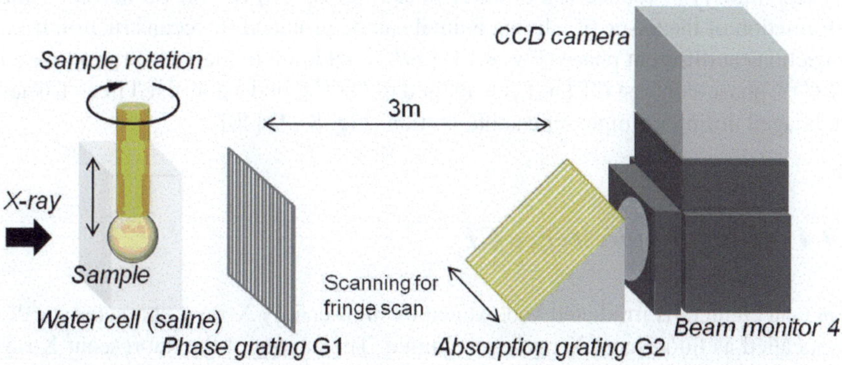

Fig. 8.10 Schematic diagram of the phase-contrast CT using Talbot interferometer showing the sample cell with sample suspended on a rotatable rod; the phase and absorption gratings (G1 and G2, respectively) and the detector (the beam monitor) [32]

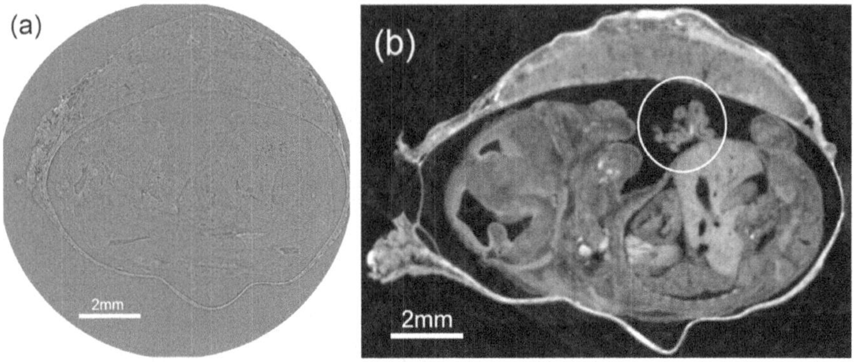

Fig. 8.11 Cross-section of a fetus **a** Refraction enhanced CT images. The sample-to-detector distance was 3 m. **b** Umbilical hernia in the fetus (circle) recorded with phase-contrast CT [30]

the image contrast is much higher by phase-contrast CT than by refraction enhanced CT (Fig. 8.11).

8.4.3 Four-Dimensional (4D) CT

A recent technological innovation in clinical CT achieves 3D-CT containing a moving structure (4D-CT) over a period of time. Projections are imaged, and CT images are reconstructed retrospectively or prospectively, resulting that a dynamic volume data set of various organs, such as musculoskeletal and cardiac dynamic imaging, is provided [37]. In particular, prospective 4D-CT is often used in small animal imaging because of the fast heartbeats and the difficulty in breathing control [8, 38]. The projections are obtained in synchrony with the physiological signals (ECG and respiratory) to reduce a motion artifact (Fig. 8.12) [8], and furthermore, the deformation of the tissue of a living animal can be provided by reconstruction from projections at different phase (Fig. 8.13) [39]. In addition to these absorption-based 4D-CTs, phase-contrast CT has been applied to 4D-CT, and a non-fixed fresh tissues are imaged during a simple repeatable motion (Fig. 8.14) [40].

8.4.4 X-Ray Fluorescence CT

When an element is irradiated and excited by high-energy X-rays, characteristic X-rays, called as fluorescent X-rays, are emitted. The energy of the fluorescent X-ray depends on the element; therefore, it is widely used for highly sensitive chemical analysis. In fluorescent CT based on these phenomena, fluorescent X-rays are imaged sequentially during sample rotation, the distribution of the element in the sample

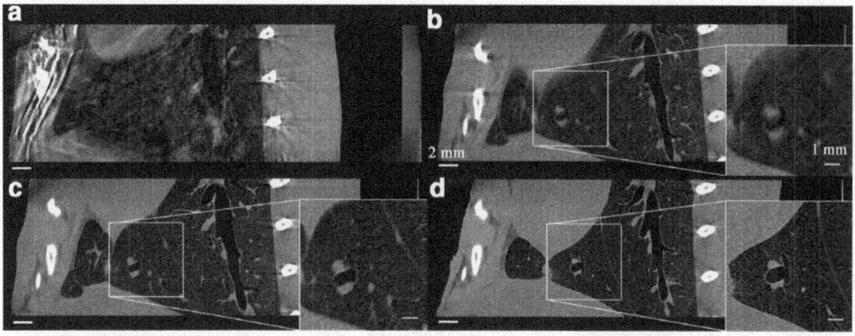

Fig. 8.12 Sagittal slices through the rat heart of without gating (**a**), respiratory only gating (**b**), respiratory and cardiac gating (**c**), and dead rat (**d**). The edge of the heart is blurred with respiratory only gating (**b**). The respiratory and cardiac gating improved the blur (**c**) so that the resolution was nearly equal to that of a dead rat (**c** and **d**). Reprinted from Ref. [8] with permission from SPIE

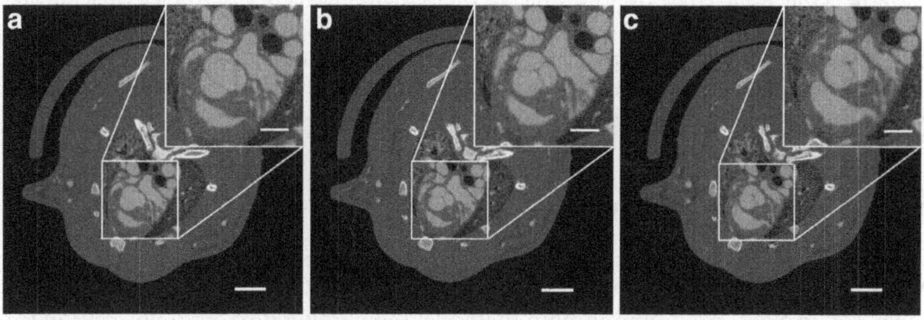

Fig. 8.13 Synchrotron CT images at the height of the aortic valve of a living rat during three phases in ECG cycle at **a** 40, **b** 125, and **c** 185 ms after the R peak. Nanoemulsions of lipid-soluble iodinated CT contrast agent (Fenestra™) are used as a contrast agent. Bars in normal image and in zoomed images: 5 and 2.5 mm. Adapted and reprinted with permission Ref. [39]

is reconstructed, and the distribution of iron in a bee is the first demonstration in fluorescent CT [41]. The advantage is simultaneous imaging of various elements, such chlorine ($Z = 17$) and copper ($Z = 29$). In experiment setup, fluorescence detector (spectrometer) is usually mounted perpendicular to the incident beam to reduce the Compton scattering from the sample and improve the signal-to-noise ratio (S/N). Therefore, this setup allows to place the detector for the projected images behind the sample on the incident beam direction, leading to the combination with anatomical information from conventional absorption CT (Fig. 8.15) [42–44]. On the other hand, the compensation of the self-absorption effects of the incident beam and the fluorescence of the sample and the long scan time are required [42].

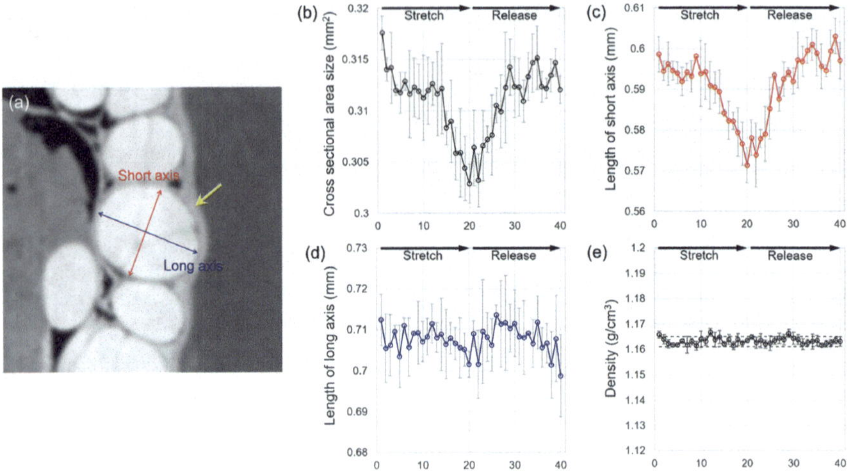

Fig. 8.14 Demonstration of 4D phase-contrast CT. **a** A magnified CT image of a tendon fascicle. Short and long axes are presented. **b** Cross-sectional area size in the stretching and releasing process. Lengths of **c** short axis and **d** long axis of the tendon fascicle. **e** Mass density averaged in the cross-section. Spacing between the dotted lines indicates density resolution. Reprinted from Ref. [40] with permission from Royal Society of Chemistry

8.4.5 X-Ray Luminescence CT

X-ray luminescence CT is based on a selective excitation similar to X-ray fluorescence CT in Sect. 8.4.4, and in both methods, the combination with absorption CT can provide the anatomical imaging simultaneously since the emitted light from the sample due to the X-ray irradiation is detected at the different axis on the transmitted direction in the absorption CT. However, while the accumulated element in a tissue is usually detected in fluorescence CT, fabricated nanophosphors, which emit near-infrared (NIR) lights, such as Gd_2O_2S:Tm and $BaYF_5$:Eu^{3+}, are used as contrast agents in X-ray luminescence CT. The tomogram reconstruction is a severely ill-conditioned inverse problem because of the high scattering and low absorption of the emitted visible and NIR light in the tissues [45, 46], resulting in low image quality. However, this technique has a potential for in vivo imaging because both X-ray and NIR lights propagate long distances in tissue [47].

8.4.6 Fluorescence Molecular CT

Nowadays, fluorescence imaging is widely used to study molecular dynamics in cell and organ, and similar to other X-ray CT techniques, tomographic reconstruction of the fluorescence images provides the three-dimensional distribution of a fluorescent probe or protein concentration, "fluorescence molecular CT". The concept is

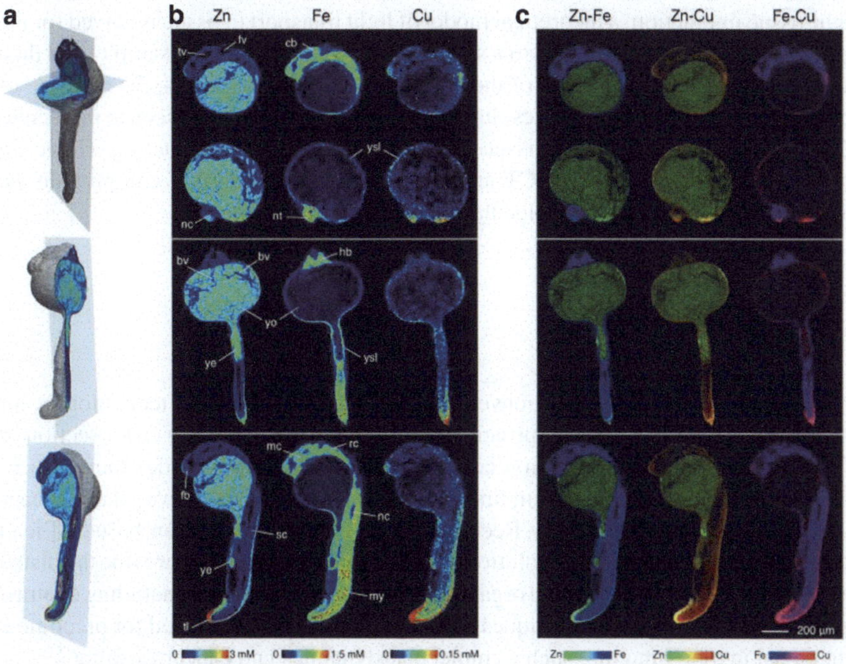

Fig. 8.15 Visualization of the elemental distribution in a zebrafish embryo by X-ray fluorescence tomography. **a** 3D-rendering of the embryo indicating the spatial orientation of the virtual slices that are displayed in panels **b** and **c**. Slices include a sagittal and transverse section (top), a coronal section (middle), and a sagittal section offset to the left (bottom). **b** Elemental distributions of Zn, Fe, and Cu for each of the 4 slices. Individual concentration scales for each element are displayed at the bottom of each column. Abbreviations: third ventricle (tv), fourth ventricle (fv), cerebellum (cb), notochord (nc), neural tube (nt), blood vessels (bv), hindbrain (hb), yolk (yo), yolk extension (ye), yolk syncytial layer (ysl), myotome (my), telencephalon (tc), mesencephalon (mc), diencephalon (dc), rhombencephalon (rc), and tail (tl). **c** False-color overlays of the elemental distributions of Zn, Fe, and Cu indicating regions of colocalization. The concentration scales of each element were normalized and color-coded as follows: Zn (green), Fe (blue), and Cu (red). Reprinted from Ref. [44] with permission from Royal Society of Chemistry

basically the same as X-ray luminescence CT described in Sect. 8.4.5, and combination with probes absorbing and fluorescing in NIR spectral range is mainly used due to the limited penetration depth of light into tissue. The emitted light is not X-ray; therefore, this technique is used for preclinical applications in small animals because consideration of radiation dose is not required [48] even though light/tissue interaction in tissues, such as tissue absorption and scattering, induced poor spatial resolution. Particularly, in NIR wavelength regions, the light in the over-thousand-nanometer NIR (OTN-NIR) reduces the scattering in biological tissues, resulting in higher spatial resolution imaging [49, 50], and furthermore, the tomogram for OTN-NIR using β-$NaGdF_4$:Yb^{3+}/Ho^{3+} as a contrast agent shows a potential for visualization of deeper areas in biological sample [51]. On the other hand, to overcome the light/tissue interaction, a theoretical model of light transport in tissue is solved for the propagation of light emitting from a source location throughout the sample, and then ART is used for reconstruction of the tomogram [52]. Nevertheless, the accurate of information in tissue at spatial resolution is still lower in the fluorescence molecular CT than X-ray CT because of tissue heterogeneity and severe scattering. However, hybrid system combing X-ray CT and fluorescence molecular CT can provide the annotation of anatomy with molecular information [53].

8.5 Summary

In this chapter, we review the conventional CT and new imaging technologies for visualizing inside of tissues. In principle, CT allows to visualize the cross-section of a sample based on the attenuation coefficients of X-ray with non-destructive techniques, and the development from first- to fifth-generation CT achieves shorter scanning time and larger field of view. Recently, micro-CT and synchrotron-based CT lead to higher spatial and contrast resolution, and new fluorescent probes provide the distribution of the protein in tissue and organ. CT with improved systems including contrast agents and reconstruction techniques will be a more attractive method for preclinical functional in vivo imaging, such as molecular dynamics and cancer imaging.

References

1. T. Flohr, CT systems. Curr. Radiol. Rep. **1**, 52–63 (2013). https://doi.org/10.1007/s40134-012-0005-5
2. S.W. Smith, *The Scientist and Engineer's Guide to Digital Signal Processing*. https://www.dspguide.com/
3. W. Stiller, Basics of iterative reconstruction methods in computed tomography: a vendor-independent overview. Eur. J. Radiol. **109**, 147–154 (2018). https://doi.org/10.1016/j.ejrad.2018.10.025
4. A.H. Andersen, A.C. Kak, Simultaneous algebraic reconstruction technique (SART): a superior implementation of the ART algorithm. Ultrason. Imaging **6**, 81–94 (1984). https://doi.org/10.1016/0161-7346(84)90008-7
5. S. Strengell, J. Keyriläinen, P. Suortti et al., Radiation dose and image quality in K-edge subtraction computed tomography of lung in vivo. J. Synchrotron. Rad. **21**, 1305–1313 (2014). https://doi.org/10.1107/S160057751401697X
6. F.A. Dilmanian, X.Y. Wu, E.C. Parsons et al., Single- and dual-energy CT with monochromatic synchrotron X-rays. Phys. Med. Biol. **42**, 371–387 (1997). https://doi.org/10.1088/0031-9155/42/2/009
7. H. Lusic, M.W. Grinstaff, X-ray computed tomography contrast agents. Chem. Rev. **113**, 1641–1666 (2013). https://doi.org/10.1021/cr200358s
8. T. Sera, H. Yokota, K. Fujisaki, et al., In vivo-CT system with respiratory and cardiac gating using synchrotron radiation, in *Medical Imaging 2007: Physiology, Function, and Structure from Medical Images* (International Society for Optics and Photonics, 2007), p. 651103. https://doi.org/10.1117/12.709619

9. H. Xing, W. Bu, S. Zhang et al., Multifunctional nanoprobes for upconversion fluorescence, MR and CT trimodal imaging. Biomaterials **33**, 1079–1089 (2012). https://doi.org/10.1016/j.biomaterials.2011.10.039
10. H. Chen, M.M. Rogalski, J.N. Anker, Advances in functional X-ray imaging techniques and contrast agents. Phys. Chem. Chem. Phys. **14**, 13469–13486 (2012). https://doi.org/10.1039/c2cp41858d
11. J.F. Hainfeld, D.N. Slatkin, T.M. Focella et al., Gold nanoparticles: a new X-ray contrast agent. Br. J. Radiol. **79**, 248–253 (2006). https://doi.org/10.1259/bjr/13169882
12. S.D. Rawson, J. Maksimcuka, P.J. Withers et al., X-ray computed tomography in life sciences. BMC Biol. **18**, 21 (2020). https://doi.org/10.1186/s12915-020-0753-2
13. J.M. de S. e Silva, I. Zanette, P.B. Noël, et al., Three-dimensional non-destructive soft-tissue visualization with X-ray staining micro-tomography. Sci. Rep. **5**, 1–7 (2015). https://doi.org/10.1038/srep14088
14. Y. Zhou, K.K.H. Chan, T. Lai et al., Characterizing refractive index and thickness of biological tissues using combined multiphoton microscopy and optical coherence tomography. Biomed. Opt. Express **4**, 38–50 (2012). https://doi.org/10.1364/BOE.4.000038
15. N. Yagi, Y. Suzuki, K. Umetani et al., Refraction-enhanced X-ray imaging of mouse lung using synchrotron radiation source. Med. Phys. **26**, 2190–2193 (1999). https://doi.org/10.1118/1.598735
16. Y. Suzuki, N. Yagi, K. Uesugi, X-ray refraction-enhanced imaging and a method for phase retrieval for a simple object. J. Synchrotron. Rad. **9**, 160–165 (2002). https://doi.org/10.1107/S090904950200554X
17. M.J. Kitchen, R.A. Lewis, N. Yagi et al., Phase contrast X-ray imaging of mice and rabbit lungs: a comparative study. Brit. J. Radiol. **78**, 1018–1027 (2005). https://doi.org/10.1259/bjr/13024611
18. T. Sera, K. Uesugi, N. Yagi, Refraction-enhanced tomography of mouse and rabbit lungs. Med. Phys. **32**, 2787–2792 (2005)
19. T. Sera, H. Yokota, G. Tanaka et al., Murine pulmonary acinar mechanics during quasi-static inflation using synchrotron refraction-enhanced computed tomography. J. Appl. Physiol. **115**, 219–228 (2013). https://doi.org/10.1152/japplphysiol.01105.2012
20. T. Sera, H. Yokota, S. Nakamura et al., Synchrotron refraction enhanced tomography of an intact common Marmoset (Callithrix jacchus). Open J. Radial. **1**, 28–37 (2011). https://doi.org/10.4236/ojrad.2011.12005
21. G.-H. Chen, J. Zambelli, N. Bevins et al., X-ray phase sensitive imaging methods: basic physical principles and potential medical applications. Curr. Med. Imaging Rev. **6**, 90–99 (2010). https://doi.org/10.2174/157340510791268533
22. M.J. Kitchen, G.A. Buckley, T.E. Gureyev et al., CT dose reduction factors in the thousands using X-ray phase contrast. Sci. Rep. **7**, 15953 (2017). https://doi.org/10.1038/s41598-017-16264-x
23. A. Snigirev, I. Snigireva, V. Kohn et al., On the possibilities of X-ray phase contrast microimaging by coherent high-energy synchrotron radiation. Rev. Sci. Instrum. **66**, 5486–5492 (1995). https://doi.org/10.1063/1.1146073
24. P. Cloetens, R. Barrett, J. Baruchel et al., Phase objects in synchrotron radiation hard X-ray imaging. J. Phys. D.: Appl. Phys. **29**, 133–146 (1996). https://doi.org/10.1088/0022-3727/29/1/023
25. R.A. Lewis, K.D. Rogers, C.J. Hall, et al., *Medical Imaging 2002: Physics of Medical Imaging* (International Society for Optics and Photonics, 2002), pp. 286–297. https://doi.org/10.1117/12.465570
26. T.E. Gureyev, S. Mayo, S.W. Wilkins et al., Quantitative in-line phase-contrast imaging with multienergy X rays. Phys. Rev. Lett. **86**, 5827–5830 (2001). https://doi.org/10.1103/PhysRevLett.86.5827
27. A. Momose, Recent advances in X-ray phase imaging. Jpn. J. Appl. Phys. **44**, 6355 (2005). https://doi.org/10.1143/JJAP.44.6355

28. A. Momose, T. Takeda, Y. Itai et al., Phase-contrast X-ray computed tomography for observing biological soft tissues. Nat. Med. **2**, 473–475 (1996). https://doi.org/10.1038/nm0496-473
29. M. Engelhardt, J. Baumann, M. Schuster et al., High-resolution differential phase contrast imaging using a magnifying projection geometry with a microfocus X-ray source. Appl. Phys. Lett. **90**, 224101 (2007). https://doi.org/10.1063/1.2743928
30. M. Hoshino, K. Uesugi, N. Yagi, Phase-contrast X-ray microtomography of mouse fetus. Biol. Open **1**, 269–274 (2012). https://doi.org/10.1242/bio.2012430
31. M. Hoshino, K. Uesugi, T. Tsukube et al., Quantitative and dynamic measurements of biological fresh samples with X-ray phase contrast tomography. J. Synchrotron. Radiat. **21**, 1347–1357 (2014). https://doi.org/10.1107/S1600577514018128
32. M. Hoshino, K. Uesugi, N. Yagi et al., Optical properties of in situ eye lenses measured with X-ray Talbot interferometry: a novel measure of growth processes. PLoS ONE **6**, e25140 (2011). https://doi.org/10.1371/journal.pone.0025140
33. G. Schulz, T. Weitkamp, I. Zanette et al., High-resolution tomographic imaging of a human cerebellum: comparison of absorption and grating-based phase contrast. J. R. Soc. Interface **7**, 1665–1676 (2010). https://doi.org/10.1098/rsif.2010.0281
34. K. Noda-Saita, A. Yoneyama, Y. Shitaka et al., Quantitative analysis of amyloid plaques in a mouse model of Alzheimer's disease by phase-contrast X-ray computed tomography. Neuroscience **138**, 1205–1213 (2006). https://doi.org/10.1016/j.neuroscience.2005.12.036
35. S.A. McDonald, F. Marone, C. Hintermüller et al., Advanced phase-contrast imaging using a grating interferometer. J. Synchrotron Radiat. **16**, 562–572 (2009). https://doi.org/10.1107/S0909049509017920
36. M. Shinohara, T. Yamashita, H. Tawa et al., Atherosclerotic plaque imaging using phase-contrast X-ray computed tomography. Am. J. Physiol. Heart Circul. Physiol. **294**, H1094–H1100 (2008). https://doi.org/10.1152/ajpheart.01149.2007
37. Y. Kwong, A.O. Mel, G. Wheeler et al., Four-dimensional computed tomography (4DCT): a review of the current status and applications. J. Med. Imag. Radiat. Oncol. **59**, 545–554 (2015). https://doi.org/10.1111/1754-9485.12326
38. S.J. Blocker, M.D. Holbrook, Y.M. Mowery et al., The impact of respiratory gating on improving volume measurement of murine lung tumors in micro-CT imaging. PLoS ONE **15**, e0225019 (2020). https://doi.org/10.1371/journal.pone.0225019
39. T. Sera, H. Yokota, K. Fujisaki et al., Development of high-resolution 4D in vivo-CT for visualization of cardiac and respiratory deformations of small animals. Phys. Med. Biol. **53**, 4285–4301 (2008). https://doi.org/10.1088/0031-9155/53/16/005
40. M. Hoshino, K. Uesugi, N. Yagi, 4D X-ray phase contrast tomography for repeatable motion of biological samples. Rev. Sci. Instrum. **87**, 093705 (2016). https://doi.org/10.1063/1.4962405
41. P. Boisseau, L. Grodzins, Fluorescence tomography using synchrotron radiation at the NSLS. Hyperfine Interact. **33**, 283–292 (1987). https://doi.org/10.1007/BF02394116
42. T. Ohigashi, N. Watanabe, H. Yokosuka et al., Combined imaging system for X-ray fluorescence and transmission X-ray microtomography. AIP Conf. Proc. **705**, 1352–1355 (2004). https://doi.org/10.1063/1.1758052
43. R. Lopes, I. Lima, G. Pereira et al., Synchrotron radiation X-ray microfluorescence techniques and biological applications. Pramana **76**, 271–279 (2011). https://doi.org/10.1007/s12043-011-0043-1
44. D. Bourassa, S.-C. Gleber, S. Vogt et al., 3D imaging of transition metals in the zebrafish embryo by X-ray fluorescence microtomography. Metallomics **6**, 1648–1655 (2014). https://doi.org/10.1039/C4MT00121D
45. C. Li, A. Martinez-Davalos, S.R. Cherry, Numerical simulation of X-ray luminescence optical tomography for small-animal imaging. J. Biomed. Opt. **19**, 046002 (2014). https://doi.org/10.1117/1.JBO.19.4.046002
46. G. Zhang, F. Liu, J. Liu et al., Cone beam X-ray luminescence computed tomography based on Bayesian method. IEEE Trans. Med. Imaging **36**, 225–235 (2017). https://doi.org/10.1109/TMI.2016.2603843

47. G. Pratx, C.M. Carpenter, C. Sun et al., X-ray luminescence computed tomography via selective excitation: a feasibility study. IEEE Trans. Med. Imaging **29**, 1992–1999 (2010). https://doi.org/10.1109/TMI.2010.2055883
48. V. Ntziachristos, C.-H. Tung, C. Bremer et al., Fluorescence molecular tomography resolves protease activity in vivo. Nat. Med. **8**, 757–761 (2002). https://doi.org/10.1038/nm729
49. G. Hong, J.C. Lee, J.T. Robinson et al., Multifunctional in vivo vascular imaging using near-infrared II fluorescence. Nat. Med. **18**, 1841–1846 (2012). https://doi.org/10.1038/nm.2995
50. E. Hemmer, N. Venkatachalam, H. Hyodo et al., Upconverting and NIR emitting rare earth based nanostructures for NIR-bioimaging. Nanoscale **5**, 11339–11361 (2013). https://doi.org/10.1039/c3nr02286b
51. M. Umezawa, T. Sera, H. Yokota, et al., Computed tomography for in vivo deep over-1000 nm near-infrared fluorescence imaging. J. Biophotonics, e202000071 (2020). https://doi.org/10.1002/jbio.202000071
52. F. Stuker, J. Ripoll, M. Rudin, Fluorescence molecular tomography: principles and potential for pharmaceutical research. Pharmaceutics **3**, 229–274 (2011). https://doi.org/10.3390/pharmaceutics3020229
53. G. Zacharakis, H. Kambara, H. Shih et al., Volumetric tomography of fluorescent proteins through small animals in vivo. Proc. Nat. Acad. Sci. **102**, 18252–18257 (2005). https://doi.org/10.1073/pnas.0504628102

Part IV
Application of NIR Light for Medical Photonics

Chapter 9
NIR Light Induced Photodynamic Therapy

Masao Kamimura

Abstract Photodynamic therapy (PDT) is an emerging cancer therapy technique which destroys cancer tissues under photoirradiation. This therapy technique uses various photosensitizers which produce reactive oxygen species (ROS) to destroy cancer cells under light irradiation. However, generally, activation of these photosensitizers requires visible light irradiation and thus, most current photosensitizes are quite difficult to kill cancer tissues in deep tissues. Alternatively, tissue penetrable near-infrared (NIR) light-induced PDT is a promising technique for cancer treatment in deep tissues. Recently, various NIR responsive materials, such as rare-earth doped ceramics nanoparticles are used for NIR PDT. This chapter reviews the current progress of NIR light-induced PDT.

Keywords Photodynamic therapy · Cancer treatment · Upconversion · Photosensitizer · Reactive oxygen species

9.1 Photodynamic Therapy (PDT)

9.1.1 Brief Outline of PDT

Photodynamic therapy (PDT) is photo irradiation-initiated cancer treatment technique by using photo-responsive materials, i.e., photosensitizers [1, 2]. Today, various photosensitizes are widely reported for PDT, and these photosensitizers absorb specific wavelength light and it generates reactive oxygen species. Because of the reactive oxygen species can attack cancer cells, based on photoirradiation light source and photosensitizes, PDT technique can kill cancer cells. Compared to other current cancer treatment techniques, such as chemotherapy, radiotherapy, or surgery, photodynamic therapy has several advantages. For example, because most of the photosensitizers are safety materials, PDT technique does not show side effects against

M. Kamimura (✉)
Department of Materials Science and Technology, Faculty of Industrial Science and Technology, Tokyo University of Science, 6-3-1 Niijuku, Katsushika, Tokyo 125-8585, Japan
e-mail: masaokamimura@rs.tus.ac.jp

K. Soga et al. (eds.), *Transparency in Biology*,
https://doi.org/10.1007/978-981-15-9627-8_9

patients. In addition, photoirradiation is performed under the limited target area, thus, PDT treatment can also inhibit additional damage for normal tissues. Furthermore, PDT therapy only requires intravenous injection of PDT agents into the patient and light irradiation from the outside of the patient's body. Therefore, PDT technique does not show side effects and physical damage.

9.1.2 Wavelength of PDT

The treatment effect of cancer PDT is strongly dependent on the property of photosensitizes. Thus, various photosensitizes are developed and applied for PDT studies. Generally, most of current PDT, including clinical trial level, mainly uses porphyrin derivatives as photosensitizes, due to price of these organic molecules are low and it effectively generates reactive oxygen species.

However, most of these photosensitizes require short wavelength light irradiation for activation. Usually, PDT uses ultraviolet (UV: <400 nm) or visible (VIS: 400–700 nm) light sources. These short wavelength photoirradiation causes serious problems. It is well known that the short wavelength light, especially UV or blue (400–500 nm) light are quite difficult to penetrate into the human tissues, due to the strong absorption of biomolecules, such as protein in the body. Therefore, UV or blue light-responsive photosensitizers are difficult to use in PDT. To address this problem, red (600–700 nm) light is mainly used as light source of PDT, due to the good permeability of red light, as compared with UV or blue light. For example, one of the porphyrin derivatives, Chlorin e6 (Ce6) is famous photosensitizer of PDT. The Ce6 can absorb 400 nm (blue) and 660 nm (red), however, 660 nm light is mainly used for light source of Ce6 activation, due to the problem of tissue permeability.

By using red light irradiation, PDT technique can treat few mm deep tumor tissues. However, this depth is not enough for effective cancer treatment technique. Thus, tissue penetrative light-initiated PDT, i.e., near-infrared (NIR) light-induced PDT have recently mush attention.

9.1.3 Application of Near-Infrared Light for PDT

NIR light (>700 nm) is well known to show excellent tissue permeability and thus, is an attractive light source for biophotonics, including fluorescence bioimaging, manipulation technique, and PDT. NIR light possesses low scattering and absorption of biomolecules and water in the human body. Therefore, NIR wavelength region is called as "biological window". In the past three decades, NIR biophotonics became hot topic and numerous researches are reported in the world.

In the case of PDT, most of photosensitizes are UV or VIS light responsive agent. Thus, NIR light cannot use directly as light source. However, by using NIR light responsive materials as transducer, NIR light-induced PDT can be achieved.

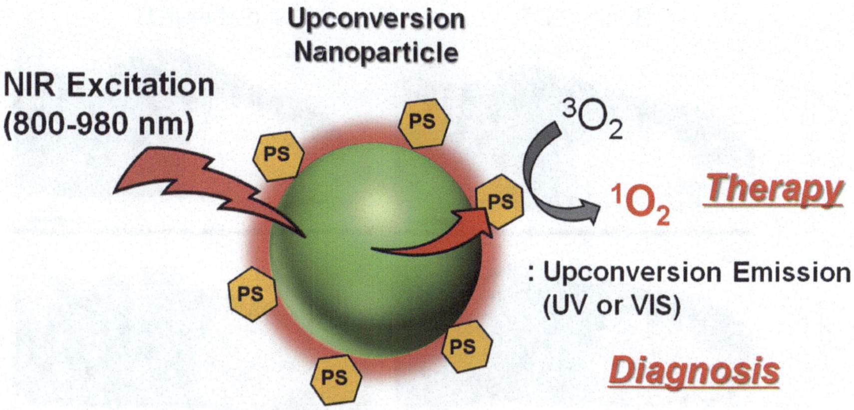

Fig. 9.1 Application of upconversion nanoparticles for PDT

For example, transduction of longer wavelength excitation light to shorter wavelength emission, photon upconversion phenomena is well used as NIR light-induced PDT. Rare-earth ion doped ceramics nanoparticles is well known to show upconversion phenomena and thus, is called as "upconversion nanoparticles" (Fig. 9.1). For example, trivalent rare-earth ions, ytterbium (Yb^{3+}), erbium (Er^{3+}), or thulium (Tm^{3+}) are doped into the crystal structure of fluoride or oxide ceramic nanoparticles, such as $NaYF_4$, $NaGdF_4$, LaF_3, or Y_2O_3, show excellent upconversion emission. These nanoparticles can be excited by NIR laser and it shows various short wavelength upconversion emission, from UV to NIR region. By using upconversion UV or VIS emission under NIR light irradiation, UV or VIS light responsive photosensitizers can be activated by tissue penetrable NIR light. Thus, PDT in deep tissue can be performed (Fig. 9.2).

9.2 Ceramic Nanophosphors for PDT and Diagnosis

9.2.1 Upconversion Nanoparticles for PDT

Recently, upconversion nanoparticles-based PDT is attracted much attention and many studies are reported. In 2007, Peng Zhang et al. report the first study of PDT by using upconversion nanoparticles [3]. In this work, Merocyanine 540 was used as photosensitizer, and it was immobilized on rare-earth doped $NaYF_4$ nanoparticles. In addition, ligand antibody was also installed on the nanoparticles surface and obtained conjugates could attach and kill MCF-7AZ breast cancer cells under 974 nm laser excitation. Since then, both of in vitro and in vivo upconversion PDT studies are reported [1, 2, 4–66]. Advantages of upconversion nanoparticles-based PDT is not only tissue permeability but also diagnosis applications. Under NIR light excitation,

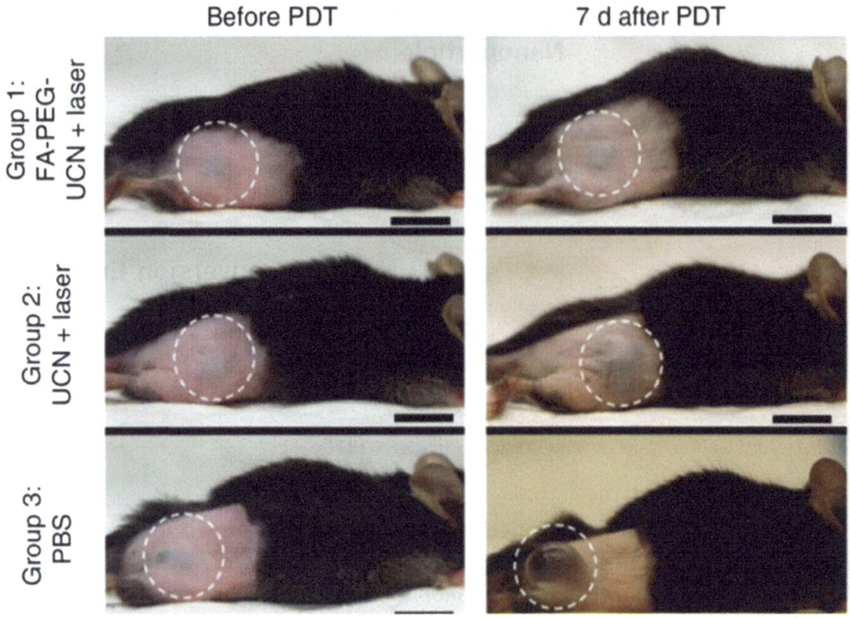

Fig. 9.2 Targeted in vivo PDT of a subcutaneous tumor model injected with upconversion PDT agent. Representative gross photos of a mouse from each group 1–3 intravenously injected with folic acid/PEG immobilized nanoparticles, unmodified upconversion nanoparticles, or PBS showing the change in tumor size (highlighted by dashed white circles) before (0 d) and 7 d after PDT treatment. Scale bars, 10 mm. (Reproduced with permission from 9) © 2012 Springer Nature

upconversion nanoparticles show short wavelength emission. By using this emission, cancer therapy and diagnosis are performed at the same time by using same probe. These combination techniques of diagnosis and therapy is so-called "theranostics", and similar studies are widely reported. For example, Hsin-Cheng Chiu and his co-workers reported photosensitizer and photothermal organic dyes immobilized upconversion nanoparticles for PDT, photothermal therapy (PTT), and fluorescence imaging [59]. This work successfully performed in vivo PDT and PTT. In addition, fluorescence imaging of brain tumor was also successfully observed.

9.2.2 Upconversion PDT with Organic Molecule Photosensitizes

Upconversion nanoparticles-based PDT uses various photosensitizers. For example, Ce6 is one of the most used photosensitizers for upconversion PDT. Shun Yang et al. reported light responsive polymer-modified upconversion nanoparticles for PDT [31]. This PDT agent contains Ce6 and anti-cancer drug, and is used for PDT and chemotherapy. Caixia Yue et al. also reported polymer/Ce6/anticancer drug conjugated upconversion nanoparticles [49]. In this work, reactive oxygen species sensitive polymer was introduced on the nanoparticles surface, and generated reactive oxygen species under NIR light irradiation was used not only for direct therapy but also for trigger of drug release. Furthermore, Jun Xu et al. reported Ce6 loaded upconversion nanoparticles for PDT and immunotherapy [53]. Moreover, Xiaogang Liu and his co-workers reported dopamine core/$NaGF_4$:Yb^{3+}, Er^{3+} shell nanoparticle and Ce6 nanoconjugates for PTT and PDT applications [64].

Rose Bengal is also a major photosensitizes for PDT. In 2012, Kai Liu et al. reported Rose Bengal conjugated upconversion nanoparticles for PDT [11]. In this study, cancer cell specific ligand, folic acid was installed on the nanoparticle surface, and specific binding against folic acid receptor positive cancer cells and successful PDT was performed. Xiangliang Yang's group also reported rare-earth doped LaF_3 nanoparticles conjugated with Rose Bengal for PDT in culture cancer cells [36]. Liuen Liang et al. also reported ligand antibody installed Rose Bengal/upconversion nanoparticle nanocomplex for PDT [47].

In addition, Zinc Phthalocyanine (ZnPC) is also typical propylene like photosensitizers. In 2010, Yong Zhang's group reported ZnPC based upconversion PDT [6]. In this study, mesoporous silica structure was prepared on upconversion nanoparticles surface, and ZnPC was introduced in the pore of nanoparticle surface. The obtained complex showed successful reactive oxygen species generation and cultured murine bladder cancer cells were effectively killed under NIR irradiation. In 2012, Meng Earn Lim et al. also reported on preparation of ZnPC introduced upconversion nanoparticles for PDT [10]. In this work, ZnPC was immobilized on polyethyleneimine modified upconversion nanoparticles and applied for PDT test in both culture cancer cells and tumor beard mouse. Yueqing Gu and his co-workers also reported immobilization of ZnPC on amphiphilic chitosan modified upconversion nanoparticles for PDT [8]. Xin Wang et al. also reported similar ZnPC introduced polymer-modified upconversion nanoparticles for PDT [20]. Meng Wang et al. also reported carboxy-ZnPC immobilization on rare-earth doped $LiFY_4$ via ligand exchange method for PDT agent [27]. In addition, target ligand molecules installed ZnPC/upconversion nanoparticles are also reported by several researchers. For example, Yueqing Gu and his co-workers reported folic acid and ZnPC introduced upconversion nanoparticles for specific tumor targeting and therapy agent [17]. Lu Xia et al. also reported biocompatible poly (ethylene glycol) (PEG)/ZnPC/folic acid installed upconversion nanoparticles for PDT [30].

9.2.3 Upconversion PDT with Inorganic/Metal Photosensitizes

Inorganic or metal nanoparticles are also used as photosensitizer of upconversion PDT. In these studies, photosensitizer nanomaterials are conjugated with upconversion nanoparticles via chemical bonding or incorporated into nanocomplex structure.

For example, titanium oxide (TiO_2) is used as UV responsive photosensitizer. Yong Zhang and his co-workers reported TiO_2/ceramics core/shell nanoparticles [41]. In this work, TiO_2 shell structure was formed on NIR-to-UV upconversion nanoparticles, $NaYF_4$:Yb^{3+}, Tm^{3+} surface. The obtained nanoparticles show reactive oxygen species generation under NIR excitation, and it can be used for cancer treatment. Zhiyao Hou et al. also reported nanocomplex formation by using TiO_2 nanoparticles and rare-earth doped ceramics nanoparticles for mitochondria-targeted PDT [38].

Carbon nanomaterials, such as fullerene or graphene are also used as photosensitizes. Xiaomin Liu et al. reported folic acid and fullerene conjugated upconversion nanoparticles for PDT [18]. In this work, the obtained nanocomplex could attach to cancer cells by folic acid ligand on nanoparticle surface, and generated reactive oxygen species form fullerene under NIR irradiation effectively killed cancer cells. In addition, Xiu-Ping Yan and his co-workers also reported fullered immobilized upconversion nanoparticles for PDT [29]. Furthermore, Hongjie Zhang's group also reported graphene conjugated rare-earth doped $NaYF_4$ nanoparticles for PDT [21].

In addition, other types of inorganic or metal nanomaterials are also reported as photosensitizer of upconversion PDT. Yong Zhang and his co-workers reported preparation of rare-earth doped $NaYF_4$ core-Zinc Oxide (ZnO) shell nanoparticle and its application for PDT [35]. Ruichan Lv et al. also reported black phosphorous photosensitizer immobilized upconversion nanoparticles for PDT [48]. In addition, same research group also reported gold nanocluster as a photosensitizer of upconversion PDT [42].

9.2.4 Upconversion PDT Agent for Multimodal Theranostics

Application of upconversion nanoparticles for both PDT and various imaging techniques, i.e., theranostics is one of the most interesting features of upconversion PDT. By using emission of the rare-earth doped ceramic nanoparticles, the prepared PDT agent also can be used for cancer diagnosis. In addition, combination of upconversion emission with various imaging techniques, such as magnetic resonance (MR) or X-ray CT are widely reported.

In 2012, Zengxia Zhao et al. reported rare-earth doped $NaGdF_4$ nanoparticles based theranostics agent [14]. In this work, photosensitizer molecule was introduced on nanoparticle surface, and the obtained nanoparticles can be used for PDT,

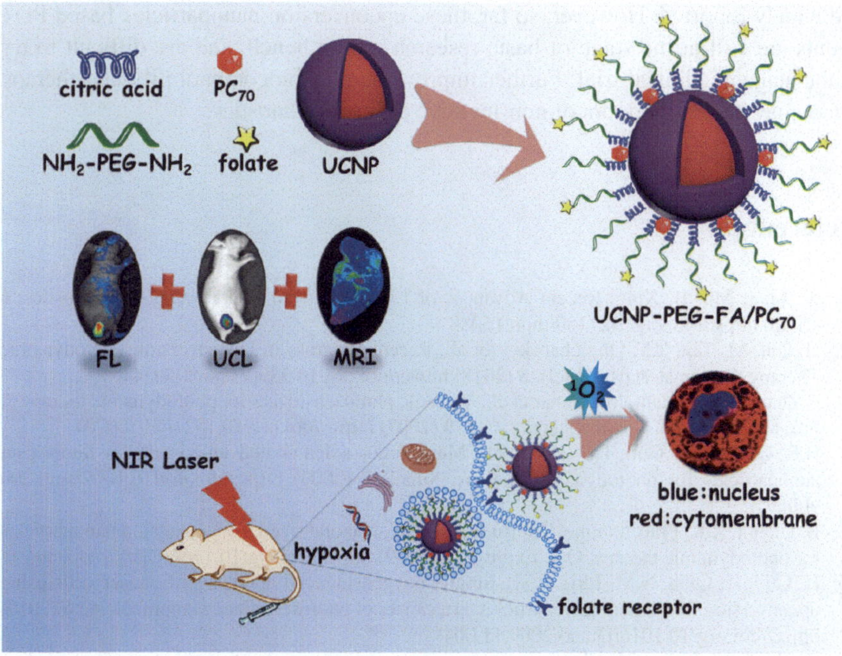

Fig. 9.3 Schematic strategies of trimodal imaging-guided upconversion PDT. (Reproduced with permission from 38) © 2015 Springer Nature

fluorescence imaging, and MR imaging. Since then, $NaGdF_4$ based theranostics agents are widely reported [24, 28, 45, 46, 60, 62]. For example, fullerene was conjugated with rare-earth doped $NaGdF_4$ nanoparticles and applied for PDT and MR/fluorescence imaging (Fig. 9.3) [37]. Furthermore, because of the rare-earth doped ceramic nanoparticle show both of short wavelength upconversion emission and over-1000 nm NIR emission under the same NIR excitation source, combination of PDT and over-1000 nm NIR emission is also promising candidate for effective theranostics [51]. Over-1000 nm NIR light, also called as NIR-II (1050–1400 nm) or NIR-III (1550–1700 nm), show excellent permeability into the human body. Thus, over-1000 nm emission and PDT-based theranostics have possible applications in future cancer medicine.

9.3 Summary

In this chapter, tissue penetrable NIR light-induced PDT based on rare-earth doped ceramic nanoparticles are reviewed. In the past decade, many researchers reported various combination of photosensitizers and rare-earth doped ceramic nanoparticles. Thus, effective therapy results of these studies including in vivo experiments are widely reported. However, so far, these upconversion nanoparticles-based PDT agents are still at the stage of basic research on the bench, and are difficult to try at the stage of clinical trial. Further improvement of biocompatibility and therapy effect may open new doors of non-invasive cancer theranostics.

References

1. X. Ai, J. Mu, B. Xing, Recent Advances of Light-Mediated Theranostics. Theranostics, **6** (2016) https://doi.org/10.7150/thno.16088
2. H. Qiu, M. Tan, T.Y. Ohulchanskyy et al., Recent Progress in Upconversion Photodynamic Therapy. Nanomaterials (Basel). **8** (2018) https://doi.org/10.3390/nano8050344
3. P. Zhang, W. Steelant, M. Kumar et al., Versatile photosensitizers for photodynamic therapy at infrared excitation. J. Am. Chem. Soc. **129** (2007) https://doi.org/10.1021/ja0700707
4. H.S. Qian, H.C. Guo, P.C. Ho et al., Mesoporous-silica-coated up-conversion fluorescent nanoparticles for photodynamic therapy. Small. **5** (2009) https://doi.org/10.1002/smll.200900692
5. B. Ungun, R.K. Prud'homme, S.J. Budijono et al., Nanofabricated upconversion nanoparticles for photodynamic therapy. Opt. Express **17** (2009) https://doi.org/10.1364/OE.17.000080
6. H. Guo, H. Qian, N.M. Idris et al., Singlet oxygen-induced apoptosis of cancer cells using upconversion fluorescent nanoparticles as a carrier of photosensitizer. Nanomedicine **6** (2010) https://doi.org/10.1016/j.nano.2009.11.004
7. W. Wang, Q. Shang, W. Zheng et al., A novel near-infrared antibacterial material depending on the upconverting property of Er^{3+}-Yb^{3+}-Fe^{3+} Tridoped TiO_2 Nanopowder. J. Phys. Chem. C **114** (2010) https://doi.org/10.1021/jp102320x
8. S. Cui, H. Chen, H. Zhu et al., Amphiphilic chitosan modified upconversion nanoparticles for in vivo photodynamic therapy induced by near-infrared light. J. Mater. Chem. **22**, 4861 (2012). https://doi.org/10.1039/c2jm16112e
9. N.M. Idris, M.K. Gnanasammandhan, J. Zhang et al., In vivo photodynamic therapy using upconversion nanoparticles as remote-controlled nanotransducers. Nat. Med. **18** (2012) https://doi.org/10.1038/nm.2933
10. M.E. Lim, Y.L. Lee, Y. Zhang et al., Photodynamic inactivation of viruses using upconversion nanoparticles. Biomaterials **33** (2012) https://doi.org/10.1016/j.biomaterials.2011.11.033
11. K. Liu, X. Liu, Q. Zeng et al., Covalently assembled NIR nanoplatform for simultaneous fluorescence imaging and photodynamic therapy of cancer cells. ACS Nano. **6** (2012) https://doi.org/10.1021/nn300436b
12. Y.I. Park, H.M. Kim, J.H. Kim et al., Theranostic probe based on lanthanide-doped nanoparticles for simultaneous in vivo dual-modal imaging and photodynamic therapy. Adv. Mater. **24** (2012) https://doi.org/10.1002/adma.201202433
13. X.F. Qiao, J.C. Zhou, J.W. Xiao et al., Triple-functional core-shell structured upconversion luminescent nanoparticles covalently grafted with photosensitizer for luminescent, magnetic resonance imaging and photodynamic therapy in vitro. Nanoscale **4** (2012) https://doi.org/10.1039/c2nr30938f
14. Z. Zhao, Y. Han, C. Lin et al., Multifunctional core-shell upconverting nanoparticles for imaging and photodynamic therapy of liver cancer cells. Chem. Asian J. **7** (2012) https://doi.org/10.1002/asia.201100879
15. A. Zhou, Y. Wei, B. Wu et al., Pyropheophorbide A and C (RGDyK) comodified chitosan-wrapped upconversion nanoparticle for targeted near-infrared photodynamic therapy. Mol. Pharm. **9** (2012) https://doi.org/10.1021/mp200590y
16. X. Chen, Z. Zhao, M. Jiang et al., Preparation and photodynamic therapy application of $NaYF_4$:Yb, Tm–$NaYF_4$:Yb, Er multifunctional upconverting nanoparticles. New J. Chem. **37**, 1782 (2013). https://doi.org/10.1039/c3nj00065f

17. S. Cui, D. Yin, Y. Chen et al., In Vivo targeted deep-tissue photodynamic therapy based on near-infrared light triggered upconversion nanoconstruct. ACS Nano. **7** (2013) https://doi.org/10.1021/nn304872n
18. X. Liu, M. Zheng, X. Kong et al., Separately doped upconversion-C_{60} nanoplatform for NIR imaging-guided photodynamic therapy of cancer cells. Chem. Commun. (Camb) **49** (2013) https://doi.org/10.1039/c3cc41013g
19. G. Tian, W. Ren, L. Yan et al., Red-emitting upconverting nanoparticles for photodynamic therapy in cancer cells under near-infrared excitation. Small **9** (2013) https://doi.org/10.1002/smll.201201437
20. X. Wang, Q. Zhang, J. Zhao et al., One-step self-assembly of ZnPc/$NaGdF_4$:Yb,Er nanoclusters for simultaneous fluorescence imaging and photodynamic effects on cancer cells. J. Mater. Chem. B **1** (2013) https://doi.org/10.1039/c3tb20533a
21. Y. Wang, H. Wang, D. Liu et al., Graphene oxide covalently grafted upconversion nanoparticles for combined NIR mediated imaging and photothermal/photodynamic cancer therapy. Biomaterials **34** (2013) https://doi.org/10.1016/j.biomaterials.2013.06.045
22. Q. Yuan, Y. Wu, J. Wang et al. Targeted bioimaging and photodynamic therapy nanoplatform using an aptamer-guided G-quadruplex DNA carrier and near-infrared light. Angew. Chem. Int. Ed. Engl. **52** (2013) https://doi.org/10.1002/anie.201305707
23. L. Zeng, L. Xiang, W. Ren et al., Multifunctional photosensitizer-conjugated core–shell Fe_3O_4@$NaYF_4$:Yb/Er nanocomplexes and their applications in T_2-weighted magnetic resonance/upconversion luminescence imaging and photodynamic therapy of cancer cells. RSC Adv. **3**, 13915 (2013). https://doi.org/10.1039/c3ra41916a
24. Q. Chen, C. Wang, L. Cheng et al., Protein modified upconversion nanoparticles for imaging-guided combined photothermal and photodynamic therapy. Biomaterials **35** (2014) https://doi.org/10.1016/j.biomaterials.2013.12.046
25. A. Punjabi, X. Wu, A. Tokatli-Apollon et al., Amplifying the red-emission of upconverting nanoparticles for biocompatible clinically used prodrug-induced photodynamic therapy. ACS Nano. **8** (2014) https://doi.org/10.1021/nn505051d
26. H. Wang, Z. Liu, S. Wang et al., MC540 and upconverting nanocrystal coloaded polymeric liposome for near-infrared light-triggered photodynamic therapy and cell fluorescent imaging. ACS Appl. Mater. Interfaces **6** (2014) https://doi.org/10.1021/am500097f
27. M. Wang, Z. Chen, W. Zheng et al., Lanthanide-doped upconversion nanoparticles electrostatically coupled with photosensitizers for near-infrared-triggered photodynamic therapy. Nanoscale **6** (2014) https://doi.org/10.1039/c4nr01826e
28. X. Wang, K. Liu, G. Yang et al., Near-infrared light triggered photodynamic therapy in combination with gene therapy using upconversion nanoparticles for effective cancer cell killing. Nanoscale **6** (2014) https://doi.org/10.1039/c4nr02495h
29. X. Wang, C.X. Yang, J.T. Chen et al., A dual-targeting upconversion nanoplatform for two-color fluorescence imaging-guided photodynamic therapy. Anal. Chem. **86** (2014) https://doi.org/10.1021/ac500060c
30. L. Xia, X. Kong, X. Liu et al., An upconversion nanoparticle—Zinc phthalocyanine based nanophotosensitizer for photodynamic therapy. Biomaterials **35** (2014) https://doi.org/10.1016/j.biomaterials.2014.01.068
31. S. Yang, N. Li, Z. Liu et al., Amphiphilic copolymer coated upconversion nanoparticles for near-infrared light-triggered dual anticancer treatment. Nanoscale **6** (2014) https://doi.org/10.1039/c4nr05305b
32. Y. Zhang, G.K. Das, V. Vijayaragavan et al., "Smart" theranostic lanthanide nanoprobes with simultaneous up-conversion fluorescence and tunable T_1-T_2 magnetic resonance imaging contrast and near-infrared activated photodynamic therapy. Nanoscale **6** (2014) https://doi.org/10.1039/c4nr01717j
33. F. Ai, Q. Ju, X. Zhang et al., A core-shell-shell nanoplatform upconverting near-infrared light at 808 nm for luminescence imaging and photodynamic therapy of cancer. Sci. Rep. **5** (2015) https://doi.org/10.1038/srep10785

34. H. Ding, Y. Lv, D. Ni et al., Erythrocyte membrane-coated NIR-triggered biomimetic nanovectors with programmed delivery for photodynamic therapy of cancer. Nanoscale **7** (2015) https://doi.org/10.1039/c5nr02470f
35. Q.Q. Dou, A. Rengaramchandran, S.T. Selvan et al., Core-shell upconversion nanoparticle—semiconductor heterostructures for photodynamic therapy. Sci. Rep. **5** (2015) https://doi.org/10.1038/srep08252
36. A.H. Elmenoufy, Y. Tang, J. Hu et al., A novel deep photodynamic therapy modality combined with CT imaging established via X-ray stimulated silica-modified lanthanide scintillating nanoparticles. Chem. Commun. (Camb) **51** (2015) https://doi.org/10.1039/c5cc04135j
37. M. Guan, H. Dong, J. Ge et al., Multifunctional upconversion–nanoparticles–trismethylpyridylporphyrin–fullerene nanocomposite: a near-infrared light-triggered theranostic platform for imaging-guided photodynamic therapy. NPG Asia Mater. **7**, e205 (2015). https://doi.org/10.1038/am.2015.82
38. Z. Hou, Y. Zhang, K. Deng et al., UV-emitting upconversion-based TiO_2 photosensitizing nanoplatform: near-infrared light mediated in vivo photodynamic therapy via mitochondria-involved apoptosis pathway. ACS Nano. **9** (2015) https://doi.org/10.1021/nn506107c
39. X. Liu, I. Que, X. Kong et al., In vivo 808 nm image-guided photodynamic therapy based on an upconversion theranostic nanoplatform. Nanoscale **7** (2015) https://doi.org/10.1039/c5nr03690a
40. S. Lu, D. Tu, P. Hu et al., Multifunctional nano-bioprobes based on rattle-structured upconverting luminescent nanoparticles. Angew. Chem. Int. Ed. Engl. **54** (2015) https://doi.org/10.1002/anie.201501468
41. S.S. Lucky, N. Muhammad Idris, Z. Li et al., Titania coated upconversion nanoparticles for near-infrared light triggered photodynamic therapy. ACS Nano. **9** (2015) https://doi.org/10.1021/nn503450t
42. R. Lv, P. Yang, F. He et al., An imaging-guided platform for synergistic photodynamic/photothermal/chemo-therapy with pH/temperature-responsive drug release. Biomaterials **63** (2015) https://doi.org/10.1016/j.biomaterials.2015.05.016
43. D. Wang, B. Xue, X. Kong et al., 808 nm driven Nd^{3+}-sensitized upconversion nanostructures for photodynamic therapy and simultaneous fluorescence imaging. Nanoscale **7** (2015) https://doi.org/10.1039/c4nr04953e
44. S. Wang, L. Zhang, C. Dong et al., Smart pH-responsive upconversion nanoparticles for enhanced tumor cellular internalization and near-infrared light-triggered photodynamic therapy. Chem. Commun. (Camb) **51** (2015) https://doi.org/10.1039/c4cc08178a
45. L. Zeng, L. Luo, Y. Pan et al., In vivo targeted magnetic resonance imaging and visualized photodynamic therapy in deep-tissue cancers using folic acid-functionalized superparamagnetic-upconversion nanocomposites. Nanoscale **7** (2015) https://doi.org/10.1039/c5nr01932j
46. L. Zhang, L. Zeng, Y. Pan et al., Inorganic photosensitizer coupled Gd-based upconversion luminescent nanocomposites for in vivo magnetic resonance imaging and near-infrared-responsive photodynamic therapy in cancers. Biomaterials. **44** (2015) https://doi.org/10.1016/j.biomaterials.2014.12.040
47. L. Liang, A. Care, R. Zhang et al., Facile assembly of functional upconversion nanoparticles for targeted cancer imaging and photodynamic therapy. ACS Appl. Mater. Interfaces **8** (2016) https://doi.org/10.1021/acsami.6b00713
48. R. Lv, D. Yang, P. Yang et al., Integration of upconversion nanoparticles and ultrathin black phosphorus for efficient photodynamic theranostics under 808 nm near-infrared light irradiation. Chem. Mater. **28**, 4724 (2016). https://doi.org/10.1021/acs.chemmater.6b01720
49. C. Yue, C. Zhang, G. Alfranca et al., Near-infrared light triggered ROS-activated theranostic platform based on Ce6-CPT-UCNPs for simultaneous fluorescence imaging and chemo-photodynamic combined therapy. Theranostics **6** (2016) https://doi.org/10.7150/thno.14101
50. M.H. Chan, Y.T. Pan, I.J. Lee et al., Minimizing the heat effect of photodynamic therapy based on inorganic nanocomposites mediated by 808 nm near-infrared light. Small **13** (2017) https://doi.org/10.1002/smll.201700038

51. M. Kamimura, A. Omoto, H.-C. Chiu et al., Enhanced red upconversion emission of $NaYF_4:Yb^{3+}$, Er^{3+}, Mn^{2+} nanoparticles for near-infrared-induced photodynamic therapy and fluorescence imaging. Chem. Lett. **46**, 1076 (2017). https://doi.org/10.1246/cl.170322
52. D. Wang, B. Liu, Z. Quan et al., New advances on the marrying of UCNPs and photothermal agents for imaging-guided diagnosis and the therapy of tumors. J. Mater. Chem. B **5** (2017) https://doi.org/10.1039/c6tb03117j
53. J. Xu, L. Xu, C. Wang et al., Near-infrared-triggered photodynamic therapy with multitasking upconversion nanoparticles in combination with checkpoint blockade for immunotherapy of colorectal cancer. ACS Nano. **11** (2017) https://doi.org/10.1021/acsnano.7b00715
54. J.Y. Zhang, S. Chen, P. Wang et al., $NaYbF_4$ nanoparticles as near infrared light excited inorganic photosensitizers for deep penetration in photodynamic therapy. Nanoscale **9** (2017) https://doi.org/10.1039/c6nr09401e
55. B. Zheng, H. Wang, H. Pan et al., Near-infrared light triggered upconversion optogenetic nanosystem for cancer therapy. ACS Nano. **11** (2017) https://doi.org/10.1021/acsnano.7b06395
56. X. Han, Z. Zhong, J. Kou et al., ROS generated by upconversion nanoparticle-mediated photodynamic therapy induces autophagy via PI3K/AKT/mTOR signaling pathway in M1 peritoneal macrophage. Cell Physiol. Biochem. **48** (2018) https://doi.org/10.1159/000492283
57. G. Ramirez-Garcia, S.S. Panikar, T. Lopez-Luke et al., An immunoconjugated up-conversion nanocomplex for selective imaging and photodynamic therapy against HER2-positive breast cancer. Nanoscale **10** (2018) https://doi.org/10.1039/c8nr01512k
58. T. Sabri, P.D. Pawelek, J.A. Capobianco, Dual activity of rose bengal functionalized to albumin-coated lanthanide-doped upconverting nanoparticles: targeting and photodynamic therapy. ACS Appl. Mater. Interfaces **10** (2018) https://doi.org/10.1021/acsami.8b08919
59. Y.C. Tsai, P. Vijayaraghavan, W.H. Chiang et al., Targeted delivery of functionalized upconversion nanoparticles for externally triggered photothermal/photodynamic therapies of brain glioblastoma. Theranostics **8** (2018) https://doi.org/10.7150/thno.22482
60. F. Xu, Y. Zhao, M. Hu et al., Lanthanide-doped core-shell nanoparticles as a multimodality platform for imaging and photodynamic therapy. Chem. Commun. (Camb) **54** (2018) https://doi.org/10.1039/c8cc05057k
61. Y. Zhang, P. Huang, D. Wang et al., Near-infrared-triggered antibacterial and antifungal photodynamic therapy based on lanthanide-doped upconversion nanoparticles. Nanoscale **10** (2018) https://doi.org/10.1039/c8nr01967c
62. S. Dong, J. Xu, T. Jia et al., Upconversion-mediated $ZnFe_2O_4$ nanoplatform for NIR-enhanced chemodynamic and photodynamic therapy. Chem. Sci. **10** (2019) https://doi.org/10.1039/c9sc00387h
63. S Liu, W. Li, S. Gai et al. A smart tumor microenvironment responsive nanoplatform based on upconversion nanoparticles for efficient multimodal imaging guided therapy. Biomater. Sci. **7** (2019) https://doi.org/10.1039/c8bm01243a
64. S. Yan, X. Zeng, Y. Tang et al. Activating antitumor immunity and antimetastatic effect through polydopamine-encapsulated core-shell upconversion nanoparticles. Adv. Mater. **31** (2019) https://doi.org/10.1002/adma.201905825
65. X. Zhang, Z. Guo, X. Zhang et al., Mass production of poly (ethylene glycol) monooleate-modified core-shell structured upconversion nanoparticles for bio-imaging and photodynamic therapy. Sci. Rep. **9** (2019) https://doi.org/10.1038/s41598-019-41482-w
66. Z. Zhang, M.K.G. Jayakumar, X. Zheng et al., Upconversion superballs for programmable photoactivation of therapeutics. Nat. Commun. **10** (2019) https://doi.org/10.1038/s41467-019-12506-w

Chapter 10
NIR Hyperspectral Imaging

Kyohei Okubo

Abstract Hyperspectral imaging (HSI) is a powerful imaging technique for biomedical applications, such as disease detection, diagnosis, and surgery assistance. HSI provides a three-dimensional dataset (two spatial and one spectral), which allows to obtain spectral curve at each pixel in acquired images. Spectral imaging data offers diagnostic information on physiology, morphology, and composition of the biological tissues. Light in the near-infrared (NIR) region, ranging from 800–2500 nm, is useful for probing deep parts of tissue due to its lower absorption and scattering, and the absorption spectra in the NIR region convey fingerprint data due to overtone or combination vibrations of chemical bonds. This enables to investigate the distribution of chemical composition within a sample, as well as to provide key information for detecting cancer that locates beneath submucosal layer. This chapter reviews the basics of HSI, HSI acquisition systems, and biomedical NIR-HSI applications.

Keywords Hyperspectral imaging · Chemical composition · Diagnostics · Non-invasiveness · Image-guided surgery

10.1 Introduction

Hyperspectral imaging (HSI) is an emerging modality that integrates conventional imaging and spectroscopy in order to simultaneously obtain spatial (localization) and spectral (identification) information from a sample. HSI, also called imaging spectrometer, was first developed for remote sensing [1] and has been employed to various applications such as archeology and art investigation, [2, 3] food quality and safety control [4–7], and medical diagnosis [8, 9]. Among various applications of HSI, medical HSI is still at an early stage. HSI provides significant potential for noninvasive and label-free diagnosis and surgical guidance. Light penetrated through the biological tissue experiences scattering from inhomogenetiy of biological structures and absorption by hemoglobin, melanin, and water as it propagates

K. Okubo (✉)
Department of Materials Science and Technology, Faculty of Industrial Science and Technology, Tokyo University of Science, 6-3-1 Niijuku, Katsushika 125-8585, Tokyo, Japan
e-mail: kyohei.okubo@rs.tus.ac.jp

K. Soga et al. (eds.), *Transparency in Biology*,
https://doi.org/10.1007/978-981-15-9627-8_10

through the biological tissues. The absorption, fluorescence, and scattering properties of biological tissues are assumed to change depending on the progression of disease, thus the reflected, fluorescent, or transmitted light originated from tissues can deliver quantitative information for diagnosis [10]. Recent progress of hyperspectral camera, image analysis methods, and machine learning algorithms is now enabling exciting applications in the biomedical field. In this chapter, we aim to introduce and describe biomedical HSI technology and provide a summary of the literature regarding HSI applications. We first begin with the basics including optical absorption of biomolecules in near-infrared (NIR) region, the mechanisms of HSI, and the development trend of HSI acquisition system. We emphasize on the advantages of the use of NIR region ranging from 800–2500 nm for biomedical HSI applications, as well as emerging image processing, feature extraction, and selection methods.

10.2 Optical Properties of Tissue and Biomolecules

The biological tissues are mostly opaque in the visible region (400–700 nm in wavelength). On the other hand, the NIR region is useful for bioimaging [11] due to its high transparency through living tissues. Furthermore, the NIR radiation excites overtone and combination vibrational modes that induce very weak absorption as compared to mid-infrared region. This absorption allows us to directly identify and quantify biomolecules in a sample without pre-treatment.

10.2.1 Tissue Properties

The light–tissue interaction mechanisms, optical processes involved in HSI, and useful diagnostic information provided by HSI are briefly reviewed in this section. Since biological tissues are heterogeneous in composition, light entering biological tissues experience scattering and absorption as it propagates through the tissue [12]. Light scatters at the microscopic biointerface with different refractive indices. The refractive index for a medium determines the speed of light in the medium, and changes in the refractive index give rise to scattering, refraction, and reflection. We should know the refractive indices for various tissue compositions or an averaged value for the whole tissue when using HSI devices. The effective refractive index of tissue is often approximated as the volume-weighted average of the values of its compositions. Water determines a significant portion of most tissues and its refractive index ($n = 1.333$ at $\lambda = 589$ nm; $n = 1.327$ at $\lambda = 1000$ nm) indicates the minimum value for fluids and soft-tissue components. Fat globules and dense protein compartment have higher refractive index, ~1.55. In addition, melanin particles located in the epidermal layer limit the high end of the refractive index in biological tissue, with reported value >1.6. Refractive indices for whole tissues, including parts of the brain, lung, stomach, kidney, and bladder, are in the range of 1.36–1.40. The indices of extracellular fluids and intracellular cytoplasm are in the range of 1.35–1.38. The refractive index of fatty tissue is around 1.45. The membrane that surround cells and subcellular organelles (100 nm–6 mm in size; i.e., mitochodria) is composed mostly of phospholipids. The refractive index mismatch between plasma and these lipid membranes is the origin of scattering in cellular tissues. The shape and size of the cells vary among different tissue types with dimensions of a few microns and larger. Although an isolated cell can cause a strong scattering within tissue, the scattering is largely subcellular in origin. The scattering properties of support tissues composed of cells and extracellular proteins (elastin and collagen, etc.) are caused by the small-scale inhomogeneities and the large-scale variations in structures they form.

10.2.2 Optical Absorption in Near-Infrared Window Based on Harmonic Oscillator Model

The degree of transparency in biological tissues depends on the absorption and scattering by tissues, as discussed in the chapter 1. NIR light is not strongly absorbed by water or hemoglobin. It can easily pass through the body without bending. NIR spectral characteristics arise from the molecular absorption of the overtone and combination modes derived from fundamental vibrational modes which occur in the mid-infrared region. This NIR absorption is 100 times weaker than that in the mid-infrared. Molecular vibrations in the NIR region comprise combination and overtone modes of stretching and bending vibrations. There are a series of overtones for fundamental vibrations with decreasing intensity as the transition number increases. Combination modes emerge as the summation of fundamental modes. Most absorption result from the harmonics and overtones of fundamental stretching and bending vibrational modes of chemical bonds such as C–H, O–H, and N=H [13].

Here, we will briefly explain the origin of fundamental modes, combination modes, and overtone modes. The classical harmonic oscillator model shows the frequency of vibration (ν) for a diatomic molecule as follows

$$\nu = \frac{1}{2\pi}\sqrt{\frac{k}{\mu}} \tag{10.1}$$

where k is the force constant for the bond, and μ is the reduced mass from the individual atoms. The total potential energy for a bond is thus given by

$$E = \frac{1}{2}kx^2 \tag{10.2}$$

where x is the change in displacement from the center of equilibrium. Using quantum theory, a mathematical relationship for the behavior of diatomic oscillator follows a set of discrete energy levels given by

$$E_n = \left(n + \tfrac{1}{2}\right)h\nu \quad (10.3)$$

where n is the oscillating dipole quantum number for discrete potential energy levels. The frequency and discrete potential energy obtained define fundamental vibration modes of diatomic molecules, and this concept can be easily extended to N-atomic molecules. The selection rule tells that a molecule allows only to absorb the light equal to the spacing between two levels. For a diatomic molecule, the energy transition can only occur from one level to the next higher level. In that sense, overtone modes and combination modes are the forbidden transition, which explains the reason why the NIR absorption is much weaker than that in the mid-NIR. More details and descriptions of NIR absorption principles can be found in ref. [14].

To understand practical cases of NIR absorption, water molecule is first taken as an example. Figure 10.1a illustrates fundamental vibration modes of the water molecule, symmetric stretch (ν_1), bending (ν_2), and asymmetric stretch (ν_3). The strong peak at 1940 nm is due to a combination of the asymmetric stretch and bending of the water molecule ($\nu_2 + \nu_3$), and the peak at 1450 nm is due to a combination of the symmetric and asymmetric stretching modes ($\nu_1 + \nu_3$), as shown in Fig. 10.1b. There are also weak and broad combination modes near 1200 nm and 1780 nm, and a second set near 740 nm and 840 nm, respectively. Another example is cholesterol,

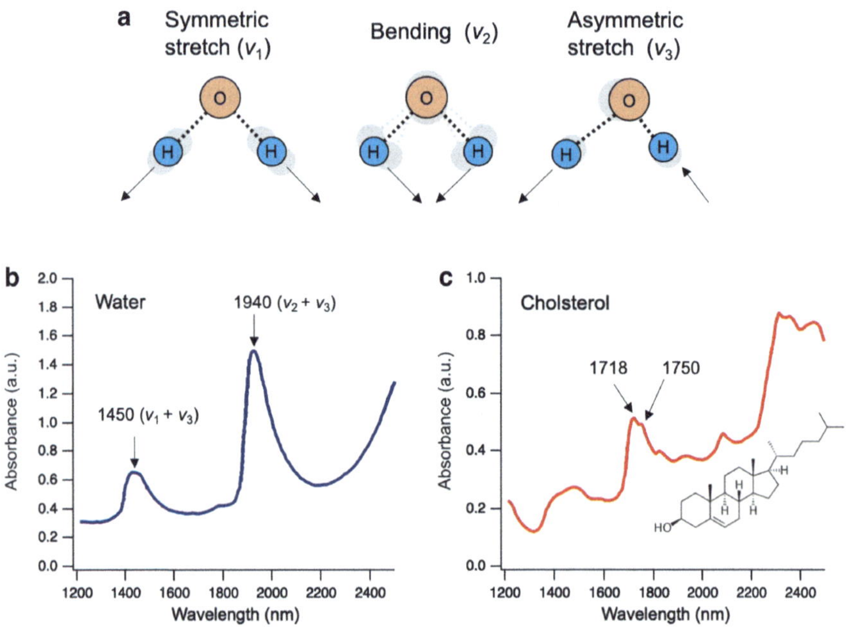

Fig. 10.1 Optical absorption of molecules in NIR region. **a** Fundamental vibration modes of water: symmetric stretch; bending; asymmetric stretch. NIR absorption spectra of **b** water and **c** cholesterol. Figure 10.1c is adapted or redrawn from reference [14] with permission from American Chemical Society Publications

a type of lipids. The absorption spectrum of crystallized cholesterol in the NIR region (1200–2400 nm) is shown in Fig. 10.1c. The data is adapted and redrawn from ref. [15]. Cholesterol has characteristic absorption peaks in the NIR region, especially at 1718–1750 nm. The absorption corresponding to these wavelengths is attributed to specific groups such as methyl and ethyl groups. This feature allows to use cholesterol as a pathological indicator for cardiovascular diseases since it is the major lipid contained in plaques in blood vessels. Oxy and deoxyhemoglobin, cytochrome c oxidase, and oxy and deoxymyoglobin have unique absorption spectra in the NIR region. These molecules have different absorption properties depending on the oxygen concentration in tissues. They are useful as in vivo oxygen indicator, which is measured by pulse oximetry.

10.3 Fundamentals of Hyperspectral Imaging

Wavelength in the visible range spectrum is appreciated as a spectral color, also known as monochromatic light. Standard camera acquires images composed of three spectral bands–red, green, and blue–while HSI systems acquire images from multiple narrow spectral bands that incorporate a wide range of electromagnetic spectrum. Each pixel in hyperspectral image represents full spectral information, which can analyze biological tissue with a detailed and accurate way. A three-dimensional (3D) data that contains spatial and spectral information is customary called hypercube. By mapping spectral information at each two-dimensional (2D) position, spatial distribution of a given chemical component and boundaries among multiple chemical components can be analyzed. HSI generally covers a neighboring portion of the light spectrum with more spectral bands (up to a few hundreds) and higher spectral resolution than multispectral imaging (such as RGB color cameras), therefore, HSI is able to capture the subtle spectral differences under different biochemical compositions while multispectral imaging may miss significant spectral information. HSI devices work in a way that the sample illuminated by the light source is projected through a front lens into entrance slit which only passes light from a narrow line. After collimation, a dispersive device (such as prism and grating) splits the light into a series of narrow spectral bands that are then focused onto a detector array. Slit width controls the amount of light entering the spectrograph. In this way, for each pixel interval along the line defined by the slit, a corresponding spectrum is projected on a column of the detector array. Thus, each line of the targeted area on the sample is projected as a 2D image onto the detector, with one spatial dimension and one spectral dimension. By scanning over a region of interest or moving the camera across the region, HSI camera collects 2D images for adjacent lines, creating a hypercube with two spatial dimensions and one spectral dimension.

10.3.1 Hardware and Systems

HSI systems comprise a monochrome sensor, an electronically controlled spatial or spectral scanner, imaging optics, and a computer for data storage, analysis, and image processing for collecting a set of spectral images. The number of the spectral bands that the imaging system can handle determines a difference between multispectral imaging and HSI. Multispectral imaging devices acquire 5–20 spectral bands, while HSI systems acquire up to a few hundreds of spectral bands. HSI systems employ monochrome detectors or sensor arrays, which can capture only two or three spectral dimensions of hypercube at one time. To capture the third dimension, spatial or spectral scanning is needed. HSI systems are classified depending on the scanning method, as shown in Fig. 10.2. The whiskbroom HSI devices, also known as point-scanning method, use a linear sensor array to obtain the λ dimension from a single point at a time while the *x, y* dimensions are obtained by spatial scanning. In the whiskbroom mode, two-axis motorized positioning tables are needed to perform the image acquisition. An advantage of this mode is the capability to control depth of field, resulting in background noise reduction. However, this mode is time-consuming due to 2D spatial scanning. The pushbroom HSI devices, also known as line scanning, use a 2D sensor array. The pushbroom mode can simultaneously acquire a line of spatial information. The first spatial dimension (x) and the λ dimension are obtained along with line, or fixed y coordinate, in each frame, and the second spatial dimension (y) is then obtained by line scanning. Staring HSI devices use a 2D sensor array coupled with an imaging monochromator, which is tuned to scan the spectral domain and in each

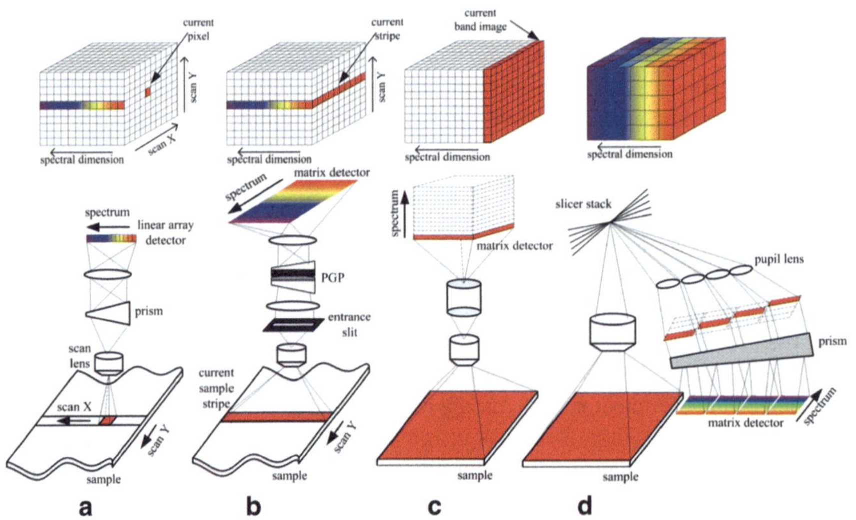

Fig. 10.2 Hyperspectral imaging acquisition systems. **a** Whiskbroom, **b** pushbroom, **c** staring, **d** snapshot. Reprinted or redrawn from reference [17] under a Creative Commons Attribution 3.0 unported license

scanning step. The pushbroom mode enables longer exposure time and higher spatial resolution than that of the whiskbroom mode. Whiskbroom and pushbroom devices relying on spatial scanning for building hypercube do not provide live display of spectral images since they are calculated and generated from the spectra after performing the spatial or spectral scan. On the other hand, staring devices make use of the tuning of the imaging wavelength, and the resulting hypercube can be generated simultaneously. As compared with other techniques, staring devices are appropriate for live imaging. In biomedical imaging, staring devices are more suitable where real-time imaging at selected spectral bands (i.e., endoscopy) is required. Wavelength scanning has the advantage of realizing high spectral resolution using monochromatic filters such as rotating filter wheels, liquid crystal tunable filters (LCTFs) and acousto-optic tunable filters (AOTFs); however, low light throughput or long image acquisition time hamper the spectral selectivity. The snapshot mode can record both spatial and spectral information simultaneously with a single exposure. The snapshot mode acquires a hypercube, with minimal postprocessing, by directly imaging the remapped and dispersed image zones on a detector [16]. Detailed comparison of advantages and disadvantages among these four different modes is briefly discussed in a review by Li et al. [17].

Si charged couple devices (CCD) and scientific complementary metal-oxide semiconductors (sCMOS) enable to capture ultraviolet (UV) visible and NIR up to 1 μm. InGaAs detectors can be used to cover the wavelength range, 1000–1700 nm. For longer infrared wavelength, HgCdTe or InSb sensors must be equipped with the system. Photomultiplier tube (PMT) arrays are another type of detector that provides an electric output after a photon strikes in photocathode in just a few nanoseconds. PMT array devices detect signals faster than CCD or sCMOS. The wavelength range of the monochromators should cover a significant part of the spectral range that the selected detector is sensitive to. In addition, the light throughput of the monochromator depends on the wavelength. The quantum efficiency of the detector also changes with the wavelength. The light throughput can be improved by using a light source with strong emission in this wavelength range or by using a back-illuminated CCD. In some applications, increasing the intensity of the light source is out of options because it might overheat the sample or initiate photochemical reaction. A calibration of the hyperspectral imaging systems is the key to achieving quantitative image analysis. Calibration can be done using a reference sample that shows a flat spectrum within a target wavelength window. Brightness can be corrected using the aid of the performed calibration data.

10.3.2 Image Analysis

HSI enables to extract diagnostic information from a vast amount of hyperspectral image data at tissue, cellular, and molecular levels. Recent development on HSI systems has realized to acquire hypercube with high spatial and spectral resolution; however, this high dimensionality also makes it difficult to automatically analyze data. Since we ought to deal with a vast amount of spatial and spectral data simultaneously, image classification methods for hyperspectral datasets are highly needed to extract and classify relevant spectral information. Therefore, the image analysis leads to investigate the relationship between spectral information and biomedical phenomena behind. The fundamentals for hyperspectral image analysis include preprocessing, feature extraction and selection, and classification.

Hyperspectral image preprocessing involves data normalization. Data normalization converts observed radiation from a sample to reflectance [18–22] or absorbance [23]. CCD arrays used in HSI systems generally have dark current even without light shining on it. Dark current is dependent on temperature and is proportional to integration time. Thus, to convert raw intensity into reflectance, reference and dark images are taken before acquiring sample images. The reference image is taken with a standard reflectance surface placed in the scene, and the dark current is measured by keeping the camera shutter closed. The raw data were then corrected using the following equation:

$$I_{\text{ref}} = \frac{I_{\text{raw}} - I_{\text{dark}}}{I_{\text{white}} - I_{\text{dark}}} \tag{10.4}$$

where I_{ref} is the calculated reflectance value, I_{raw} is the raw intensity of a given pixel, and I_{dark} and I_{white} are the dark current and the white reference intensity of the given pixel, respectively. The absorbance (A) or optical density (OD) is calculated by taking the ratio of the sample images (I_{raw}) with respect to a reference image (I_{ref}).

$$A = OD = -\log\left(\frac{I_{\text{raw}}}{I_{\text{ref}}}\right) \tag{10.5}$$

10.3.3 Feature Extraction and Selection, and Classification

Feature extraction and feature selection from diverse HSI data are one of the critical steps for various goals. Although HSI data have hundreds of spectral bands, all the bands might not contain same portion of information and few bands might contain discriminatory information. Thus, the reduction of dimensionality should be applied to select relevant bands and to extract the intrinsic features. Principal component analysis (PCA) is the most widely used feature extraction method in NIR spectral analysis and HSI applications. In NIR spectrum, some of wavelength peaks contain critical information and other spectral regions mostly contribute primarily noise. One of the merits in PCA is data compression, meaning that higher dimensional spectral data is projected to a lower dimensional underlying variable space. Usually, only the first few principal components are required to explain most signal variance. PCA employed to HSI data has enabled to reveal relative distribution of different molecular mixtures, [18] or identify keratinized tissues from cancer tissue section

[24]. Note that PCA converts the original data to a subspace spanned by principal directions, making it difficult to interpret the underlying biological meaning.

Pixel-wise classification methods can be categorized as supervised classifiers and unsupervised classifiers. Supervised classification methods make use of training a classification model by a set of fingerprint spectra with class labels, and the trained model is used to attribute new labels to unknown spectral fingerprints in HSI data. Statistical machine learning methods, including decision trees, random forest, artificial neural networks (ANNs), and kernel-based methods, have been used to classify hyperspectral images. Among various kernel-based methods, the support vector machine (SVM) classifier has been widely accepted as a prominent classification method due to its well-defined statistical learning theory that separates the linearly separable feature space into two classes with the maximum margin [25]. For medical NIR-HSI application, for example, Chihara et al. used SVM classifier in NIR hyperspectral images to create pixel-wise images that predict atherosclerotic plaques within a vessel of rabbits [26]. It was found that the location of plaques was precisely visualized with an accuracy of 83%. Jiang et al. took a Gaussian-kernel-based SVM approach to classify walnuts shell and pulp with a better recognition rate on 6257 samples than that of PCA method [27]. ANN is another type of supervised classification approach. ANNs have been used in medicine for cancer diagnostics and other illnesses. Blanco et al. performed ANN method combined with NIR hyperspectral images for detecting a difference among kidney stones [28]. They optimized the ANN model by decreasing a number of hidden layers and wavelengths, which provided almost 100% accuracy of classification using only 30 wavelengths selected. Recent advances enabled us to take deep learning approaches to classify HSI data, and some researches showed excellent classification performance as compared with machine learning methods [29]. Both machine learning and deep learning approaches have been facing difficulty due to the high dimensionality and the limited number of samples to be analyzed, as well as spectral variations among samples.

10.4 Biomedical Applications of NIR Hyperspectral Imaging

Compared with imaging technologies based on other detection principles, HSI offers advantages in terms of less invasiveness, label-free, and identification of chemical components in tissue. In this section, biomedical applications in fields such as food quality and safety, disease diagnosis, and image-guided surgery are reviewed. A wide variety of HSI applications used only visible spectral range, and thus the author selected papers that specifically employ NIR light (wavelength, >700 nm) as critical spectral information.

10.4.1 Food Quality and Safety Control

Food quality and safety control are one of the earliest application fields in NIR spectroscopy. Karl H. Norris, often regarded as the father of modern NIR spectroscopic analysis, invented NIR spectrophotometer for quantitative analysis of agricultural samples in early 1950s. In 1962, Noris and others first published an NIR application, the moisture determination in extracts of seeds [30]. Associated with a rapid development of instrumentations during 1980s and 1990s, NIR spectroscopy has been remaining one of the leading investigative methods in the area of agricultural raw materials, food ingredients, and food products [4, 31]. Except for remote sensing, most NIR-HSI applications mainly focused on food quality detection [6] since it is suitable to noninvasively visualize amounts of water and lipids in foods by the use of NIR reflectance spectra. One of the first NIR-HSI food safety applications include fecal contamination on fruit, vegetables, chicken, fungal contamination of cereal grains, and parasites in fish [32, 33]. After the NIR-HSI technology emerged in this field, it is increasingly being used for a wide variety of commodities including meat, fruit and vegetables, [34] cereals, [7] and beverages [35] for ensuring the quality and safety of food products while avoiding liability issues [9].

Williams et al. used NIR-HSI to perform both object-wise and pixel-wise classification of hardness among maize kernels [36]. Results obtained by pixel-wise classification achieved sensitivity and specificity of 0.75 and 0.97. Object-wise classification using score histogram performed better for hard kernels while that using mean spectra (wavelength, 1100–2500 nm) gave better results for medium kernels. They concluded that this classification methods could be extended to large production scale. Ma and Sun used a line-scanning NIR-HSI system (wavelengths, 1000–2000 nm) to assess the contents of monosaturated and poly-saturated fatty acids in pork meat [37]. Aheto et al. integrated NIR-HSI with an electrochemical sensor (electric nose) to quantify intramuscular fat and peroxide in pork meat. NIR-HSI and electric nose can respectively monitor non-volatile and volatile compounds of samples, and the merged dataset achieved rapid detection of those compounds [38]. Zhang et al. identified and visualized a variation of coffee beans by NIR hyperspectral images [39]. Their SVM models using pixel-wise NIR spectra (wavelengths, 950–1650 nm) provided good prediction results with over 80% classification accuracy.

10.4.2 Disease Diagnostics

Imaging modalities used routinely in clinical care are X-ray computed tomography (CT), magnetic resonance imaging (MRI), ultrasonography, single-photon emission computed tomography (SPECT), and positron emission tomography (PET). HSI is potentially useful for biomedical and clinical applications where slight spectral differences in biological tissue can be a pathological indicator. It is being widely demonstrated for individual biochemical species detection in in vitro samples and organs in patients. The use of NIR region enables to acquire biological information from deep parts of tissue, such as characteristic absorption in NIR range. Most researches that employed medical NIR-HSI used relatively short NIR range (700–1000 nm), sometimes called very near infrared (VNIR), for discriminating different materials/biomolecules and thus distinguishing malignant tissues from healthy tissues. Diagnostic applications of NIR-HSI include label-free detection of cancer [40–43] and biomarkers [44–46], blood vessel visualization enhancement [26, 47, 48], functional brain cortex mapping [49–53], and various diagnosis (i.e., shock and artery disease) [55–57].

A study by Akbari et al. demonstrated the effective use of NIR hyperspectral images for cancer detection, [40] which is the earliest representative among various diagnostic NIR-HSI applications. Spectral processing such as SVM, first derivatives, and integral of spectral diagram successfully led to delineation of gastric tumors in human subjects. They emphasized that the first derivatives in spectral region between 1226–1251 nm and 1288–1379 nm were useful for determining cancerous tissue, as they called "cancer index" (Fig. 10.3). The group of Sakaida measured a difference of VIS–NIR spectral reflectance (wavelength, 400–800 nm) between gastric tumors and normal mucosa in respected samples ex vivo and found the possibility to selectively enhance tumor regions by image processing at 770 nm wavelength [42, 43]. They concluded that the vascular density in the tumor or an increased hemoglobin concentration might affect the reflectance spectra at 770 nm because the absorption band of hemoglobin locates between 700–900 nm. Chihara et al. demonstrated in vivo detection of plaque in rabbit's vessels by label-free NIR-HSI (wavelength, 1000–2350 nm), as shown in Fig. 10.4, [26]. The hypercube analyzed by SVM showed that the increase in absorption at 1200 and 1750 nm, derived from cholesterol in plaques, was useful for identifying plaques, paving the way to less invasively evaluate the precise location of lesions during vascular surgery. Hadoux et al. recently reported in vivo retinal NIR-HSI to correlate reflectance spectra with the load of amyloid beta (Aβ) accumulated in human brain, which has wavelength dependence on light scattering (Fig. 10.5), [44]. Their study was based on in vitro evaluation of retinal Aβ deposition measured by microscopic HSI system [46]. Fabelo et al. established an intraoperative system to acquire VIS–NIR hyperspectral database (wavelength, 400–1000 nm) of in vivo human brain tissues [51, 53]. The obtained hyperspectral images with class labels were used to distinguish among normal brain tissue, brain tissue affected by cancer (both primary and secondary), blood vessels, and background elements. The database achieved to generate classification maps where the tumor boundaries are identified and delineated. The same group also reported microscopic NIR-HSI to automatically detect human brain tumor tissues, [53] and their developed NIR-HSI system was demonstrated for image-guided neurosurgery by combining both supervised and unsupervised machine learning methods [52].

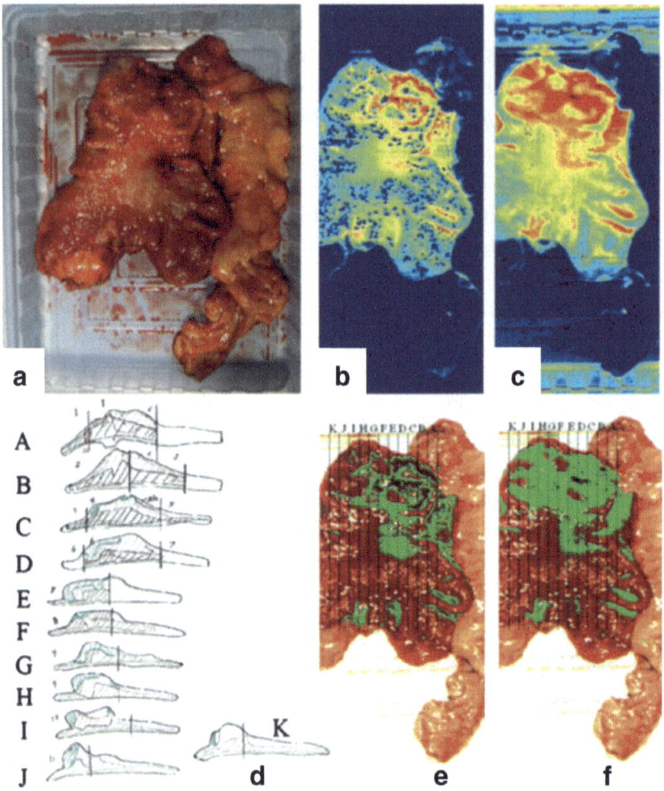

Fig. 10.3 In vitro cancer detection. **a** the RGB image; **b** the cancer enhanced regions using integral filter in the hyperspectral image (1057–2440 nm), the tissues are shown in a blue to red spectrum and the red regions represent the tumor; **c** the cancer enhanced regions using the normalized cancer index (NDCI); **d** pathological results; **e** detected tumor using integral filter **f** detected tumor using NDCI. Reprinted from reference [40] under a Creative Commons Attribution 3.0 unported license

10.4.3 Surgical Guidance

Recent advances of imaging technology enabled an extension of surgeon's vision, providing real-time biological and physiological information on targets of choice. During surgery, the surgeons process a tremendous amount of image data to make a rapid judgement to detect and delineate lesion and its margins. They have also been seeking for alternative ways of providing visualization without exposing anatomy. The use of VNIR fluorescence light (700–900 nm) has been taking increasing attention in image-guided surgery due to the availability of Food and Drug Administration (FDA)-approved NIR fluorophores, indocyanine green and methylene blue [58, 59]. However, the fluorescence imaging is often associated with an administration of such contrast agents. Therefore, HSI, as a label-free modality, has

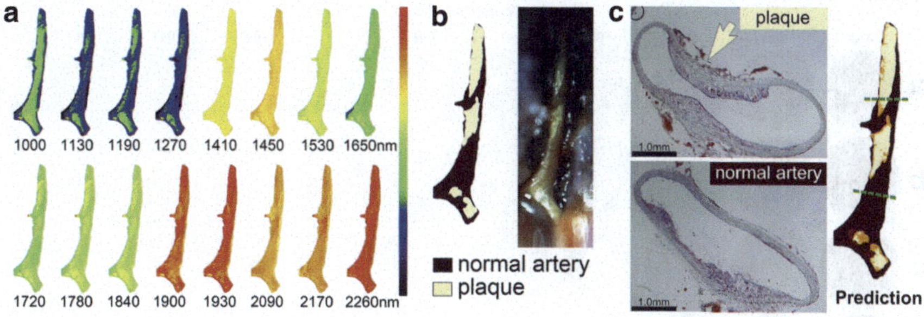

Fig. 10.4 Plaques in blood vessels. **a** The arbitrarily selected montage of a representative hyperspectral cube composed of 215 channels. Hotter colors indicate higher absorption, **b** regions of interest in a rabbit both for "normal artery" and "plaque" regions manually defined on the hyperspectral images in accordance with the histological classification, **c** the predictions by a SVM and pathological findings. The upper green dotted line shows a cutting line of a section that contains plaques. In the "plaque" section, atherosclerotic plaques are distributed on the ventral aorta wall (pale yellow arrows). Plaques, which are thinner than vessel wall, are found in some sections (pale yellow arrow head). The lower green dotted line shows a cutting line of a section with a normal artery. In the "normal artery" section, ventral aorta walls have no atherosclerotic change. Reprinted from reference [26] with permission from Elsevier Ireland Ltd.

a great potential to support minimally invasive surgery techniques by extending the surgeon's vision at molecular, cellular, and tissue levels [8, 60].

Early in 2004, NIR-HSI was first applied to the visualization of tissues that were submerged in a blood layer [61]. ANN approach achieved to produce clean images under the cloak of blood-water solution. Akbari et al. reported NIR-HSI as a visual aid to detect intestinal ischemia and anatomy of organs in pigs [57, 62]. NIR range (765–830 nm) in hyperspectral images had a significant clue for distinguishing ischemic intestine. Also, spleen, colon, small intestine, urinary bladder, and peritoneum were segmented on the basis of their identical signatures in spectra. Some notable works have attempted to integrate NIR-HSI and traditional medical devices. Zuzak et al. developed a surgical laparoscopic NIR-HSI platform to generate images with chemically different anatomical structures [18]. This NIR-HSI laparoscope, as shown in Fig. 10.6a, demonstrated intraoperative biliary imaging of live pig to observe frontal biliary system through hepatoduodenal ligament [63]. The duct-reflected spectra had characteristic peaks at 760 nm (deoxyhemoglobin), 800 nm (oxyhemoglobin), 930 nm (lipid), and 970 nm (water), all of which could be used as signature to distinguish among arterial vessels, venous structures, and bile ducts. (Figure 10.6b) Carr et al. developed an NIR-HSI-based otoscope for investigating middle ear pathologies [64]. Deeper tissue penetration of NIR light enabled good visualization of middle ear anatomy, and human middle ear fluid could be detected using strong absorption in 1400–1550 nm due to water content (Fig. 10.7). Urade et al. proposed a concept of hyperspectral enhanced reality (HYPER), allowing to overlay NIR hyperspectral images that predict tissue oxygenation on the visible operative field [65]. The superimposed images provide an effective surgical navigation tool, for example, in discriminating ischemic area in liver tissue.

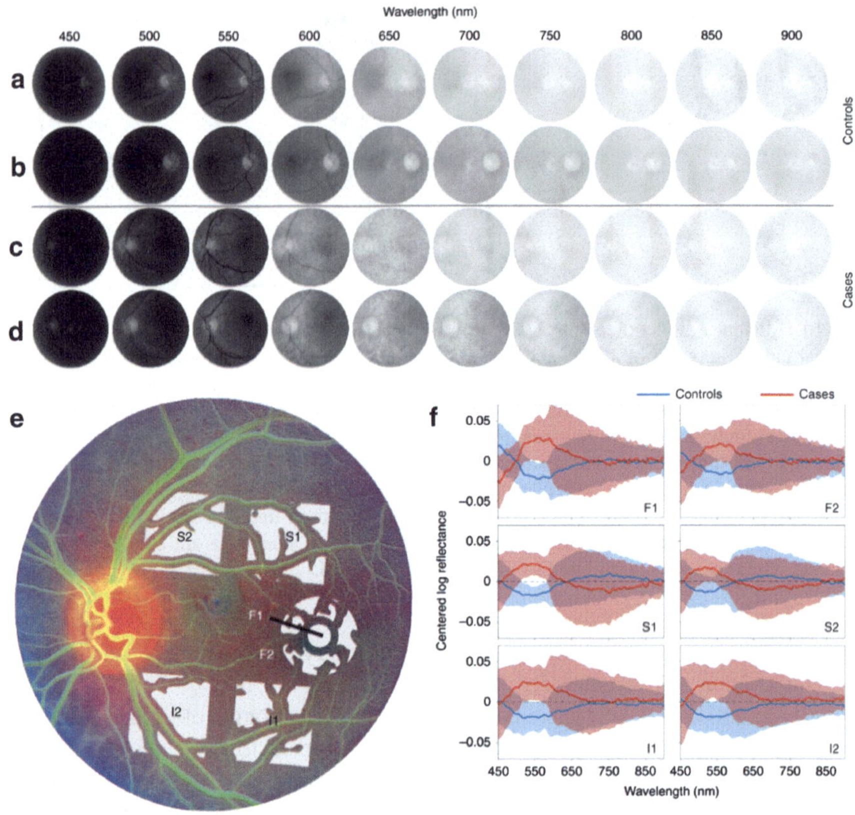

Fig. 10.5 Retina biomarker detection. Spectral variation between eyes precludes discrimination between cases and controls. (**a–d**) Representative hyperspectral montages of four eyes ($n = 2$ controls, $n = 2$ cases from 450 to 900 nm in 50 nm steps) showing the inherent variability within and between eyes owing to the retinal and choroidal vasculature, ocular pigment, and ocular media. **e** Systematic sampling method used to analyze HS images including foveal locations (F1, F2) as well as locations superior (S1, S2) and inferior (I1, I2) to the temporal vascular arcades with segmentation of the major inner-retinal blood vessels. (**e**) Centered reflectance spectra at the different sampling locations for controls ($n = 20$, blue) and cases (n = 15, red) highlighting the large degree of inter-subject variability using uncorrected spectral data, which precludes discrimination between cases and controls. Centered spectra were obtained by subtracting the average spectrum from the spectrum measured for individual participants in the principal cohort (Aβ PET+ and PET−). Reprinted from reference [44] with permission from Springer Nature

Most HSIs for image-guided surgery in the state-of-art use images in the VNIR region, rather than a spectral range over 1000 nm. The optimal spectral ranges vary depending on purposes and targets. The identification and detection of diseases might be improved by the use of over-1000 nm NIR range due to plentiful amount of available fingerprint spectra. There are also engineering challenges in integrating NIR-HSI systems with medical inspection devices such as endoscope and laparoscope, though some of the works achieved a proof-of-concept [18, 66–68]. For this

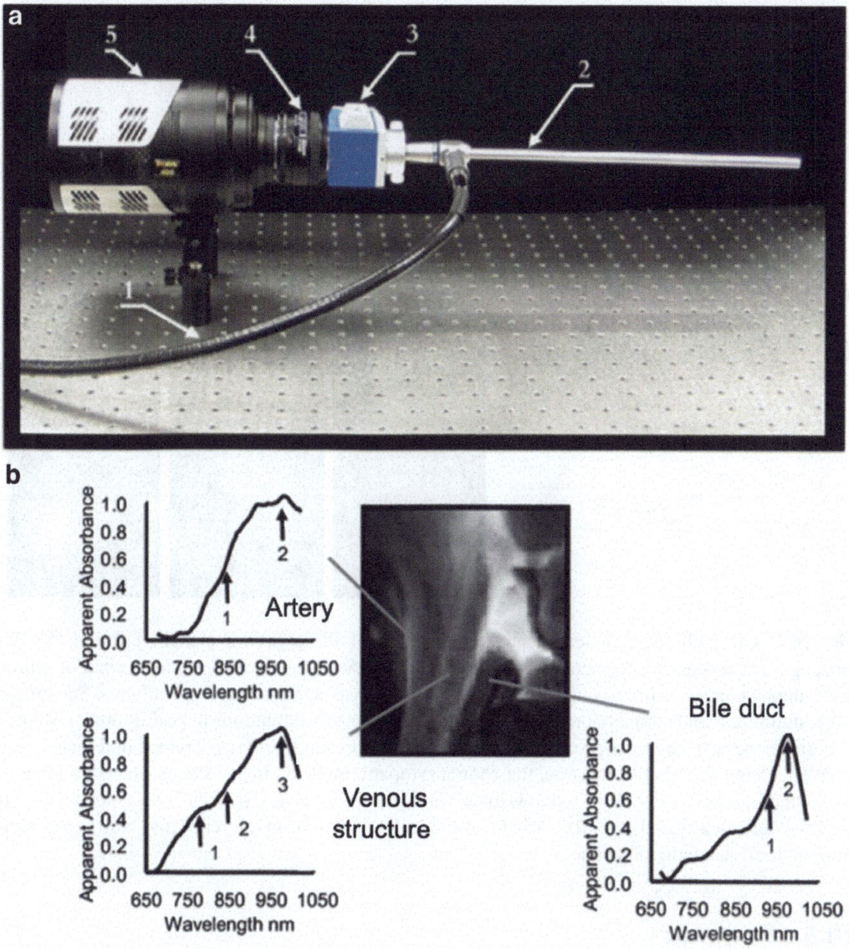

Fig. 10.6 NIR-HSI-based laparoscopy. **a** The laparoscopy coupled to NIR camera and LCTF, **b** NIR laparoscopic image of hepatoduodenal ligament with absorption spectra obtained from artery, venous structure, and bile duct. Reprinted from reference [63] with permission from Elsevier Inc

purpose, pushbroom cameras should be used to ensure high spectral resolution in order to detect weak absorption peaks derived from the overtone and combination vibration of biomolecules. To develop the NIR-HSI applications in the medical field, an accessible library of medical NIR hyperspectral images with spectral database of biomarkers will be highly needed for evaluating various HSI systems.

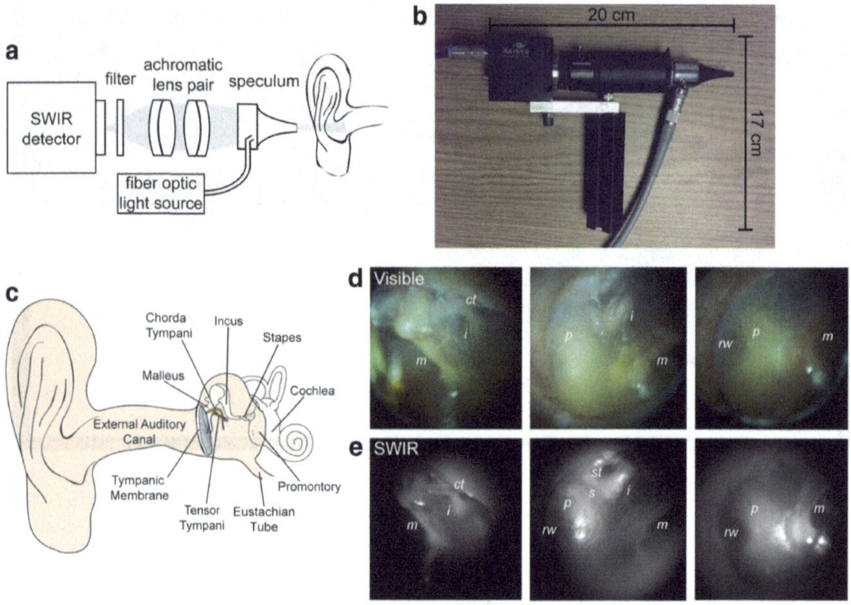

Fig. 10.7 Ear pathology. Schematic **a** and image **b** of the short-wavelength infrared (SVIR) otoscope. The SWIR otoscope comprises a compact InGaAs detector, a filter holder, a pair of achromatic doublet lenses, a fiber-coupled light source, and a disposable medical speculum. **c** Schematic of the middle ear and comparison between SWIR and visible examinations in healthy adults. Representative images from otoscope examinations of 10 healthy adult humans obtained in **d** visible and **e** SWIR. Using the SWIR otoscope, the chorda tympani, malleus, incus, stapes, stapedial tendon, cochlear promontory, and round window niche (indicated by ct, m, i, s, st, p, and rw, respectively, and shown schematically in **d**) are all clearly identifiable. Reprinted from reference [64] with permission from national academy of sciences

10.5 Summary

HSI provides 3D image cube with two spatial dimensions (position) and one spectral dimension (wavelength), also called hypercube, in label-free and less-invasive manner. Each pixel in the hypercube is analyzed by a spectral curve ranging from ultraviolet to infrared region. Spectral data acquired by HSI can provide diagnostic information on the tissue physiology, morphology, and composition. The use of NIR region enables HSI to probe deeper part of tissues as well as to obtain fingerprint for detection and identification of chemical components and even malignant lesions. In addition, HSI can be combined with various conventional imaging systems and medical devices including microscopy and endoscopy. Medical NIR-HSI technology has been rapidly developed for clinical trials, demonstrating its great potential in disease detection, diagnosis, monitoring, and image-guided surgery assistance.

References

1. A.F.H. Goetz, G. Vane, J.E. Solomon et al., Imaging spectrometry for earth remote sensing. Science **228**, 1147 (1985). https://doi.org/10.1126/science.228.4704.1147
2. H. Liang, Advances in multispectral and hyperspectral imaging for archaeology and art conservation. Appl. Phys. A **106**, 309 (2012). https://doi.org/10.1007/s00339-011-6689-1
3. C. Cucci, J.K. Delaney, M. Picollo, Reflectance hyperspectral imaging for investigation of works of art: Old master paintings and illuminated manuscripts. Acc. Chem. Res. **49**, 2070 (2016). https://doi.org/10.1021/acs.accounts.6b00048
4. A. Gowen, C. Odonnell, P. Cullen et al., Hyperspectral imaging—an emerging process analytical tool for food quality and safety control. Trends Food Sci. Tech. **18**, 590 (2007). https://doi.org/10.1016/j.tifs.2007.06.001
5. G. Elmasry, M. Kamruzzaman, D.W. Sun et al., Principles and applications of hyperspectral imaging in quality evaluation of agro-food products: A review. Crit. Rev. Food Sci. **52**, 999 (2012). https://doi.org/10.1080/10408398.2010.543495
6. M. Manley, Near-infrared spectroscopy and hyperspectral imaging: Non-destructive analysis of biological materials. Chem. Soc. Rev. **43**, 8200 (2014). https://doi.org/10.1039/c4cs00062e
7. K. Sendin, P.J. Williams, M. Manley, Near infrared hyperspectral imaging in quality and safety evaluation of cereals. Crit. Rev. Food Sci. **58**, 575 (2018). https://doi.org/10.1080/10408398.2016.1205548
8. G.L. Lu, B.W. Fei, Medical hyperspectral imaging: A review. J. Biomed. Opt. **19**, 010901 (2014). https://doi.org/10.1117/1.Jbo.19.1.010901
9. M.J. Khan, H.S. Khan, A. Yousaf et al., Modern trends in hyperspectral image analysis: A review. IEEE Access **6**, 14118 (2018). https://doi.org/10.1109/ACCESS.2018.2812999
10. M.C. Pierce, R.A. Schwarz, V.S. Bhattar et al., Accuracy of in vivo multimodal optical imaging for detection of oral neoplasia. Cancer Prev. Res. **5**, 801 (2012). https://doi.org/10.1158/1940-6207.CAPR-11-0555
11. A.M. Smith, M.C. Mancini, S. Nie, Second window for in vivo imaging. Nat. Nanotechnol. **4**, 710 (2009). https://doi.org/10.1038/nnano.2009.326
12. M.S. Patterson, B.C. Wilson, D.R. Wyman, The propagation of optical radiation in tissue i. Models of radiation transport and their application. Laser Med. Sci. **6**, 155 (1991). https://doi.org/10.1007/BF02032543
13. B.G. Osborne, T. Fearn, P.H. Hindle, *Practical NIR Spectroscopy with Applications in Food and Beverage Analysis* (Longman Scientific and Technical, Harlow, 1993)
14. D.A. Burns, E.W. Ciurczak, *Handbook of Near-Infrared Analysis*, 3rd edn. (CRC Press, 2007)
15. E. Peuchant, C. Salles, R. Jensen, Determination of serum cholesterol by near-infrared reflectance spectrometry. Anal. Chem. **59**, 1816 (1987). https://doi.org/10.1021/ac00141a017
16. L. Weitzel, A. Krabbe, H. Kroker et al., 3d: The next generation near-infrared imaging spectrometer. Astron. Astrophys. Suppl. Ser. **119**, 531 (1996). https://doi.org/10.1051/aas:1996266
17. Q. Li, X. He, Y. Wang et al., Review of spectral imaging technology in biomedical engineering: Achievements and challenges. J. Biomed. Opt. **18**, 100901 (2013). https://doi.org/10.1117/1.JBO.18.10.100901
18. K.J. Zuzak, S.C. Naik, G. Alexandrakis et al., Characterization of a near-infrared laparoscopic hyperspectral imaging system for minimally invasive surgery. Anal. Chem. **79**, 4709 (2007). https://doi.org/10.1021/ac070367n
19. C. Li, G.S. Mitchell, J. Dutta et al., A three-dimensional multispectral fluorescence optical tomography imaging system for small animals based on a conical mirror design. Opt. Express **17**, 7571 (2009). https://doi.org/10.1364/OE.17.007571
20. S. Bégin, B. Burgoyne, V. Mercier et al., Coherent anti-stokes raman scattering hyperspectral tissue imaging with a wavelength-swept system. Biomed. Opt. Express **2**, 1296 (2011). https://doi.org/10.1364/BOE.2.001296

21. U. Maeder, K. Marquardt, S. Beer et al., Evaluation and quantification of spectral information in tissue by confocal microscopy. J. Biomed. Opt. **17**, 1 (2012). https://doi.org/10.1117/1.JBO.17.10.106011
22. R.A. Schultz, T. Nielsen, J.R. Zavaleta et al., Hyperspectral imaging: A novel approach for microscopic analysis. Cytometry **43**, 239 (2001). https://doi.org/10.1002/1097-0320(20010401)43:4%3c239::AID-CYTO1056%3e3.0.CO;2-Z
23. M.G. Sowa, J.R. Payette, M.D. Hewko et al., Visible-near infrared multispectral imaging of the rat dorsal skin flap. J. Biomed. Opt. **4**, 474 (1999). https://doi.org/10.1117/1.429957
24. J.R. Duann, C.I. Jan, M. Ou-Yang et al., Separating spectral mixtures in hyperspectral image data using independent component analysis: Validation with oral cancer tissue sections. J. Biomed. Opt. **18**, 126005 (2013). https://doi.org/10.1117/1.Jbo.18.12.126005
25. B. Park, R. Lu, *Hyperspectral Imaging Technology in Food and Agriculture (Food Engineering Series)* (Springer, New York, 2015)
26. H. Chihara, N. Oishi, A. Ishii et al., In vivo detection of atherosclerotic plaque using non-contact and label-free near-infrared hyperspectral imaging. Atherosclerosis **250**, 106 (2016). https://doi.org/10.1016/j.atherosclerosis.2016.04.029
27. L. Jiang, B. Zhu, X. Rao et al., Discrimination of black walnut shell and pulp in hyperspectral fluorescence imagery using gaussian kernel function approach. J. Food Eng. **81**, 108 (2007). https://doi.org/10.1016/j.jfoodeng.2006.10.023
28. F. Blanco, M. Lopez-Mesas, S. Serranti et al., Hyperspectral imaging based method for fast characterization of kidney stone types. J. Biomed. Opt. **17**, 076027 (2012). https://doi.org/10.1117/1.Jbo.17.7.076027
29. A. Signoroni, M. Savardi, A. Baronio et al., Deep learning meets hyperspectral image analysis: A multidisciplinary review. J. Imaging **5**, 52 (2019). https://doi.org/10.3390/jimaging5050052
30. J.R. Hart, K.H. Norris, C. Golumbic, Determination of the moisture content of seeds by near-infrared spectrophotometry of their methanol extracts. Cereal Chem. **39**, 94 (1962)
31. C. Scotter, Use of near infrared spectroscopy in the food industry with particular reference to its applications to on/in-line food processes. Food Control **1**, 142 (1990). https://doi.org/10.1016/0956-7135(90)90006-X
32. K.C. Lawrence, W.R. Windham, B. Park et al., A hyperspectral imaging system for identification of faecal and ingesta contamination on poultry carcasses. J. Near Infrared Spec. **11**, 269 (2003). https://doi.org/10.1255/jnirs.373
33. Y.Z. Feng, D.W. Sun, Application of hyperspectral imaging in food safety inspection and control: A review. Crit. Rev. Food Sci. **52**, 1039 (2012). https://doi.org/10.1080/10408398.2011.651542
34. I. Chandrasekaran, S.S. Panigrahi, L. Ravikanth et al., Potential of near-infrared (NIR) spectroscopy and hyperspectral imaging for quality and safety assessment of fruits: An overview. Food Anal. Method **12**, 2438 (2019). https://doi.org/10.1007/s12161-019-01609-1
35. C. Nansen, K. Singh, A. Mian et al., Using hyperspectral imaging to characterize consistency of coffee brands and their respective roasting classes. J. Food Eng. **190**, 34 (2016). https://doi.org/10.1016/j.jfoodeng.2016.06.010
36. P.J. Williams, S. Kucheryavskiy, Classification of maize kernels using NIR hyperspectral imaging. Food Chem. **209**, 131 (2016). https://doi.org/10.1016/j.foodchem.2016.04.044
37. J. Ma, D.W. Sun, Prediction of monounsaturated and polyunsaturated fatty acids of various processed pork meats using improved hyperspectral imaging technique. Food Chem. **321**, 126695 (2020). https://doi.org/10.1016/j.foodchem.2020.126695
38. J.H. Aheto, X.Y. Huang, X.Y. Tian et al., Multi-sensor integration approach based on hyperspectral imaging and electronic nose for quantitation of fat and peroxide value of pork meat. Anal. Bioanal. Chem. **412**, 1169 (2020). https://doi.org/10.1007/s00216-019-02345-5
39. C. Zhang, F. Liu, Y. He, Identification of coffee bean varieties using hyperspectral imaging: Influence of preprocessing methods and pixel-wise spectra analysis. Sci. Rep. **8**, 2166 (2018). https://doi.org/10.1038/s41598-018-20270-y
40. H. Akbari, K. Uto, Y. Kosugi et al., Cancer detection using infrared hyperspectral imaging. Cancer Sci. **102**, 852 (2011). https://doi.org/10.1111/j.1349-7006.2011.01849.x

41. H. Akbari, L.V. Halig, D.M. Schuster et al., Hyperspectral imaging and quantitative analysis for prostate cancer detection. J. Biomed. Opt. **17**, 076005 (2012). https://doi.org/10.1117/1.JBO.17.7.076005
42. A. Goto, J. Nishikawa, S. Kiyotoki et al., Use of hyperspectral imaging technology to develop a diagnostic support system for gastric cancer. J. Biomed. Opt. **20**, 016017 (2015). https://doi.org/10.1117/1.Jbo.20.1.016017
43. S. Kiyotoki, J. Nishikawa, T. Okamoto et al., New method for detection of gastric cancer by hyperspectral imaging: A pilot study. J. Biomed. Opt. **18**, 026010 (2013). https://doi.org/10.1117/1.JBO.18.2.026010
44. X. Hadoux, F. Hui, J.K.H. Lim et al., Non-invasive in vivo hyperspectral imaging of the retina for potential biomarker use in alzheimer's disease. Nat. Commun. **10**, 4227 (2019). https://doi.org/10.1038/s41467-019-12242-1
45. T. Wang, Q. Li, X. Li et al., Use of hyperspectral imaging for label-free decoding and detection of biomarkers. Opt. Lett. **38**, 1524 (2013). https://doi.org/10.1364/OL.38.001524
46. S.S. More, R. Vince, Hyperspectral imaging signatures detect amyloidopathy in alzheimer's mouse retina well before onset of cognitive decline. ACS Chem. Neurosci. **6**, 306 (2015). https://doi.org/10.1021/cn500242z
47. A. Bjorgan, M. Milanic, L.L. Randeberg, Estimation of skin optical parameters for real-time hyperspectral imaging applications. J. Biomed. Opt. **19**, 066003 (2014). https://doi.org/10.1117/1.Jbo.19.6.066003
48. H. Akbari, Y. Kosugi, K. Kojima et al., Blood vessel detection and artery-vein differentiation using hyperspectral imaging, in *The 31st Annual International Conference of the IEEE Engineering in Medicine and Biology Society*. (Minnesota, Minneapolis, 2009), p. 1461
49. M. Ferrari, K.H. Norris, M.G. Sowa, Medical near infrared spectroscopy 35 years after the discovery. J. Near Infrared Spec. **20**, 7 (2012). https://doi.org/10.1255/jnirs.982
50. L. Giannoni, F. Lange, I. Tachtsidis, Hyperspectral imaging solutions for brain tissue metabolic and hemodynamic monitoring: Past, current and future developments. J. Opt. **20**, 044009 (2018). https://doi.org/10.1088/2040-8986/aab3a6
51. H. Fabelo, S. Ortega, R. Lazcano et al., An intraoperative visualization system using hyperspectral imaging to aid in brain tumor delineation. Sensors (Basel) **18**, 430 (2018). https://doi.org/10.3390/s18020430
52. H. Fabelo, S. Ortega, D. Ravi et al., Spatio-spectral classification of hyperspectral images for brain cancer detection during surgical operations. PLoS ONE **13**, e0193721 (2018). https://doi.org/10.1371/journal.pone.0193721
53. H. Fabelo, S. Ortega, A. Szolna et al., In-vivo hyperspectral human brain image database for brain cancer detection. IEEE Access **7**, 39098 (2019). https://doi.org/10.1109/ACCESS.2019.2904788
54. S. Ortega, H. Fabelo, R. Camacho et al., Detecting brain tumor in pathological slides using hyperspectral imaging. Biomed. Opt. Express **9**, 818 (2018). https://doi.org/10.1364/Boe.9.000818
55. L.C. Cancio, A.I. Batchinsky, J.R. Mansfield et al., Hyperspectral imaging: A new approach to the diagnosis of hemorrhagic shock. J. Trauma Acute Care Surg. **60**, 1087 (2006). https://doi.org/10.1097/01.ta.0000217357.10617.3d
56. J.A. Chin, E.C. Wang, M.R. Kibbe, Evaluation of hyperspectral technology for assessing the presence and severity of peripheral artery disease. J. Vasc. Surg. **54**, 1679 (2011). https://doi.org/10.1016/j.jvs.2011.06.022
57. H. Akbari, Y. Kosugi, K. Kojima et al., Detection and analysis of the intestinal ischemia using visible and invisible hyperspectral imaging. IEEE Trans. Bio-Med. Eng. **57**, 2011 (2010). https://doi.org/10.1109/Tbme.2010.2049110
58. S. Gioux, H.S. Choi, J.V. Frangioni, Image-guided surgery using invisible near-infrared light: Fundamentals of clinical translation. Mol. Imaging **9**, 237 (2010). https://doi.org/10.1038/nrclinonc.2013.123
59. A.L. Vahrmeijer, M. Hutteman, J.R. van der Vorst et al., Image-guided cancer surgery using near-infrared fluorescence. Nat. Rev. Clin. Oncol. **10**, 507 (2013). https://doi.org/10.1038/nrclinonc.2013.123

60. J. Freeman, F. Downs, L. Marcucci et al., Multispectral and hyperspectral imaging: Applications for medical and surgical diagnostics, in the 19th annual international conference of the IEEE engineering in medicine and biology society. Chicago, Illinois **2**, 700 (1997)
61. S.T. Monteiro, Towards applying hyperspectral imagery as an intraoperative visual aid tool, in *The 4th International Conference on Visualization.* (Marbella, Spain, Imaging and Image Processing, 2004), p. 438
62. H. Akbari, Y. Kosugi, K. Kojima et al., Hyperspectral imaging and diagnosis of intestinal ischemia, in *The 30th Annual International Conference of the IEEE Engineering in Medicine and Biology Society.* (Canada, Vancouver, 2008), p. 1238
63. K.J. Zuzak, S.C. Naik, G. Alexandrakis et al., Intraoperative bile duct visualization using near-infrared hyperspectral video imaging. Am. J. Surg. **195**, 491 (2008). https://doi.org/10.1016/j.amjsurg.2007.05.044
64. J.A. Carr, T.A. Valdez, O.T. Bruns et al., Using the shortwave infrared to image middle ear pathologies. Proc. Natl. Acad. Sci. U.S.A. **113**, 9989 (2016). https://doi.org/10.1073/pnas.1610529113
65. T. Urade, E. Felli, M. Barberio et al., Hyperspectral enhanced reality (HYPER) for anatomical liver resection. Surg. Endosc. (2020). https://doi.org/10.1007/s00464-020-07586-5
66. R.T. Kester, N. Bedard, L. Gao et al., Real-time snapshot hyperspectral imaging endoscope. J. Biomed. Opt. **16**, 056005 (2011). https://doi.org/10.1117/1.3574756
67. H.T. Lim, V.M. Murukeshan, A four-dimensional snapshot hyperspectral video-endoscope for bio-imaging applications. Sci. Rep. **6**, 24044 (2016). https://doi.org/10.1038/srep24044
68. R. Kumashiro, K. Konishi, T. Chiba et al., Integrated endoscopic system based on optical imaging and hyperspectral data analysis for colorectal cancer detection. Anticancer Res. **36**, 3925 (2016)

Chapter 11
Multimodal Imaging with NIR Light

Thi Kim Dung Doan

Abstract Multimodal imaging with NIR light is of great interest in medical research as its feasibility in combining with various techniques such as magnetic resonance imaging (MRI), X-ray computed tomography (CT), positron emission tomography (PET), single-photon emission computed tomography (SPECT), ultrasound (US), and photoacoustic (PA) imaging. This chapter gives an overview and backgrounds of diverse contrast agents for NIR mulimodal imaging associated with MRI, CT, PET/SPECT, and PA as a primal modality. It highlights potential applications in therapeutic practice using NIR light.

Keywords Multimodality · Magnetic resonance imaging · X-ray computed tomography · Photoacoustic imaging · Photon emission tomography · Single-photon emission computed tomography

11.1 Multimodal Imaging

The human body is a natural complex system that contains multiplex parts together with numerous biochemical processes. Imaging technologies have been developed for decades to visualize different organs or molecules inside the body. "Multi-modalities" refers to the integration of different imaging techniques with principal merits to compensate the intrinsic limitation of each modality. Such integration has demonstrated the efficacy in diagnostics and treatment by improving the screening, surveillance, prognosis process, and guiding for therapy [1–9]. The idea of multi-modalities includes the design and development of integrated hardware platforms and multimodal probes. Multimodal platforms, such as CT/MRI, PET/SPECT/CT,

---

T. K. D. Doan (✉)
Research Institute for Biomedical Science, Tokyo University of Science, 2669, Yamazaki, Noda, Chiba 278-0022, Japan
e-mail: doan@rs.tus.ac.jp

Division of Functional Imaging, Exploratory Oncology Research & Clinical Trial Center, National Cancer Center, 6-5-1 Kashiwanoha, Kashiwa 277-8577, Japan

K. Soga et al. (eds.), *Transparency in Biology*,
https://doi.org/10.1007/978-981-15-9627-8_11

PET/SPECT/MRI, US/MRI, and PA/NIR/MRI [10–16], have been applied to identify or describe the structural, functional, and molecular variations in cancer tumors. Among them, PET-CT is the most popular one, which widely applied in medical researches [17, 18]. MRI and CT provide images with high structural information of tissue compared to high functional techniques such as SPECT or PET [19–23]. MRI is well-known for supplying 3-dimensional (3D) high spatial resolution and soft-tissue contrast that are good enough to discriminate morphological changes in a small structure, but it has insufficient sensitivity for accurate visualization of the molecular and functional information compared to SPECT or PET. Meanwhile, SPECT and PET use ionizing radiation to provide more functional and physiological information of tissue [18, 24].

Optical imaging is a versatile tool to obtain real-time high spatial-temporal resolution images as well as to monitor the dynamic events from the molecular to whole body levels and high instrument sensitivity. It greatly contributes to the expansion of clinical multimodality techniques because of the ability to image structural, functional, and molecular information at different spatial resolutions and scales. However, this technique often uses visible (VIS) and ultraviolet (UV) light sources which are, due to the scattering and absorption of light by water and tissue components, limited imaging depth of less than one millimeter [25, 26]. Further problems restricting their temporal use is the autofluorescence of the biological tissue and phototoxicity [26–28]. Ongoing efforts of overcoming the restriction at the very superficial regions by utilizing NIR light has been well discussed in chapters I and II [29–31]. In brief, at the swallow surfaces, optical imaging creates shadow images of the target regions, by going centimeter deeper, the images of whole volume regions can be recognized. The deep penetration is governed by the interaction between photon and media, particularly the nature of media corresponding to the photon wavelengths [32]. Light is an electromagnetic wave, and its interaction to transmitted media obeys Lorentz and Maxwell's law. According to this law, dielectric constants of media varied with light frequencies; it means the refractive index of the tissue change due to wavelengths of the incident light. As the wavelength increases, the absorption by water and scattering by fat becomes significant. Fortunately, the tissue-optical-transparent windows at 650–950, 1000–1350, and 1550–1870 nm, provides a pathway for deeper penetrating of light into the biological object, also called NIR-I, -II, and -III, respectively [33]. The NIR region over 1000 nm is sometimes called OTN-NIR. NIR imaging, consequently, has tremendously emerged as a potent technique for biomedical researches and applications. The yield of light transmission through the skin was proved relatively good in the range of 810 nm [34], the near-infrared light over 1000 nm can penetrate as deep as centimeter-level under the skin, and the penetration of light longer than 1500 nm through muscle tissues were also recorded [35]. A novel strategy to utilize OTN-NIR light by capitalizing CT approach has been discussed in previous chapters, permits obtaining cross-sectional images (tomograms) by projecting of the incident electromagnetic wave from multiplex sources and reconstruct into 3D tomographic images using reconstruction algorithms [36].

Herein, the ongoing research and the achievement of NIR based multimodal imaging will be introduced with the focus on the compensation that each modality

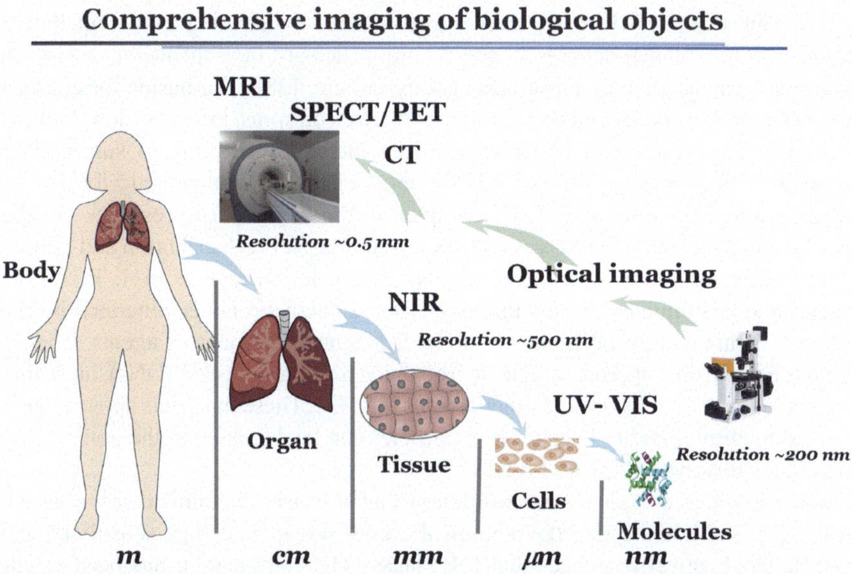

Fig. 11.1 An illustration of the overall vision from human bodies providing by a combination of magnetic resonance and optical imaging

contributes to overall platforms. The design and composition of imaging probes that play a crucial role in the combinations will also be discussed together with the potential applications in medical research and practices (Fig. 11.1).

11.2 Combination Modalities

11.2.1 Magnetic Resonance Imaging

MR image formation originates from the presence of protons that comprises up to 70% of human bodies. Water and fat in the bodies are the rich sources of protons [37]. Magnetic moments of proton nuclei orient randomly until being exposed to an external magnetic field B_0. If a radio wave (RF) with a particular frequency, called Larmor frequency, is applied perpendicular to the main magnetic field, a local B_1 field is produced. Then the resonance occurs. The distribution of the spin states is perturbed and again precess to the B_1 orientation. When the RF wave is off, all of the spins return to their primary position in two different approaches called "spin-lattice" and "spin-spin" relaxation. The times of relaxation are also called T_1-longitudinal or T_2-transverse relaxation, respectively. Because proton distributes variously among tissues and organs, the relaxation times are also varied, as a stem in the dissimilarity in the contrast of obtained images.

The contrast of an MR image derives from various contributing factors, including intrinsic proton relaxation times T_1 and T_2, proton density, flow, diffusion, and perfusion and extrinsic such as pulse sequence types, etc. [38]. The basics for contrast enhancement effects depend on the number of electron spins, in case of longitudinal T_1, where the presence of paramagnetic ions increased visibility of surrounding organs or body regions [39]. Gd^{3+} is the most effective paramagnetic ion for T_1 enhancement until now [40]. Gd^{3+} complex in the bodies easily escapes out the capillary and gets into the intercellular space for nonspecific distribution and rapidly reaches the equilibrium state in normal tissues and lesion regions [41]. T_2, on the other hand, is stimulated by fluctuations of local magnetic fields generated by the contrast agents and proportional to the total magnetic moment of agents. The T_2 agent is made from superparamagnetic iron oxide nanoparticles (SPIONs), including magnetite, Fe_3O_4, γFe_2O_3, and other ferrites [42, 43]. These materials show a lower contrast medium toxicity because iron particles can be degraded in the normal iron recycling pathways [41] (Fig. 11.2).

MRI provides a high spatial resolution and intrinsic anatomical information, while NIR is able to capture the biological events in real-time, high sensitivity, and specific biochemical responses that MRI lacks. The combination has been extensively performed and reported [44–46]. Bardhan et al. utilized NIR fluorescence

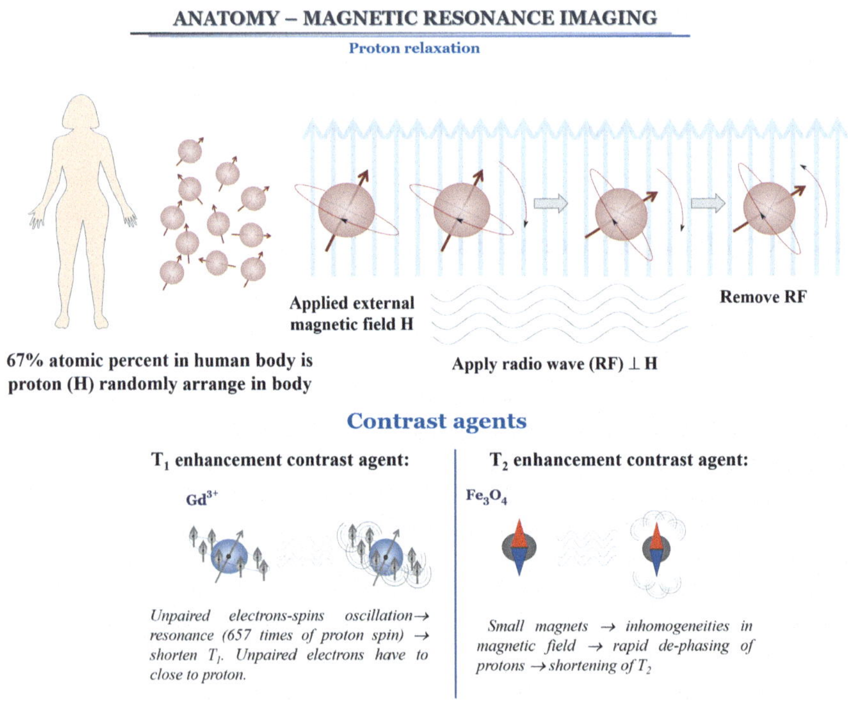

Fig. 11.2 Principle of magnetic resonance of proton spins and contrast enhancement agents

and MRI to track human epidermal growth factor receptor 2 (HER2) targeting nano complexes from SPION and ICG in vivo and examine the nanocomplex distribution and fate over 72 h [47]. As shown in Fig. 11.3, the agent circulated throughout the body, the differences in tumor uptake quantities between HER2 overexpressing and HER2 low expressing tumors were observed. The results showed that the combination of NIR and MRI could reveal a detailed picture of the geographic distribution of nanoparticles (NPs) in tumors and internal organs. A discrepancy between the intensity distribution due to time between MR and fluorescent imaging was clearly observed. The discrepancy was attributed to the intrinsic differences between two techniques in terms of sensitivity, properties of signal releasing of probes, and the distinction of techniques [48, 49]. For instance, the MRI contrast enhancement effect is greatly dependent on the proton density of the surrounding environment. However, inside the tumor core, the vasculature is weird, and there may be necrotic which limit the delivery of oxygen and glucose lead to the low sensitivity of MRI signal. More importantly, the contrast of MRI is derived from the proton density in a pixel or voxel of images; therefore, the detailed structure or morphology of the region can be lost. In contrast, fluorescent images provide a real-time signal which is recorded from the actual distribution of the nanostructure; it sequentially displays the complex structures inside the tumors. The discrepancy between MR and fluorescence in VIS

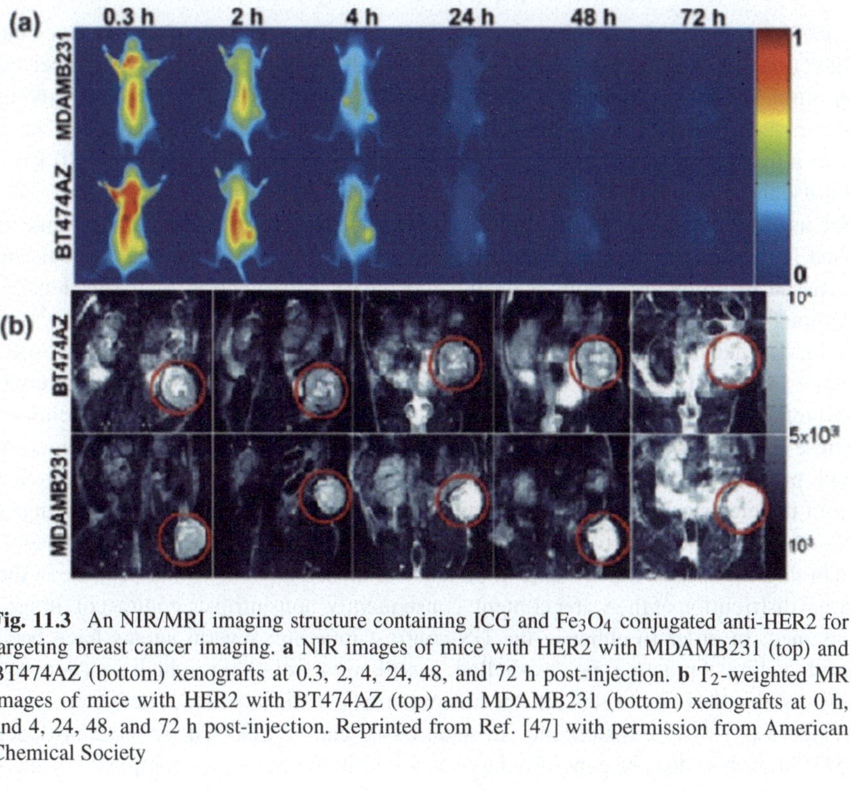

Fig. 11.3 An NIR/MRI imaging structure containing ICG and Fe_3O_4 conjugated anti-HER2 for targeting breast cancer imaging. **a** NIR images of mice with HER2 with MDAMB231 (top) and BT474AZ (bottom) xenografts at 0.3, 2, 4, 24, 48, and 72 h post-injection. **b** T_2-weighted MR images of mice with HER2 with BT474AZ (top) and MDAMB231 (bottom) xenografts at 0 h, and 4, 24, 48, and 72 h post-injection. Reprinted from Ref. [47] with permission from American Chemical Society

or NIR-I imaging in vivo is not only due to the low sensitivity of MRI but also can be attributed to the characteristic discordance of NIR and MRI. Fluorescence shorter than 1000 nm cannot penetrate the deep tumor and preferentially images as the probe accumulated near the tumor surface, but with MRI, complete cross-sections of the tumor are imaged. The probe distributing characteristics are similar; however, the signals detected at a different time according to the difference in energy-probe interaction mechanism between techniques. The distribution of the probe can be obtained from the surface to the deep tumor core if the light can penetrate deeper. The ongoing effort on using NIR over 1000 nm is expected to overcome the short-coming of routine fluorescence imaging using VIS and NIR-I and offers exceptional detection sensitivity but is limited by photon penetration depth. Thus, combining MRI will render the merits of high spatial resolution and the capability to obtain detailed anatomical structure information at significant probe depths of MRI and highly sensitive time-dependent fluorescence distribution in deep tumors as well as detailed spatial profiling of the tumors. This combination also enables internal details of tumor margin in real-time.

11.2.2 Acoustic and Photoacoustic Imaging

US/NIR imaging. Acoustic imaging or ultrasound imaging typically capitalizes the reflection and scattering of acoustic waves from a source across the tissue and received by a transducer and generates the anatomic images, [50, 51] or velocity of moving tissues such as blood in Doppler based types. An ultrasound is defined as a sound wave with a frequency higher than the audible threshold of humans, which is >20 kHz. Ultrasound is produced from a vibrating source made of piezoelectric materials. For use in medical such as diagnosis, typical frequencies up to 30 MHz range is used [52, 53]. At this frequency range, sound waves are transmitted from tissue to tissue resulting in various acoustic impedance as a function of tissue density. Although tissue impedance is a product of transmission velocity of sound waves and tissue density, the velocity is negligible as it is nearly identical in most of soft-tissue and blood. As ultrasound wave travels in the body, it is attenuated, scattered, and reflected at the tissue interfaces and resulting in impedance mismatch (echoes). These reflected sound waves can be detected by the transducer, and under interaction with piezoelectric crystals, a small current is produced. The images are constructed from the electrical signal by the computers. The contrast in ultrasound imaging is obtained from the acoustic impedance mismatch at the interfaces among soft tissues. In biological tissues, the acoustic impedance of most tissues is similar because of the equal distribution of the water content; consequently, non-intrinsic contrast of images obtained. In order to enhance the US contrast imaging, several agents have been used in clinical practice such as shelled, gas-filled microbubbles which are routinely injected into the body to generate the mismatch in acoustic impedance between tissues of interest. Thus the disturbance in focal tissue can be detected by the transducer [54]. Microbubbles are generally large 1–4 μm in diameter and composed of inert gas core stabilized by a shell surrounding made from phospholipids, proteins, or polymer. The structure is used to increase circulation time and reduce gas dissolution in blood. The gases used in the core included air, nitrogen, or heavy inert gases such as perfluorocarbons. Targeting moieties such as antibodies can be bound to the surface of microbubbles through covalent and noncovalent binding and allows ultrasound to be used as a molecular imaging modality.

Many studies have developed nanobubbles from US contrast agents to include other imaging agents such as NIR, PET, and MRI [55, 56]. US and NIR nanobubbles have been developed by encapsulating fluorophores such as Cyanine 5.5, DiI, or upconversion NPs. Sciallero et al. created a poly(vinyl alcohol) (PVA) surface-based microbubbles embedded SPIONs used as the dual-modal US and MRI contrast agents [57]. Later, a NIR/US multicolor imaging was introduced by Zhang et al., in which nanobubbles encapsulated three different cyanine fluorophores DiI, DiD, and DiR to achieve FRET. The consecutive spectra are enabling a sequent donor to excite the efficient FRET [58]. DiI is a donor for DiD and DiD following donate energy to excite DiR. As showed in Fig. 11.4

The use of US multimodal nanobubbles resulted in a combination signal obtained in solution or a tissue/tumor simulator. However, these structures did not demonstrate US contrast enhancement in vivo generally had poor contrast because the main contribution of microbubbles to US contrast enhancement is attributed to their micron size diameter [59]. The larger size microbubbles enhance the US scattering cross-section in tissues. The nanoscale agents do not show a good US enhancement effect, but the effect is due to the high concentration of nanobubbles, resulting in the coalescence of the nanobubbles into a larger US scattering microbubble.

PA/NIR imaging. PA has emerged over the last decade as a novel imaging technique by capitalizing acoustic detection of optical absorption in the tissues based on the photoacoustic effect [60, 61]. In general, as a laser pulse irradiates into biological tissues, it delivers energy into the tissues. The absorption of energy by tissues will be converted into kinetic energy that produces transient thermoelastic expansion resulting in an ultrasonic emission. The received signal from the transducer is then processed and generated into images as ultrasound imaging. The ultrasound imaging is the most widely used in diagnosis because of its simplicity, portability, and safety. However, most biological tissues exhibit a comparable acoustic impedance, due to the similarity in water content, resulting in a non-specific contrast of images [51, 54, 62].

The contrast agents from microbubbles are often applied, but with the size at several micrometers, microbubbles have restricted at vascular compartments. Since the PA was applied, more specific and high-resolution contrast images in biological tissues up to centimeters depths are expected for various clinical imaging applications, as showed in Fig. 11.5(b–d) [63]. Human body components such as blood, hemoglobin, melanin, lipid, and water are intrinsic chromophores in the body with their own characterized absorption. For example, blood has a strong absorption at ~500 nm, while cellular nuclei, DNA, and RNA strongly absorb UV light [64]. PA can discern this variation and measure the oxyhemoglobin saturation (sO_2) in blood vessels, which is attributed to ischemia-hypoxia level in the body by measure hemoglobin (Hb) absorption changes when it bounces to oxygen to form oxygenated

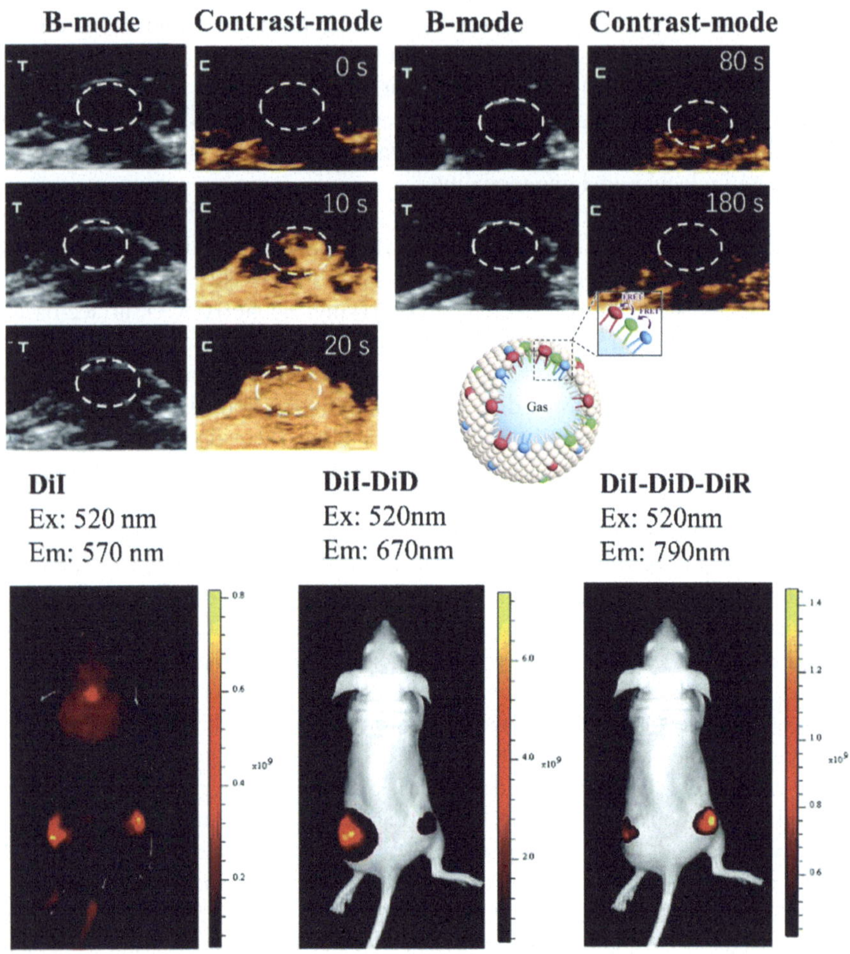

Fig. 11.4 Bimodal ultrasound and NIR imaging by using triple NIR molecules for inducing FRET. The upper part is ultrasound images of tumor, the bottom is the NIR images. Reprinted from Ref. [58] with permission from Royal Society of Chemistry

hemoglobin (HbO_2) [65, 66]. However, most of the tissue components do not provide sufficient distinction in absorption; various exogenous contrast agents are often employed to increase the specificity of the tissue imaging such as ICG or methylene blue [67–69]. Photoacoustic effect is initiated by high-energy laser pulse for returning sound waves with sufficient detectable amplitude. However, the NIR-I region has low maximum permissible exposure (MPE), a value that limits the greatest exposure of energy possible in each pulse that most individuals can tolerate without sustaining the injury. MPE limit is low in NIR-I, resulting in weak return ultrasound and a spatial resolution, but increases in NIR-II [64, 70]. NIR-II is an open window for PA to gain advantages in imaging such as high MPE, high irradiating intensity, high energy

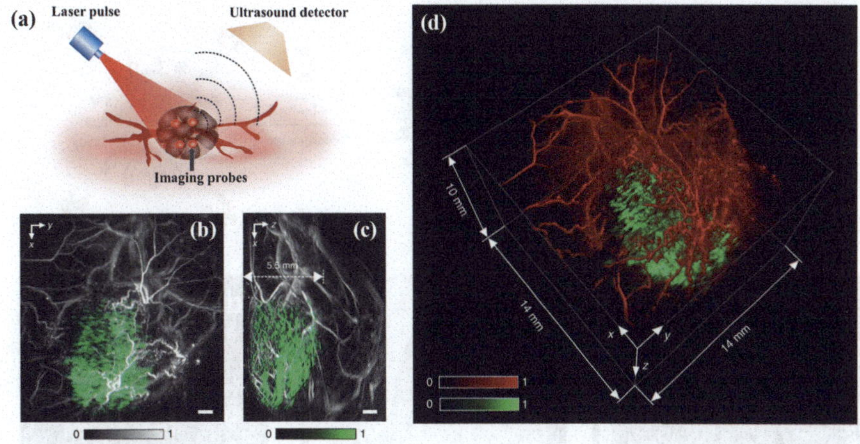

Fig. 11.5 **a** Illustration of photoacoustic imaging. In vivo PA imaging of subcutaneous tumors of AGP1-expressing HT29 cells in mice in 3D with **b** is xy plane, **c** xZ plane, and **d** whole volume-rendered, fused 3D image of the data set shown in (**a**, **b**). Reprinted from Ref. [63] with permission from Springer Nature

deposited on the tissues, maximizing oxyhemoglobin absorption presenting around 1064 nm allowing measure oxygen level in deep tissue. In OTN-NIR, tissues are more transparent, have minimal autofluorescence and background signal. Intrinsic PA signal generated by NIR-II irradiation required contrast agents for enhancing the signals.

A combination of PA and NIR-II imaging can address multiple clinical questions [71]. Although PA provides high structural information, it has a relatively lower temporal resolution than NIR-II imaging. Current PA imaging can provide volumetric images for noninvasive diagnosis and assessment of treatment outcomes. The real-time NIR-II imaging at a fast video frame rate makes it possible for highly sensitive and specific detection of tumor margins. Cheng et al. reported a combination of dual-modal NIR-II fluorescence and PA imaging by using CH1000 dye encapsulated with amphiphilic phospholipids to form a donor-acceptor chromophore-based NP [71]. The structure has an absorption in the NIR-I window and a fluorescence peak in the NIR-II region. The dual PA/NIR-II was used to differentiate the neovasculature from normal blood vessels and the abnormal angiogenic blood vessels (Fig. 11.6) associated with thyroid carcinoma FTC-133, which generally required a significant time over hours or days to detect the minimum variation in the microvasculature. The combination was possible to show the quantitative blood vessels density across the tumor, and the overlap image shows the correlative blood vessels in terms of density or vessel widths.

PA/NIR combination is of great interest because of the similarity between two techniques in terms of spatial resolution and deep penetration by using the same imaging agent. It is vital to provide detailed information about minor variation structure in vivo with high-quality spatial resolution and deep tissue penetration such as

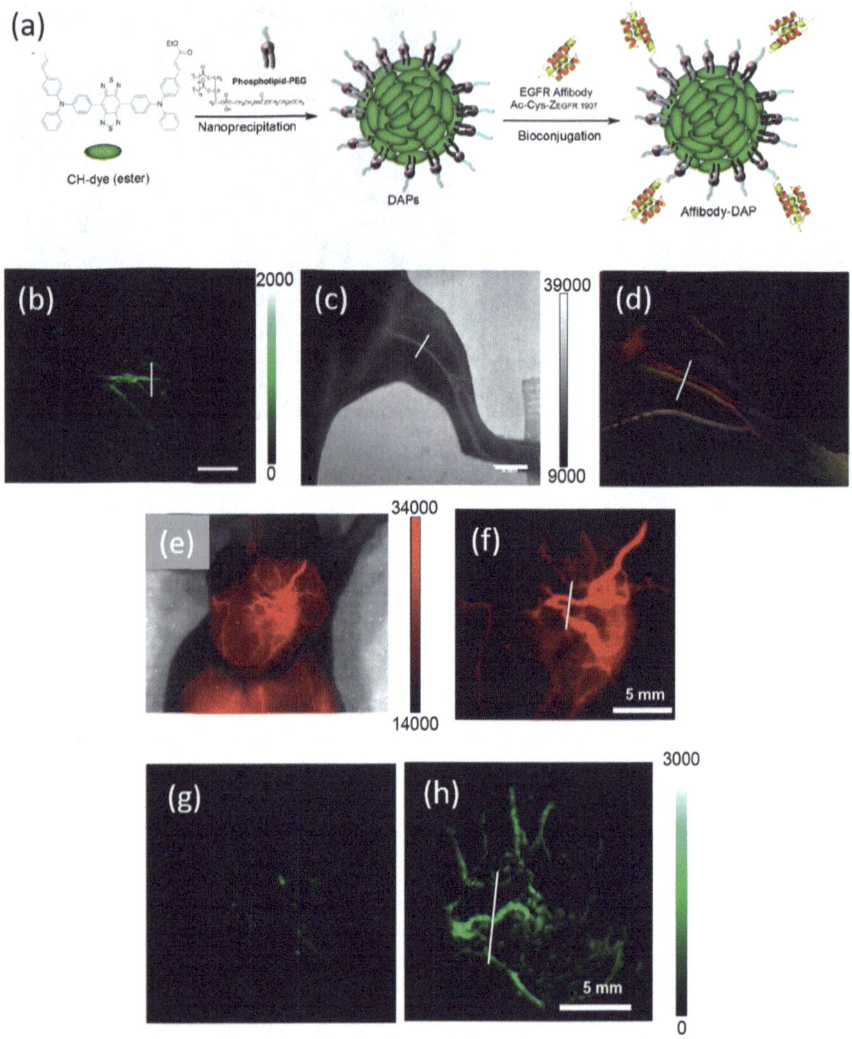

Fig. 11.6 Dual NIR and PA imaging of blood vessels of mice using Affibody-DAPs. **a.** Synthesis route of the imaging agent. **b**. Photoacoustic imaging of mouse hind limb vasculature 30 min after the injection. **c.** NIR image of mouse hind limb vasculature in the NIR-II (ex.808 nm, 1000 nm long-pass filter) window. **d.** Overlaid image based on the first 150 frames. **e.** Color-coded fluorescence merged image of the FTC-133 thyroid tumor in the NIR-II window, and **f.** high magnification image of thyroid tumor. **j**, **k.** 3D volume rendering of PA of the thyroid tumor in a mouse before (**j.**) and after (**k.**). Reprinted from Ref. [71] with permission from American Chemical Society

feeding arteries and draining veins highly correlated with each other. PA imaging is possible to provide real-time cross-sectional, but it is labor and time-consuming to process the volumetric images. NIR fluorescence is able to track rapid biological processes at a molecular level and higher sensitivity toward molecular probes for detecting tumor biomarkers. There is still a shortcoming for PA/NIR imaging is the discrepancy between probe sensitivity that may significantly interfere with the PA signals. An optimal structure of imaging agent should be considered before translating in clinical.

11.2.3 Nuclear Imaging

Imaging involving to radionuclide modalities is the most ubiquitous in medical practice and therapeutic research. The techniques refer to the detection of high-energy photons from the decay of radioisotopes in a target.

CT/NIR imaging. X-ray CT refers to a whole body X-ray imaging procedure from a narrow X-rays tube, which is aimed at a target being rapidly rotated 360° around the body in sync with multiplex detectors from many different angles to detect signals. The produced X-ray intensity profiles generate successive cross-sectional images of the body by computational algorithms, called tomographic images [72, 73]. Because the X-rays absorption is different between electron-dense bone tissues and electron-poor soft tissues creating the apparent contrast in specific areas of the body. Therefore, X-ray CT is particularly used for visualizing bone structures [74]. High-contrast images and distinct anatomical visualization of several tissues, such as blood vessels, stomach, and gastrointestinal organs, could be obtained because of the inherent distinctions in X-ray absorption and attenuation by different components of the body. By contrast, soft tissues with similar inherent CT contrast cannot be distinguished clearly [72]. Consequently, CT contrast agents with large atomic weight (high Z) elements are usually required to delineate regions of interest better. CT is also often used as exogenous agents possess high electron densities to enhance the contrast. Nanostructure of heavy metal or high electron density entities are among the general CT agents, including gold, bismuth, etc. [74, 75]. A multiplexing of CT to other anatomical techniques such as MRI or PET has been widely employed to overlay structures that cannot be obtained intrinsically by other techniques [76]. More detailed information about CT was introduced in Chap. 8. The advantages of CT are to provide supreme contrast in the body; however, the spatial resolution and sensitivity are shortage and insufficient to imaging soft tissues and cellular or molecular imaging [76–78]. X-ray CT has been studied in combining to rapid and high sensitive fluorescence techniques. for providing real-time visualization of the tumor margin, clarifying the borderline between tumor and adjacent normal tissue [79].

The advantages of CT and NIR bimodality with high-resolution/sensitivity toward molecules, tissue, and cells can be used to detect molecular events or tumor imaging and treatment aid [79, 80]. Fig. 11.7 illustrated an experiment designed by Dai

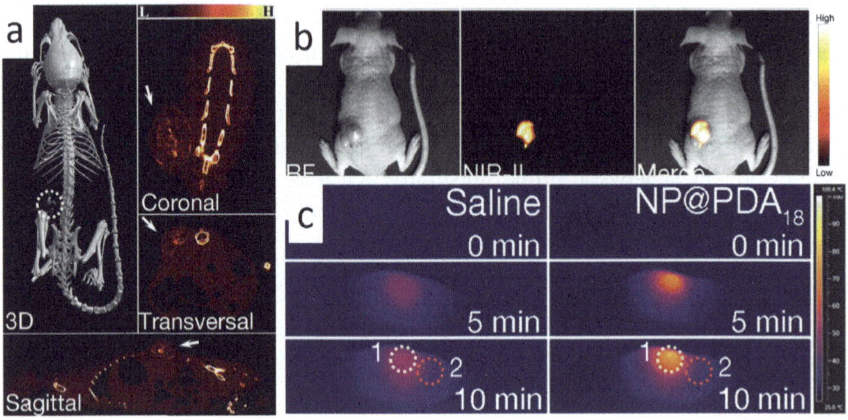

Fig. 11.7 **a** X-ray CT images of HeLa tumor-bearing nude mice after intratumoral injection of 100 μL of NPs solution, **b** NIR-II imaging of nude mice bearing HeLa tumors after injection of 100μL NPs, **c** photothermal imaging of tumor-bearing mice with radiation. After 10 min of treatment, the temperature of the tumor rising high while the normal tissue remains unchanged. Reprinted from Ref. [80] with permission from American Chemical Society

et al. using a theranostic platform from $NaYF_4$:Nd^{3+}@$NaLuF_4$@PDA nanocomposites for NIR/CT spatial- and temporal-specific imaging and photothermal therapy toward the tumor treatment [80]. The NIR imaging of the HeLa tumor-bearing nude mice after intratumoral injection was performed. It was found that NIR-II luminescence is clearly observed in tumor sites with minimal autofluorescence, a high signal/noise ratio, and maximal penetration depth. X-ray CT was used to obtain further physiological and anatomical details of the same mice. CT images showed significant X-ray attenuation signals within the tumor sites after the intratumoral injection of NPs. These results demonstrate the successful construction of NIR-II plus CT imaging agent for mapping tumor regions and adjacent normal regions. Then, the in vivo photothermal temperature elevation profiles of tumors (region 1) and adjacent normal tissues (region 2) were recorded, and photothermal therapy was performed on the previous mapping region by NIR/CT imaging with negligible effect on normal tissues around the tumor. NIR-II/CT was showed an outstanding capability in spatial- and temporal-specific tumor imaging with high sensitivity and structural details. A combination of CT and NIR has been proven as an effective detection of hepatocellular carcinoma (HCC) [78]. This lung cancer is high mortality rates and poor treatment responses due to the lack of highly sensitive techniques to diagnose in the early stage of the diseases. A novel bimodal CT/NIR contrast molecule highly sensitive with liver specificity reported by Shuai et al. has been successfully applied to in vivo CT/NIR imaging with high sensitivity and high selectivity to the HCC region of the HepG2 tumor-xenografted mice model and LM3 orthotopic hepatoma mice model [78]. Furthermore, NIR imaging revealed the blood vessels nearby the tumor for the CT image acquisition in the VX2 orthotopic hepatoma rabbit model. This combination showed a potential application of the bimodal contrast agent to diagnose HCC and successfully resolved the lack of quantitation to determine the location of HCC.

PET/NIR imaging. PET was developed to quantitatively measure biochemical and physiological processes in vivo by using positron-emitting radionuclides and by measuring the annihilation radiation using a coincidence technique [81].

Insight of biological processes also includes the pharmacokinetics of labeled drugs and its effects on metabolism, as well as disease relating processes [81]. The primary idea of measurement of biological or biochemical processes in vivo was already mentioned as the artificially produced radionuclides from the elements abundant in biology such as carbon, nitrogen, oxygen, phosphates. In principles, PET imaging involves a displacement of information from one place to another by means of electromagnetic radiation in the form of photons. The photon is an outcome of a radioactive decay process from a nucleus. Positron-emitting radionuclide' nucleus is a neutron-deficient and the conversion of a proton into a neutron leading to the emission of a positron (β^+) and a neutrino (ν). The decay product positron can travel up to several millimeters inside the body, and collision to another electron results in the annihilation and emit two photons of 511 keV simultaneously in two opposite directions. These two photons can be detected in coincidence by γ-ray scintigraphy in a very short time. In practice, multiple events in the body can be acquired by using a ring of detectors and using the computer capacity for image reconstruction. The radioactive reagents with short lifetime mainly employed by PET, the most frequently comprise from ^{11}C(t ½ = 20 min), ^{13}N(t ½ = 10 min), ^{15}O(t ½ = 2 min), ^{18}F(t ½ = 110 min). Besides, there are still long-lived positron-emitting radionuclides available such as ^{44}Sc, ^{62}Cu, ^{64}Cu, ^{68}Ga, to ^{82}Rb, ^{86}Y, ^{89}Zr, and ^{124}I [82]. The advantage of γ-radiation is weakly absorbed by the tissue; therefore, it is beneficial for imaging of organs deep inside the body. Many novel contrast agents from nanostructures have been proposed to avoid the rapid clearance and direct exposure of high ionizing radiation to humans from the traditional agents. For instance, Marik et al. reported a structure from ^{18}F-labeled PEGylated liposomes and analyzed their biodistribution in PET [83]. The structure showed an improved biodistribution and target site accumulation. Another report from Andreozzi et al. used ^{64}Cu on lipid NPs from the copper specific chelator 6-[p(bromoacetamido) benzyl]-1,4,8,11-tetraazacyclotetradecane-N,N′,N″,N‴ tetra-acetic acid declared a dose decrease of contrast agent [84]. PET is a quantitative and sensitive technique that can monitor various biological events or metabolic and characteristics of cancer cells in vitro and in vivo. However, it hurdled by insufficient anatomical information and relatively low spatial resolution, together with the short half-life of imaging reagent [10, 21]. It is commonly combined with MRI and/or CT for better anatomical and resolution imaging [49, 76] (Fig. 11.8).

Recently, PET/NIR combination was applied to study the application in image-guided surgery in head and neck cancer, or imaging cancers [85]. Y Sun et al. used dual-modal PET/NIR-II for image-guided surgery to clearly delineate the localization tumor by combination NIR-II molecule CH1055 and PET agent ^{68}Ga-1,4,7-triazacyclononane-triacetic acid (NOTA) [86]. The tumor could be clarified by both PET and NIR-II images with a high signal-to-background ratio. Figure 11.9. The PET imaging agent ^{68}Ga-CHS$_2$ were gradually increased from 0.5 to 1 h and then

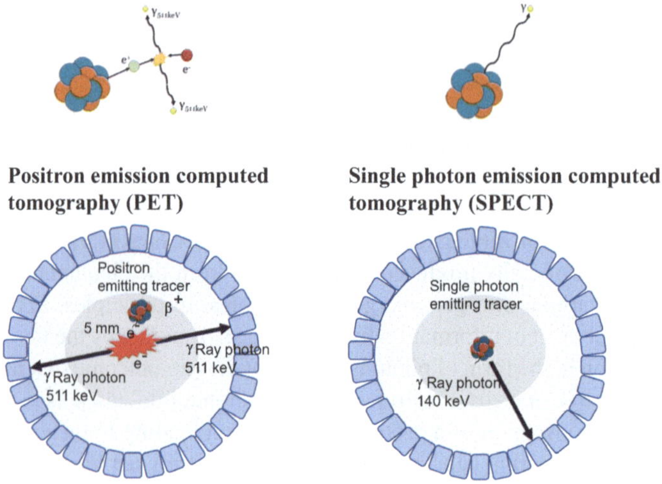

Fig. 11.8 Basic principles of PET and SPECT nucleotide activities and technical configuration for detection of the emission

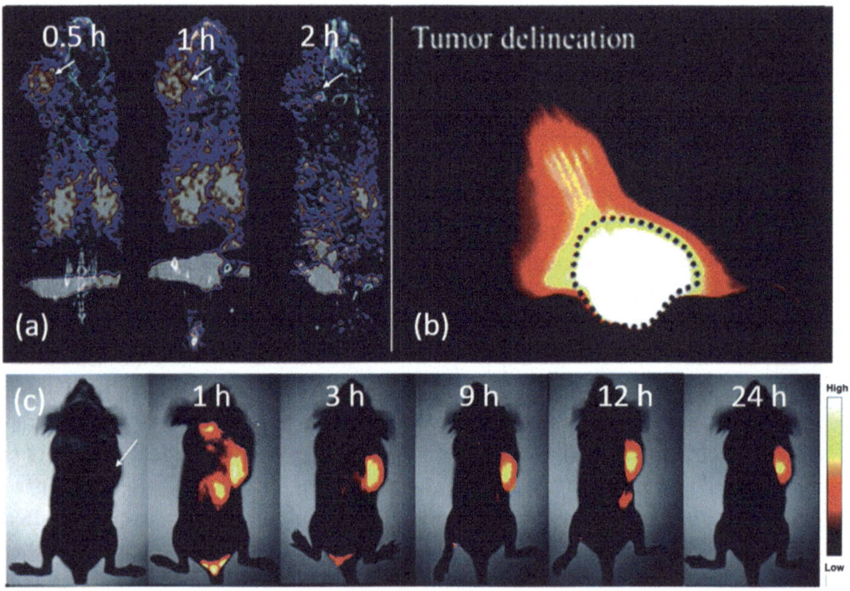

Fig. 11.9 **a** The PET/CT images of U87MG tumor-bearing mice (which arrows indicate the tumor sites after tail vein injection of 68 Ga-CHS2, **b** U87MG tumor delineation and image-guided surgery a 12 h after the injection, **c** NIR-II images of U87MG tumor mice at 1, 3, 9, 12, and 24 h after injection. Reprinted from Ref. [86] with permission from Royal Society of Chemistry

reduced 2 h after injection. The tumor contrast was obtained the maximum intensity at the 12 h time point. The tumor lesion delineation for resection margins and image-guided surgery was performed by using U87MG tumor-bearing mice. As NIR-II fluorescence image-guided surgery was carried out, the tumor delineation and resection margins can be achieved. The combination of measuring gamma probe activity and NIR imaging could lead to a reduction of examining time, which may be a useful aid for the surgery.

SPECT/NIR imaging. Analogous to PET, SPECT imaging is also based on the detection of γ-rays generated by radionuclides. SPECT, together with PET, is not just merely a picture of anatomical structures, but they render the biological and biochemistry process in a 3D vision by visualizing the distribution of radionuclides [76, 87, 88]. SPECT uses an image acquisition system similar to PET as gamma cameras integrated from multi-angles to form a ring to acquire a series of 2-D SPECT images. The computer is then used to apply a tomographic reconstruction algorithm to the projection images, yielding a 3-D image compose of multi-slides.

In the same manner to PET, SPECT uses radioactive tracer for releasing detectable gamma rays [89–91]. However, the SPECT radiation-emitting is direct events and can be detected by the camera without any delay. In contrast, PET tracers travel up to a few millimeters away, and can coincident provide higher spatial resolution than SPECT. SPECT scans are able to use in a longer time than PET. Its radioisotopes are also easier to obtain. SPECT is sensitive and robust in providing functional information of biochemical and physiologic processes with high penetration depth [76]. Radioisotope contrast agents used for SPECT are ^{99}mTc, ^{111}In, ^{123}I, ^{125}I, and $^{201}T_l$. SPECT also limited at anatomical information and low spatial resolution for in vivo applications [21, 89, 90].

PET/SPECT is sensitive and has a depth penetration; therefore, they are suitable for monitoring the pharmacokinetics, biodistribution, and target site accumulation. However, the demand for more anatomical quantification severely restricts these techniques. Thus, a combination of CT or MRI to improve anatomical information and spatial resolution, or optical imaging for further obtaining detailed information on the lesion location is usually added [10, 73, 92, 93].

Figure 11.10 illustrates comprehensive images that can be obtained by the combination of MRI, SPECT, fluorescence imaging, and CT [94]. One strategy for the combination of PET/SPECT and MRI is to utilize multimodal imaging probe composing of two reporters: Gd^{3+} chelates incorporating to PET/SPECT moieties such as radionuclide ^{18}F to develop bimodal probes [95]. The other way is to use NPs or nano-sized structures because they provide more sites for more imaging reporters and targeting biomolecules on the surface [94]. Nanostructures have a high surface area, controlled shape, and size, and tuneable chemical reactivity inner or outer the surface. Kryza et al. reported using Gd_2O_3 NPs for SPECT, MR, and optical imaging modalities [94]. The Gd_2O_3 is encapsulated in the cores within a polysiloxane shell, carrying Cy5 molecules deriving by a hydrophilic carboxylic layer. ^{111}In-DTPA complex was labeled on the surface of oxide NPs. The imaging probe enables a correlation observation among MRI, optical imaging, and SPECT

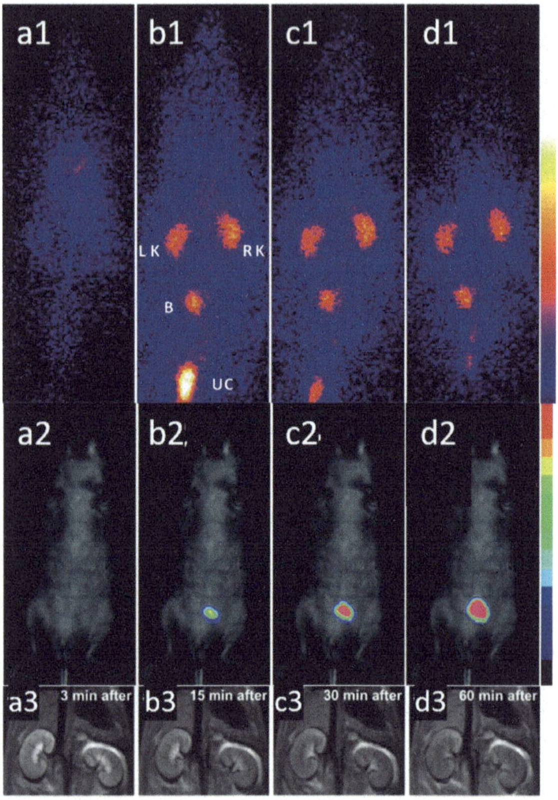

Fig. 11.10 SPECT in vivo imaging (a-d1): planar sequential scintigraphy images during injection (a1), 15 (b1), 30 (c1), 60 (d1) min after injection of imaging agents. NIR imaging (a-d2): Ventral planar fluorescence images acquired at the end of injection (a2), 15 (b2), 30 (c2), and 60 (d2) post-injection of imaging agents. MR imaging (a-d3): T1-weighted sequence corona images of kidney acquired at 3 (a3), 15 (b3), 30 (c3), and 60 (d3) min after injection. Reprinted from Ref. [94] with permission from American Chemical Society

regarding renal elimination. Use γ counting in the organ permits evaluating quantitative biodistribution of the probe and confirm that these NPs circulated freely in the blood vessel and were rapidly cleared by renal excretion without accumulation in liver and RES uptake. In general, the biodistribution and clearance of NPs in vivo depend on their physical or biochemical characteristics such as size, surface functionality, and charges, which regulate the fate of NPs to be taken up by the RES system and leading to the clearance of NPs from systemic circulation. Optical fluorescence signal could be observed in the urinary bladder and kidneys showed minimal particle retention in major organs (especially RES tissue), and the primary route of elimination was renal excretion. The trace of NPs can also be confirmed in various tissues by using γ-counting from PET or SPECT. By using multimodal imaging, the presence, the absence of significant uptake in tissues, biodistribution, and pharmacokinetics of NPs was well observed and monitored [96].

11.2.4 Other Techniques

There are significant demands for visualizing molecular process, chemical-physical properties of microenvironment or molecular structures in vivo in order to elucidate the mechanism of disease development. These questions can be addressed by combination imaging techniques at different scales. Hereby, a brief introduction of over-scale multimodal imaging will be discussed with the focus including cathodoluminescence, chemiluminescence, or visualization of viscosity.

Cathodoluminescence (CL)/NIR imaging. At the cellular level, correlation of light and electron microscopy (EM) is by far the most interest paradigm in identifying specific structures of sub-cellular life [31, 97–99]. However, it hampers at monochromic images and is unable to discriminate multiple complex molecules. Correlation of multicolor CL and deeply penetrative NIR light has been exploited to take the advantages of superior resolution near-atomic-level from EM and deep tissue observation from NIR without any damage to biological organisms [31]. Fukushima et al. reported a correlative technique of dual-color CL and NIR by using rare-earth-doped NPs. Figure 11.11 shows images of NPs sequentially obtained by a NIR microscope and a CL integrated to a field emission scanning electron microscopy (FESEM-CL). CL images are also registered to scanning transmission electron microscopy (STEM). The results suggested the ability of simultaneous imaging at different scale and differentiation among various NPs by the proper selection of the luminescent colors.

Chemiluminescence. Visualizing and quantifying physiological processes in vivo can clarify multiple problems about states of diseases, therapy outcomes, therapeutic effects of drug in the region of interest, or physio-biochemical homeostasis in abdominals. The use of NIR light for activation of the sensing process has been focused on pH sensing, monitoring reactive oxygen and nitrogen species, control enzyme reaction, drug delivery, or activation of chemiluminescence. pH sensing can reveal acidic status in cells. The pH of solid tumors is often lower than that of normal tissues due to elevated anaerobic metabolism. Because of the discrepancy in pH values intra- and extra tumors, it is possible to differentiate the metabolic status of cancer tumor by using appropriate pH indicator molecules [100–107]. The pH indicators are made of fluorophores with a responsive pKa in the physiologically relevant range in vivo. By using such indicators, optical imaging has been demonstrated for sensing pH [100–102], temperature [103, 104], biochemical reactions at the cellular level. Enhancement or quenching of fluorescence intensity, or a shift of emission peak are correlated to pH [105]. Some fluorophores are able to switch their properties by interaction with the change of microenvironment (rising pH, temperature) and the ongoing reaction of molecular counterparts [106]. Zhao et al. designed a DNA nanomachine for pH sensing in vivo and NIR imaging [108]. DNA responsive to pH change has been used to measure pH inside living cells or organs [108]. It was incorporated into UV-emitting $NaGdF_4$ co-doped with Yb^{3+}/Tm^{3+} as an activator to render the pH response of DNA. The designed DNA nanomachine was delivered to a target site without activating the pH functions of the machine. Upon the target areas,

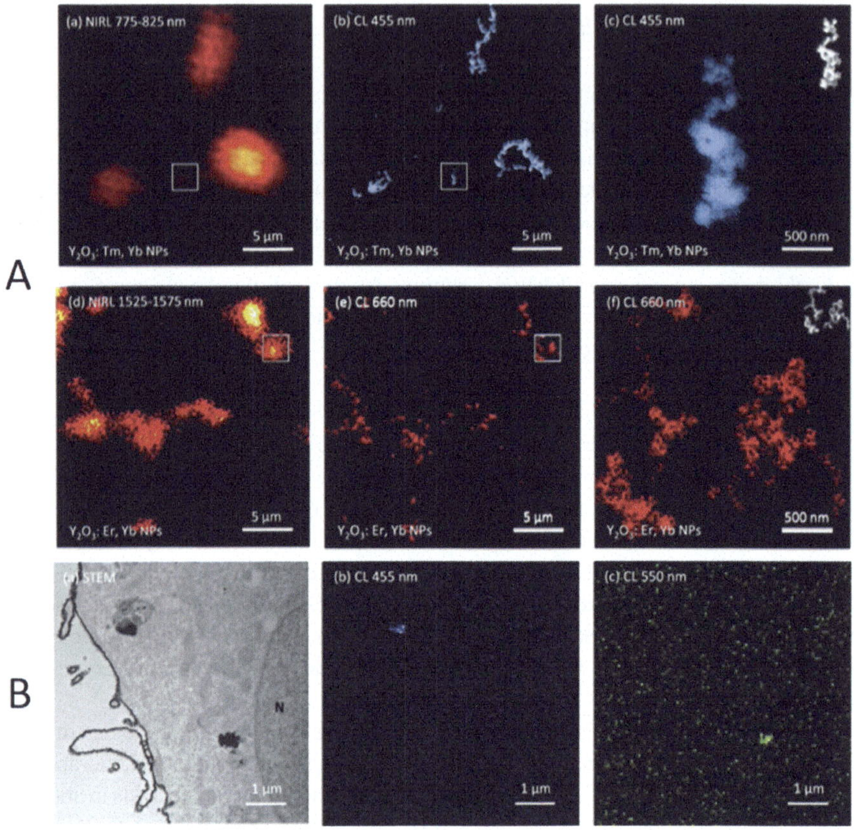

Fig. 11.11 A: The co-registration CL-NIR images of NPs: NIR emissive images at 775–825 nm (Tm) and 1525–1575 nm (Er) at the same region CL images with central peaks of 455 nm (Tm) and 660 nm (Er). B: Dual colors Hela cellular STEM image (a) and CL images (b, c) of Y_2O_3: Tm, Yb/Er, Yb NPs. Excitation source: electron beam. Acquisition wavelength: 455 nm (b), 550 nm (c). Reprinted from Ref. [31] with permission from Springer Nature

under NIR irradiation, the pH measuring function is activated in favor of the upconversion process. Upconversion is resulted in releasing UV light, selectively activated DNA machine for sensing pH with high precision both temporally and spatially. The pH response was established via the FRET effect involving to quenching agent in DNA binding fluorophore Cy5, as showed in Fig. 11.12.

Likewise, sensing metabolism products such as reactive oxygen and nitrogen species is also important because they are indicators for many physiological and pathological processes or drug-induced toxicities. It is also useful for early diagnosis of drug-induced toxicities. Various NIR activating molecular structures have been developed for monitoring drug-induced hepatotoxicity [109], detections of OH in a mouse model [110]. In some cases, the reaction release energy under the form of

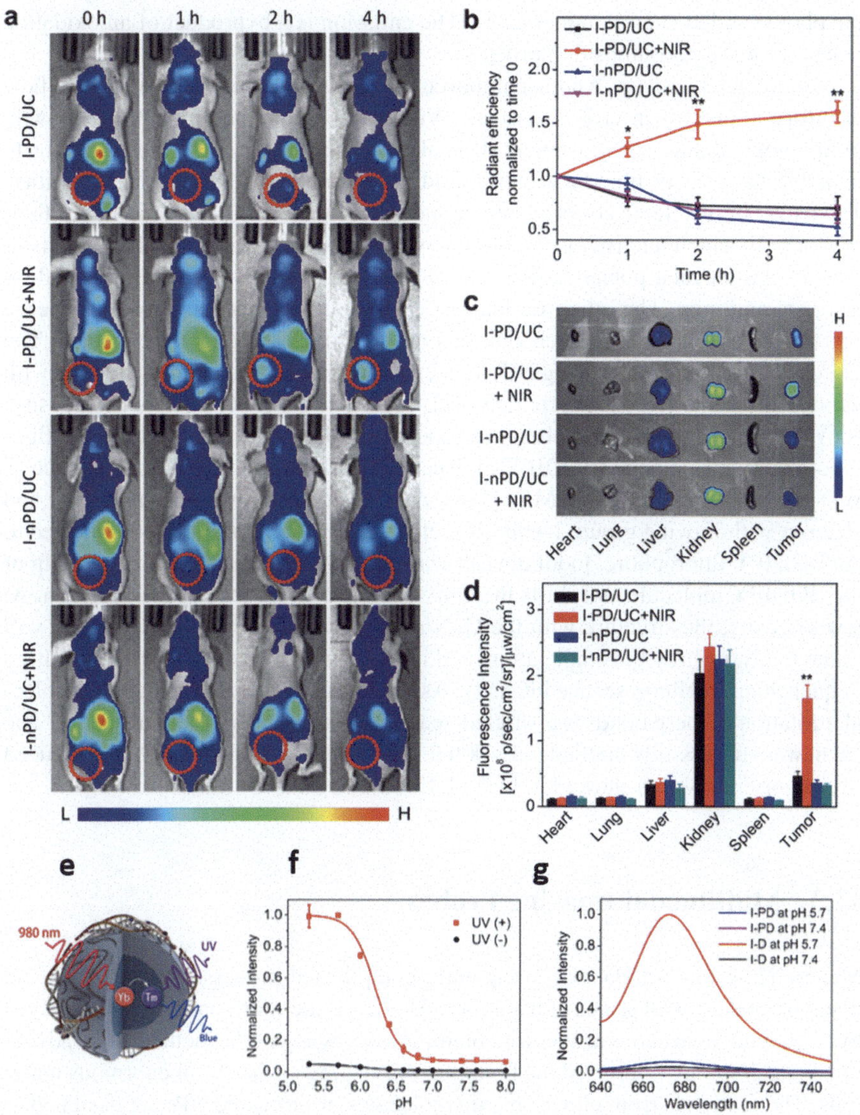

Fig. 11.12 NIR-activated pH sensing and imaging in vivo when DNA nanomachines were administered intravenously. **a** Fluorescence images of HeLa tumor-bearing mice at different times after injection **b** Quantification of in vivo fluorescence of tumor as a function of post-irradiation time, **c** ex vivo imaging and **d** quantification of signal intensities, **e** imaging agent structure, **f**, **g** correlation of emission intensity and pH. Reprinted from Ref. [108] with permission from American Chemical Society

the photon called chemiluminescence. The emission is a byproduct of an oxidation reaction for the decomposition process.

Visualizing Viscosity. Another important factor that governs the materials flow or diffusion process in vivo is viscosity. Viscosity is accounted for cellular growth, morphology change, or movement. It is also an important environmental factor to maintain the flow of blood and other fluid in a number of physiological activities. The native intracellular viscosity can be measured by imaging the variation of fluorophores, in which intensity is monitored by a twisted intramolecular charge transfer (TICT) of molecular planarization. The TICT effect caused a quenching to the fluorescence emission. The other mechanism is based on the Förster resonance energy transfer (FRET) process, which can lead to the Stokes shifts of fluorophores. The viscosity measurement by using NIR molecules has gained tremendous attention with various molecular structures reported [111, 112]. The design of a series of viscosity-activated NIR-II emissive probes was reported by Dou et al., based on the difluoroboron dipyrromethene (BODIPY) dye namely WD with functional groups such as WD-NO_2, $-CH_3$, $-OCH_3$, $-NME_2$ [113]. A viscosity-sensitive molecular rotor and electron-withdrawing group, 1-ethyl-2-methyl-benz iodolium salt, was connected to the BODIPY fluorophore. Iodolium salt was used to enlarge the conjugated chain of the BODIPY molecule and shifts the emission wavelength to the NIR-II region. At low viscosity, the structure with the C–N single bond of the BODIPY moiety will rotate freely, destroy the conjugation, and increase nonradiative transition, resulting in quenching the fluorescence intensity. As the viscosity increasing, the intramolecular rotation of these bonds is restricted, leading to the recovery of fluorescence. The agent was successfully applied to track the variation of viscosity in diabetes-induced liver injury in vivo, as showed in Fig. 11.13.

11.3 Multimodal Imaging Probes

Materializing multimodal is consequently complicated processes by adapting and assembling data at all scales of resolution. However, the shortcomings can be solved by employing exogenous probes for obtaining images across the methods. The advent of nanotechnology flourished the booming of imaging probes from enormous materials. The materials from either organic molecules or inorganic NPs are designed to specifically target a region of interest in the biological body and response to distinct stimuli. Table 11.1 summarizes some typical materials used as multimodal imaging probes. For the aim of developing an NIR based multimodal imaging probe, there are some requirements that materials have to be achieved to be applicable.

1. Ability to respond to the multi-stimuli: the interactions between materials and excitation sources must be comparable with each technique. The signal's characteristics must depend less on the nature of the environment.

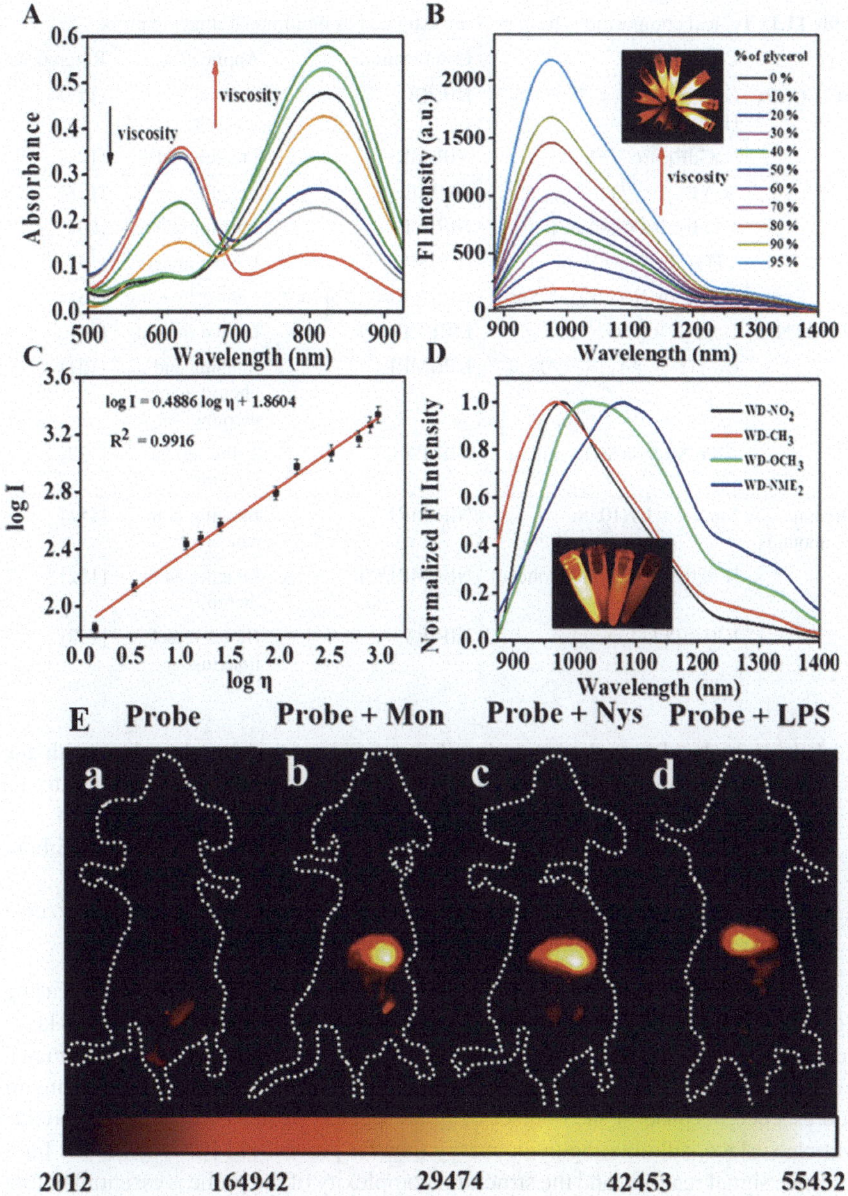

Fig. 11.13 **A** Absorption and **B** fluorescence spectra of viscosity agents WD-NO_2 in ethanol–glycerol mixture with varying ratios. **C** The linear relationship of fluorescence intensity and log η. **D** Normalized fluorescence spectra of series agents in 95% glycerol. **E** Fluorescence imaging of viscosity-ty changes by exogenous drug stimulation. (a) only injected with WD-NO_2, treated with monensin and nystatin (b, c), LPS (d) accompanied with an intraperitoneal injection of WD-NO_2. Reprinted from Ref. [113] with permission from American Chemical Society

Table 11.1 Typical compositions have been investigated as multimodal imaging probe

Types	Compositions	Combination	Applications	References
RED-CNPs	Gd_2O_3: Tm^{3+}; Er^{3+}; Ho^{3+}/Yb^{3+}	NIR/CL	Cellular imaging	[117]
	$NaGdF_4$: Er^{3+}/Yb^{3+}	NIR/MRI	Imaging, DDS	[123]
	$NaYF_4$: Er^{3+}/Yb^{3+}	PA/NIR	Imaging	[128]
	$NaYF_4$: Nd@$NaLuF_4$	NIR/CT	Imaging-guided	[80]
	YPO_4: Er, Yb@^{111}In	NIR/SPECT	In vivo imaging	[132]
	$NaLuF_4$@$NaGdF_4$: ^{153}Sm	NIR/MRI/CT//SPECT	Cancer imaging	[131]
Hybrid NPs	Gd_2O_3@Cy5@^{111}In	MRI/NIR/CT	In vivo imaging	[86]
	Gd-DTPA@Ag_2Se QDs	NIR/MRI	Imaging and photodynamic therapy	[131]
	AlPc-MoS_2@SiO_2-chitosan	NIR/PA/CT	Three-modal imaging	[132]
Organic compounds	Mn2 + @CH1055	NIR/MRI	Imaging brain cancer	[151]
	IR820@Gd-DOTA@Iohexol	NIR/MRI/CT	Imaging and therapy	[152]
	ICG@PEG Liposome	NIR/PA	Blood vessel imaging	[153]

2. Is able to be detected: the probe characteristics must be accordant with the imaging modality. The emission or effect output must be sufficient to be detectable by the instruments in a complex nature of the biological system.
3. Stability and durability: the emission signals must be stable and sustain without fading to the chemical environment.
4. Is able to modify and functionalize: the surface of materials must be able to coat or contain functional groups for further conjugations to biomolecules.

Controlling inherent dissimilarities in the sensitivity or mechanism of releasing signals of each system contrast agents to meet stringent regulations also required laborious effort [7]. The other challenge is involved in the physical property mismatch between multi-agents. For example, iron oxide is an agent that causes quenching of fluorescence because of the contact of fluorescent molecules with the metal oxide particle surface, culminating in an energy transfer process [114]. Consider the fluorescence signal readout, and the structural complexity of the probe is essential before adding targeting groups and biocompatible modifications [115] (Fig. 11.14).

RED-CNPs with their long-circulating times and facile functionalization become a potential NIR-based multimodal imaging agents with a numerous variety ranging from oxides [116, 117], to fluorides [118, 119], and sulfides [120], as well as Gd oxysalts [121]. RED-CNPs render stable platforms that are potent for multimodal imaging by tailoring dopants and the core-shell structures to integrate with contrast agents from MRI, CT, PET/SPECT. $NaYF_4$ is extensively used as a host material for

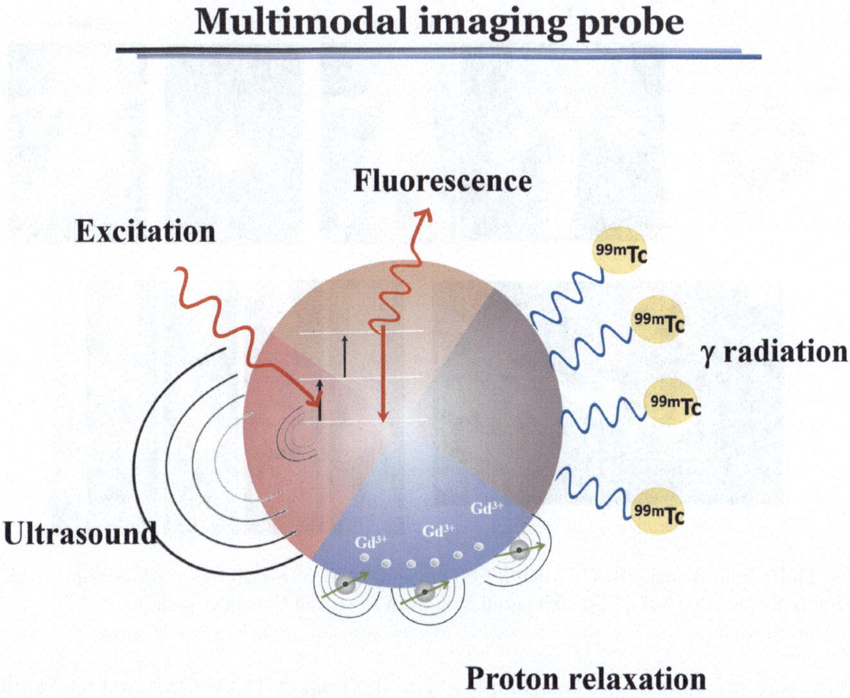

Fig. 11.14 A schematic of the NIR/MRI imaging probe from RED-CNPs using Gd^{3+} ions as a T_1 contrast enhancement agent

optical application due to its low phonon relaxation, less quenching, high luminescence intensity together with a T_1 relaxation enhancement effect when Gd^{3+} ions are co-doped in the matrix [122–124]. T_1 agents are made from non-lanthanide paramagnetic metals such as manganese with five unpaired electrons in case of bivalent manganese some time is also used in some cases [125]. For extended applications in Xray CT, dopants from heavy elements such as traditional iodine (I), bismuth (Bi), lutetium (Lu), bari (Ba) [76, 78, 126, 127]. RED-CNPs possess highly controllable absorption behavior based on the f-f transitions leading to a consequently independent of absorption to particle size or morphology. This favors an opportunity for multi-wavelength PA/NIR imaging [128]. Lanthanide-based core-shell nanocomposite $NaLuF_4$: Yb,Tm@$NaGdF_4$(^{153}Sm) used for CT, MRI, SPECT, and UCL was firstly developed by Sun et al. [129], as shown in Fig. 11.15. Er^{3+}/Yb^{3+} ions co-doped YPO_4 NPs also demonstrated the capability for over 1000 nm near-infrared fluorescence and SPECT/CT imaging [92].

Hybrid NPs are the most widespread strategy to develop multimodal imaging probes because it is facile in combining any forms of materials such as inorganic NPs to organic molecules or biomaterials [130, 131]. Because the presence of multi-agents imaging and therapeutic agents triggers the mismatch of each other, the agent platforms need to be designed for the trade-off efficacy in imaging and treatment [132].

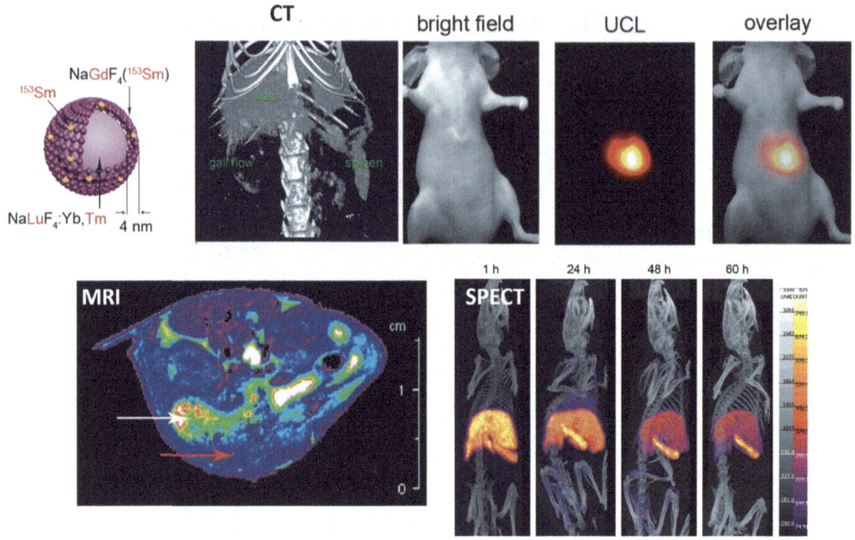

Fig. 11.15 Four modal NIR/CT/MRI/SPECT imaging with core-shell rare earth doped ceramic NPs. Reprinted from Ref. [129] with permission from American Chemical Society

Various combinations have been employed used CT agents [133–135] combined with organic dyes, QDs [136, 137], and surface modification with PET or SPECT agents [138–140]. Some combinations can be performed with PA agents [141, 142].

As discussed in the section above, the substances used as potential NIR/MRI imaging probes are abundant ranging from inorganic NPs such as rare-earth-doped NPs, the hybrid structure of metals NPs, semiconductor, carbon, to metal-ion-derived complexes (Gd or Mn, etc.) or iron oxide NPs in order to be used in MRI. However, a great part of the researches focuses on the fundamental multi-functionalities of materials with less concrete proofs of biocompatible or bio clearance. Critical concerns arose from patient safety when NPs were found retained, accumulated long-term in the body without clearance [143–145]. Organic molecular dyes are well-defined architectures with rapid metabolism, low toxicity, and are preferable for multimodal imaging. The combination of Gd-chelates and NIR molecular probes are a favorite for medical research and clinical practice [146–148]. NIR organic dyes were well discussed in the previous chapter, including several families [148–150]. The development of molecular multimodal imaging agents for MRI, PET, or SPECT to NIR can be rendered by metal ion chelates conjugation or co-assembly in macromolecules [151–153].

11.4 Applications

Multimodal imaging can provide further information in diagnosis, treatment, and monitoring than individual ones. At present, it has been widely investigated in medical research and clinical applications, for instance, in the detection of diseases such as cardiovascular diseases [154], neuropsychiatric diseases [155], and other clinical diseases [156]. In addition, it can effectively guide the surgical resection of the tumor [157]. NIR based multimodality is a novel paradigm that is expected to solve multiple problems in the current combination.

11.4.1 Intraoperative Imaging

Intraoperative imaging is a method using imaging techniques to aid or guiding the delineation and localization of the regions of interest during surgery [158]. While pre- or post-surgery practice predominantly uses X-ray CT, MRI and US, intraoperative guiding technique mainly depends on surgeon interpretation and pathology for confirmation. In recent years, surgical aid by imaging has evolved significantly as favors of novel techniques such as fluorescence, Raman, PA imaging. The imaging guiding surgery works as an on-site navigating system of the lesions, sectioning, or margining the tumor regions for rapidly and accurately deciding the treatment method during surgery. Fluorescent imaging is highly compatible with the intraoperative technique because it provides real-time imaging with a highly sensitive and excellent resolution that helps surgeons differentiate the precise borders between normal and lesion areas. The intraoperative settings using NIR-I light have been developed and demonstrated the potential for accurate pathological detection. Currently, various NIR imaging systems utilized ICG or cyanide family work as on-site surgical navigation systems [159, 160]. Some are already commercially available, as shown in Fig. 11.16. However, due to the lack of deep tissue penetration in NIR-I, 3D precision tumor detection cannot achieve, and detection depth is the main limitation for this method in clinical applications. In the case of using NIR-II or -III, this problem can be solved. A 3D NIR imaging could allow for tumor margin detection during operation, which would be enormously meaningful to the treatment. When the limitation of the penetration depth is overcome, such an outcome as 3D concurrent anatomical and functional information of the body can be obtained. In an effort to demonstrate the benefit of multimodal in tumor treatment, Xie et al. designed NPs for PET/NIR/MRI triple modality [85]. The experiment reveals the possibility that a tumor U87MG xenograft mouse model could be visualized in vivo by this triple modality probe. Another combination has been reported by Ewelt et al. to visualize diffuse gliomas, which are not always detected even by MRI contrast agents [161]. The method used ^{18}F-FET and Gd-chelate for pre-operative diagnosis, and 5-aminolevulinic derived targeting tumor for NIR image-guided malignant glioma dissection. The multimodal imaging can provide valuable information pre-diagnostic with 70.5% sensitivity and

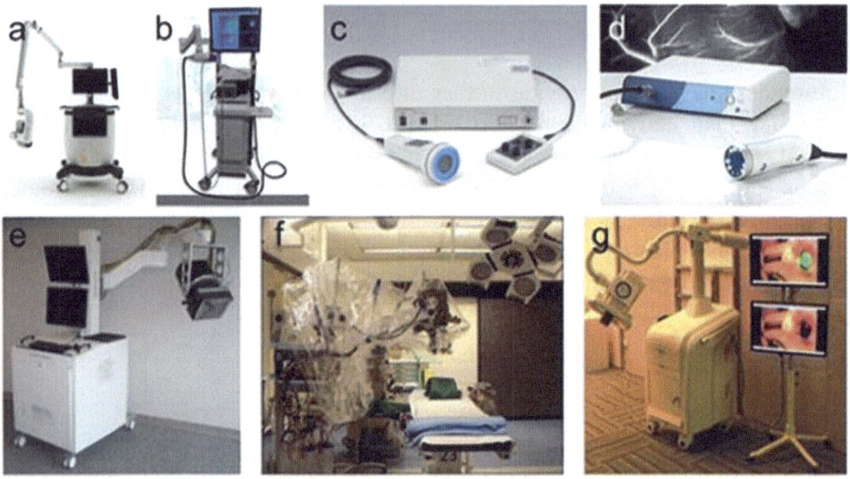

Fig. 11.16 Currently available systems for Imaging-Guided Cancer Surgery: **a** The Novadaq SPY™ system, **b** ArtemisTM, **c** Hamamatsu's Photodynamic Eye (PDE™), **d** Fluoptics' Fluobeam®. Functional intraoperative FMI systems: **e** FLARE™ imaging system, **f** Multispectral FMI system from Technische Universität München & Helmholtz Zentrum, **g** Surgical navigation system GXMI Navigator from the Institute of Automation, Chinese Academy of Science. Reprinted from Ref. [159] with permission from Ivyspring International

92.3% specificity in detection. It is also proved the role of a guiding technique in tumor resection. The example of applying NIR/MRI in surgery-guiding used Gd-ICG liposome is shown in Fig. 11.17 [162]. ICG molecules emit fluorescence in 810–830 nm, and Gd^{3+} chelates were encapsulated by a bilayer of phospholipids. The MRI was used to localize preclinically ovarian tumors and plan for surgery. NIR imaging was used for tumor resection during surgery. The results showed that the dual-modal co-registration of ovarian tumors was obtained in mice.

Multimodal NIR/MRI can also apply to detect tumor microfoci, a highly sensitive site, where the treatment relies heavily on the accurate localization of tumor lesions [163]. Many other anatomical imaging techniques such as CT or MRI alone cannot effectively detect the tumor microfoci (<2 mm) of satellite tumor nodules. Using NIR-I imaging was reported to offer sufficient optical contrast between the lesion and its surrounding normal tissue, accurate delineation of tumor borders, and sensitive detection of tumor margin delineation, which may help during liver cancer surgery. By combination with MRI, valuable pre-operative diagnostic information, or post-operational residues, remains in bodies can also be obtained.

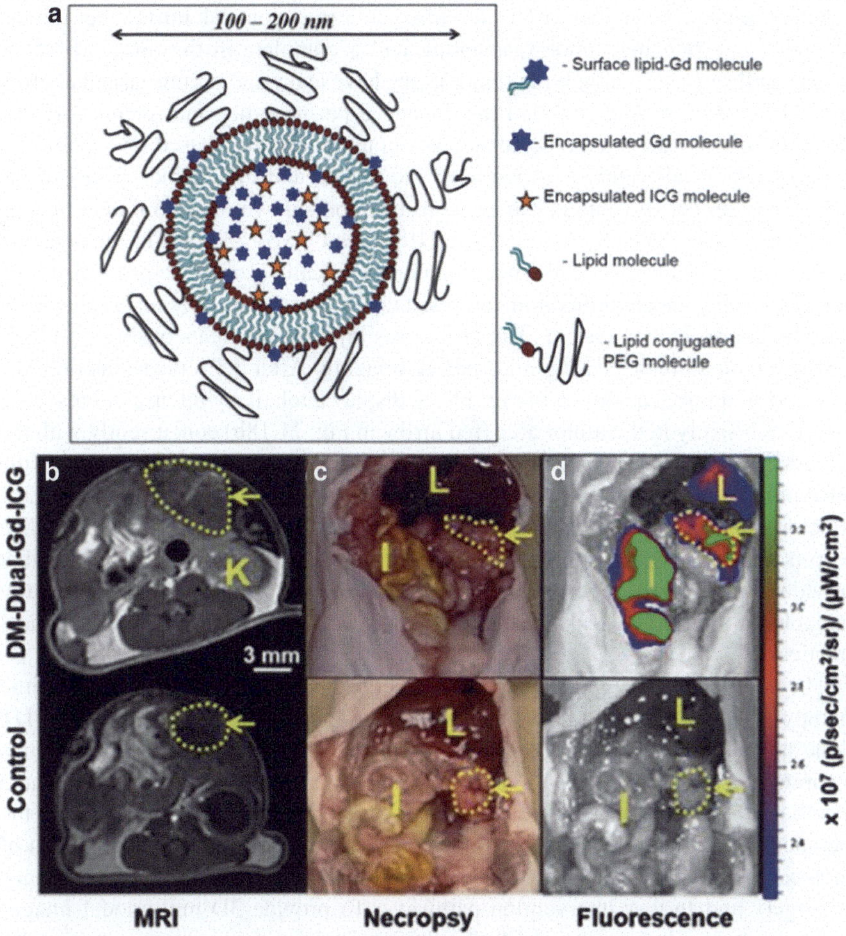

Fig. 11.17 An application of NIR/MR imaging probe in ovarian cancer Image-Guiding: **a** Schematic structure of the dual NIR/MRI probe from Gd-ICG liposome. **b** Cross-sectional image obtained by MRI. **c** A coronal opening abdomen and **d** NIR fluorescence images of a nude mouse. Similar MR and NIR findings were noted with a second intraperitoneal tumor model. Arrow, tumor; I, intestine; K, kidney; L, Liver. Reprinted from Ref. [162] with permission from Springer Nature

11.4.2 Imaging and Cancer Therapy

Cancer tumors are stiff and often develop drug resistance due to their microenvironment by imposing barriers that limit drug delivery to the tumor [164]. A comprehensive understanding of the mechanisms, microenvironment, and barriers are sequentially needed to address to enhance the efficacy of cancer therapies. Imaging cancer tumors are performed in two ways, passive or active targeting. Passive targeting uses non-specific imaging probes, the structures with their intrinsic surface without any targeting groups. Molecules or NPs circulate after injection leak into cancer tumors through the permeable tumor vasculature and accumulate in the tumors because blood capillary vessels supplying tumor tissue have leaky and porous vascular structures. This phenomenon is called the enhanced permeability and retention (EPR) effect. NIR/MRI can extend the advantages in tumor imaging based on capabilities to accurately localize the distribution of the probes and the characteristics of the microenvironment surrounding them with the depths down to several centimeters. Lee's group used SPION conjugated to NIR dyes Cy5.5 (25.6 nm in diameter) to perform in vivo cancer NIR/MR imaging nonspecifically [165]. The probe had high stability in physiological medium and resistance to being taken by macrophages of the reticuloendothelial system. The probe was injected subcutaneously into Lewis lung carcinoma tumor-bearing mice before imaging. After 3.5 h post-injection, the NIR optical images, as shown in Fig. 11.18, display a relatively intense fluorescence signal exclusively in the tumor area (red arrow in Fig. 11.18b) concurrently with the MR images. Without any specific targeting ligands on the surface of the probe, the MR/optical dual imaging of cancer in vivo by passive targeting based on the EPR effect was successfully obtained. However, this Cy5.5-based probe cannot visualize deep-positioned tissues due to the use of NIR-I. It can actively target tumors that provide tracking capabilities before, during, and after treatment. This is potentially valuable for the next critical phase of cancer research when transferred into clinical practice.

The multimodal imaging is often used to delineate tumor areas for photodynamic therapy (PDT). PDT is a promising approach to cancer treatment; however, PDT alone is insufficient to attain the contemporary medical diagnosis, which requires accurate and detailed information of the cancer tumors such as size, shape, and specificities. The uniqueness of PDT is that the same molecule may serve as both the therapeutic agent and the imaging contrast agent. This embarks the development of photosensitizer (PS) multimodal probe for treatment and imaging probe in guidance of surgery and further intervention pathways. To provide 3D multimodal images for a comprehensive insight of PS distribution in the tumor heterogeneity, multimodal imaging in conjunction with PDT can be PA, MRI, CT, PET/SPECT, and NIR imaging itself. Moreover, multi-parametric morphological and functional information, the information of PS uptake and vascular destruction, oxygenation dynamics, and the overall evaluation of tumor response can also be obtained. PDT involving in cell death mechanisms following treatment in a dynamic process in that molecular imaging-based real-time monitoring becomes a key to monitoring the outcome. It is a particular challenge to identify tumor structural and morphological changes in tumor volume due to PDT or tumor vasculature post-PDT. In many cases, photoacoustic PA is useful for monitoring of PDT-induced oxygen saturation change and evaluating the content of the whole tumor volume, thus predicting the therapeutic efficacy [166].

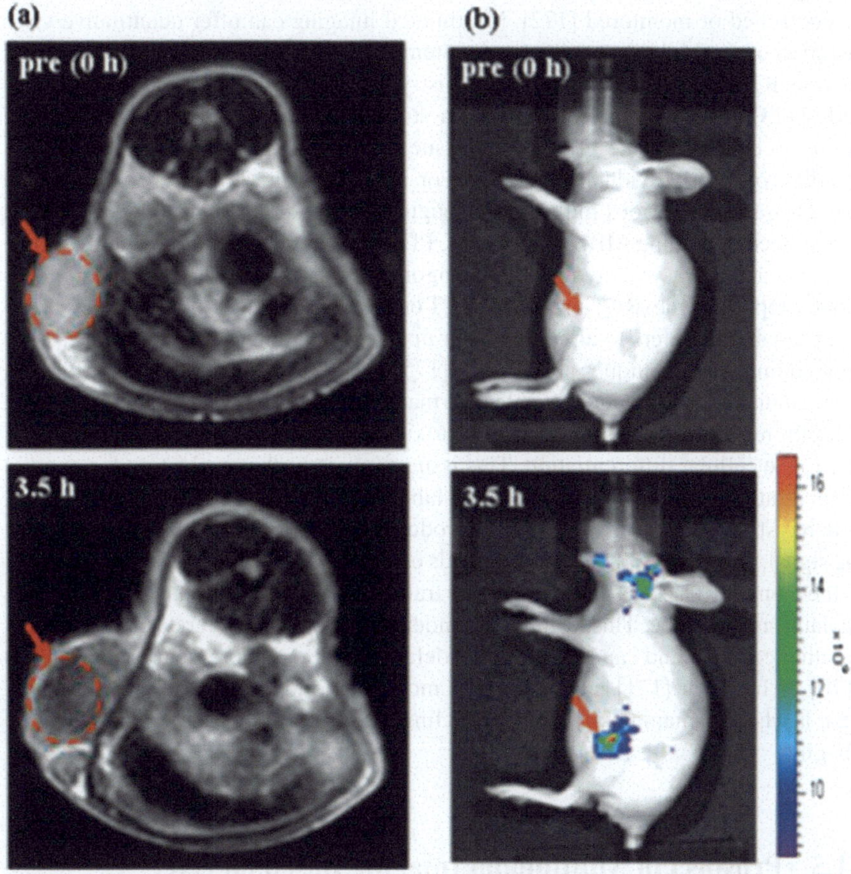

Fig. 11.18 Applying NIR/MR imaging probe made from iron oxide combining to Cy 5.5 for non-specific lung cancer targeting: **a** T2-weighted fast spin-echo images taken at 0 and 3.5 h post-injection. The dashed circle with the red arrow indicates the allograft tumor region. **b** NIR images of the same mouse taken at 0 and 3.5 h post-injection. The red arrows indicate the position of the allograft tumor. Reprinted from Ref. [165] with permission from American Chemical Society

11.4.3 Toward Regenerative Medicine Applications

The current growing field of regenerative medicine prompts toward clinical trials in various prospected applications [167, 168]. The transplanting of stem cells sequentially seeks for the innovative and breakthrough techniques for tracking the cell migration and its reciprocity with microenvironment targets, evaluating the cell fate, distribution, viability, or differentiation noninvasively and longitudinally [169–171]. Tracking and monitoring cell fate are highly needed, especially for induced pluripotent stem cells (iPSC) that can differentiate into any cell types. Stem cell therapy latently possesses safety risks and low efficacy since the translational pathway cannot be controlled or monitored [172]. Multimodal imaging can offer quantitative visualization of the biological processes of stem cells in vivo, where invasive methods such as RT-PCR and immunohistochemistry are mostly used for monitoring stem cells [173]. The drawback in long-term studies of these approaches reduces the ability to check cell fate in vivo and tissue regeneration. It also limits our understanding of interindividual variability. For addressing the limitation of long-term cellular tracking, various multimodal imaging configurations have been developed and applied, including MRI, CT, SPECT, PET, optical imaging, etc., with their own merits and demerits widely reviewed. Ongoing research involves pre-labeling techniques, reporter gene (RG) imaging by PET or optical techniques [168, 174, 175]. The cell pre-labeling refers to an intracellular uptake or loading of imaging probes. The applied imaging techniques are SPECT, [175] PET, [176] MRI, [177] etc. Although these studies above provide valuable information on migrating and distribution of the cells, the remaining issues are the cellular toxicity and the reduction of imaging probes due to the cellular differentiation. This issue requires further evaluation for optimal concentration and sensitivity used in pre-labeling. RG imaging uses a reporter gene, which is stably expressed in cells and introduces to imaging probe the RG to produce the signal. RG imaging sensitivity depends on reporter concentration in cells, as well as the number RG expressing cells, probe transport to the reporter, the intensity of the signal from the probe. This technique is independent of the proliferate, and the signal is rather constant and can be repeated indefinitely. Green fluorescent protein (GFP) [170] or luciferin [171] are fluorescent molecules often used as imaging probes. That is why the imaging in deep tissue is limited and cannot be used for whole body imaging.

11.5 Prospect of Multimodal Imaging Based on Over 1000 nm NIR Light

With the intense growth in recent years, despite its early age, NIR imaging techniques herald a prospected outlook. NIR based multimodal imaging is expected to bring a breakthrough in medical imaging and molecular biology. In particular, it facilitates the accurate visions of biological processes, molecules, cellular, and tissue architecture simultaneously to the anatomical details of dynamic processes or morphological changes. As optical imaging can go deeper into the body, a potential 3D deep tissue imaging can be obtained. Most of the optical imaging techniques currently employed, due to the limitation in deep penetration, provide only a shadow image of the target. The NIR-I provides a deeper penetration imaging; however, a large 3D volume image at centimeter-level can be obtained if the OTN-NIR applied. 3D fluorescent imaging allows quantitative imaging signals from the probe at different depths, colocalization of tissue biologic heterogeneity, and co-registration imaging to localize the tissue in the bodies. More importantly, photothermal OTN-NIR imaging is very promising as it forms a volumetric thermometry NIR/MR multimodal imaging for comparison of absolute and relative thermometry, 3D-3D imaging techniques. On the other hand, local micro-viscosity detection in cells was proposed based on Förster resonance energy transfer (FRET) of several molecular probes whose fluorescence intensity and/or fluorescence lifetime properties depend on the solvent viscosity, which enabling imaging viscosity in living cells. With the rapid advance, NIR molecules using for FRET in NIR can be developed in the near future, and the detection of micro-viscosity in deep tissue also can be elucidated. The NIR imaging also initiates pH detection, microenvironment alteration by tracking the oxygen level at the local sites. The combination based on NIR with other modalities also embarks a static/dynamic combo of molecules and functional imaging in vivo and comprehensive insight of human bodies.

11.6 Summary

Individual combinations of NIR and traditional imaging techniques inherit distinct compensating fragments; for example, the combination of NIR to MRI yields a co-registration of 3D functional or molecular images to a 3D anatomical map, promote a co-comprehensive insight of molecular role on the structure distribution. NIR/PA images are beneficial in coincident detectable signals to obtain the images of dynamic targets at the same scale and spatial resolution. In X-ray CT imaging, the structural distribution in hard tissue is possible to visualize because of the abundance of electron density. An additional NIR technique permits the enhancement of the soft-tissue distribution with high sensitivity. NIR/PET/SPECT provides sensitive biochemical reactions in deep tissue with high temporal-spatial resolution. NIR fluorescent imaging succeeds preeminent characteristics of traditional fluorescent imaging techniques such as high temporal-spatial resolution, excellent sensitivity together with high signal-to-noise ratios in deep tissue when the longer wavelength over 1000 nm is used. Fluorescent molecular probes are highly sensitive to the change of physiological and biochemical environments in the bodies. Employing NIR light paves the ways for various process sensing in vivo. The materials used for NIR based multimodal are also diverse from rare earth doped ceramic NPs, hybrid nanostructures, or incorporation of organic molecules. The multimodal based on NIR light is useful for medical and therapy practice, such as imaging guiding surgery, regenerative medicine, and cancer treatment aid.

References

1. J. Chul Paeng, D. Soo Lee, Multimodal molecular imaging in vivo. Open Nucl. Med. J. **2**, 145 (2010). https://doi.org/10.2174/1876388x01002010145

2. P. Danhier, G. De Preter, J. Magat et al., Multimodal cell tracking of a spontaneous metastasis model: Comparison between MRI, electron paramagnetic resonance and bioluminescence. Contrast Media Mol. Imaging **9**, 143 (2014). https://doi.org/10.1002/cmmi.1553
3. M. Moseley, G. Donnan, Multimodality imaging: Introduction. Stroke **35**, 2632 (2004). https://doi.org/10.1161/01.STR.0000143214.22567.cb
4. F. Kiessling, Z. Liu, J. Gätjens, Advanced nanomaterials in multimodal imaging: design, functionalization, and biomedical applications. J. Nanomater. **2010**, 1 (2010). https://doi.org/10.1155/2010/894303
5. X. Li, X.N. Zhang, X.D. Li et al., Multimodality imaging in nanomedicine and nanotheranostics. Cancer Biol. Med. **13**, 339 (2016). https://doi.org/10.20892/j.issn.2095-3941.2016.0055
6. P. Wang, T. Kim, M. Harada et al., Nano-immunoimaging. Nanoscale Horiz **5**, 628 (2020). https://doi.org/10.1039/c9nh00514e
7. A. Louie, Multimodality imaging probes: Design and challenges. Chem. Rev. **110**, 3146 (2010). https://doi.org/10.1021/cr9003538
8. L. Frullano, T.J. Meade, Multimodal MRI contrast agents. J. Biol. Inorg. Chem. **12**, 939 (2007). https://doi.org/10.1007/s00775-007-0265-3
9. D.E. Lee, H. Koo, I.C. Sun et al., Multifunctional nanoparticles for multimodal imaging and theragnosis. Chem. Soc. Rev. **41**, 2656 (2012). https://doi.org/10.1039/c2cs15261d
10. R. Thomas, I.K. Park, Y.Y. Jeong, Magnetic iron oxide nanoparticles for multimodal imaging and therapy of cancer. Int. J. Mol. Sci. **14**, 15910 (2013). https://doi.org/10.3390/ijms140815910
11. G. Gebhart, P. Flamen, E.G.E. De Vries et al., Imaging diagnostic and therapeutic targets: Human epidermal growth factor receptor 2. J. Nucl. Med. **57**, 81S (2016). https://doi.org/10.2967/jnumed.115.157941
12. C. Kim, Z. Chen, Multimodal photoacoustic imaging: systems, applications, and agents. Biomed. Eng. Lett. **8**, 137 (2018). https://doi.org/10.1007/s13534-018-0071-6
13. L. Deng, X. Cai, D. Sheng, et al., A laser-activated biocompatible theranostic nanoagent for targeted multimodal imaging and photothermal therapy. Theranostics **7** (2017). https://doi.org/10.7150/thno.21283
14. S. Biju, J. Gallo, M. Bañobre-López et al., A magnetic chameleon: biocompatible lanthanide fluoride nanoparticles with magnetic field dependent tunable contrast properties as a versatile contrast agent for low to ultrahigh field MRI and optical imaging in biological window. Chem. A Eur. J. **24**, 7388 (2018). https://doi.org/10.1002/chem.201800283
15. R. Yan, Y. Hu, F. Liu et al., Activatable NIR fluorescence/MRI bimodal probes for in vivo imaging by enzyme-mediated fluorogenic reaction and self-assembly. J. Am. Chem. Soc. **141**, 10331 (2019). https://doi.org/10.1021/jacs.9b03649
16. R. Lv, P. Yang, F. He et al., A yolk-like multifunctional platform for multimodal imaging and synergistic therapy triggered by a single near-infrared light. ACS Nano **9**, 1630 (2015). https://doi.org/10.1021/nn5063613
17. A.V.F. Massicano, B.V. Marquez-Nostra, S.E. Lapi, Targeting HER2 in nuclear medicine for imaging and therapy. Mol. Imaging **17**, 1 (2018). https://doi.org/10.1177/1536012117745386
18. S. Mattsson, M. Söderberg, Radiation dose management in CT, SPECT/CT and PET/CT techniques. Radiat. Prot. Dosimetry **147**, 13 (2011). https://doi.org/10.1093/rpd/ncr261
19. K. Rajamanickam, Multimodal molecular imaging strategies using functionalized nano probes. J. Nanotechnol. Res. **01**, 119 (2019). https://doi.org/10.26502/jnr.2688-85210010
20. P.J. Slomka, R.P. Baum, Multimodality image registration with software: State-of-the-art. Eur. J. Nucl. Med. Mol. Imaging **36**, 44 (2009). https://doi.org/10.1007/s00259-008-0941-8
21. D. Chen, C.A. Dougherty, D. Yang et al., Radioactive nanomaterials for multimodality imaging. Tomography **2**, 3 (2016). https://doi.org/10.18383/j.tom.2016.00121
22. Y. Guo, J. Suri, R. Sivaramakrishna, Image registration for breast imaging: A review. Annu. Int. Conf. IEEE Eng. Med. Biol. Proc. **7 VOLS**, 3379 (2005). https://doi.org/10.1109/iembs.2005.1617202

23. H.A. Mohammed, M.A. Hassan, The image registration techniques for medical imaging (MRI-CT). Am. J. Biomed. Eng. **6**, 53 (2016). https://doi.org/10.5923/j.ajbe.20160602.02
24. P.K. Nguyen, J.C. Wu, Radiation exposure from imaging tests: Is there an increased cancer risk? Expert Rev. Cardiovasc. Ther. **9**, 177 (2011). https://doi.org/10.1586/erc.10.184
25. F. Leblond, S.C. Davis, P.A. Valdés et al., Pre-clinical whole-body fluorescence imaging: Review of instruments, methods and applications. J. Photochem. Photobiol. B Biol. **98**, 77 (2010). https://doi.org/10.1016/j.jphotobiol.2009.11.007
26. P.P. Ghoroghchian, M.J. Therien, D.A. Hammer, In vivo fluorescence imaging: A personal perspective. Wiley Interdiscip. Rev. Nanomed Nanobiotechnol **1**, 156 (2009). https://doi.org/10.1002/wnan.7
27. P.P. Laissue, R.A. Alghamdi, P. Tomancak et al., Assessing phototoxicity in live fluorescence imaging. Nat. Meth. **14**, 657 (2017). https://doi.org/10.1038/nmeth.4344
28. J. Icha, M. Weber, J.C. Waters et al., Phototoxicity in live fluorescence microscopy, and how to avoid it. BioEssays **39**, 1 (2017). https://doi.org/10.1002/bies.201700003
29. A.M. Smith, M.C. Mancini, S. Nie, Bioimaging: Second window for in vivo imaging. Nat. Nanotechnol. **4**, 710 (2009). https://doi.org/10.1038/nnano.2009.326
30. K. Welsher, S.P. Sherlock, H. Dai, Deep-tissue anatomical imaging of mice using carbon nanotube fluorophores in the second near-infrared window. Proc. Natl. Acad. Sci. U. S. A. **108**, 8943 (2011). https://doi.org/10.1073/pnas.1014501108
31. S. Fukushima, T. Furukawa, H. Niioka et al., Correlative near-infrared light and cathodoluminescence microscopy using Y_2O_3:Ln, Yb (Ln = Tm, Er) nanophosphors for multiscale, multicolour bioimaging. Sci. Rep. **6**, 1 (2016). https://doi.org/10.1038/srep25950
32. K. Soga, M. Kamimura, K. Okubo, et al., Near-infrared biomedical imaging for transparency. J. Imaging Soc. Japan **58**, 602 (2019). https://doi.org/10.11370/isj.58.602
33. E. Hemmer, A. Benayas, F. Légaré et al., Exploiting the biological windows: Current perspectives on fluorescent bioprobes emitting above 1000 nm. Nanoscale Horiz. **1**, 168 (2016). https://doi.org/10.1039/c5nh00073d
34. T.A. Henderson, L.D. Morries, Near-infrared photonic energy penetration: Can infrared phototherapy effectively reach the human brain? Neuropsychiatr. Dis. Treat. **11**, 2191 (2015). https://doi.org/10.2147/NDT.S78182
35. L.A. Sordillo, Y. Pu, S. Pratavieira et al., Deep optical imaging of tissue using the second and third near-infrared spectral windows. J. Biomed. Opt. **19**, 056004 (2014). https://doi.org/10.1117/1.jbo.19.5.056004
36. S. Sekiyama, M. Umezawa, S. Kuraoka et al., Temperature sensing of deep abdominal region in mice by using over-1000 nm near-infrared luminescence of rare-earth-doped NaYF4 nanothermometer. Sci. Rep. **8**, 1 (2018). https://doi.org/10.1038/s41598-018-35354-y
37. P. Armstrong, S.F. Keevil, Magnetic resonance imaging—1: Basic principles of image production. Br. Med. J. **303**, 35 (1991). https://doi.org/10.1136/bmj.303.6793.35
38. R.B. Lauffer, Paramagnetic metal complexes as water proton relaxation agents for NMR imaging: Theory and design. Chem. Rev. **87**, 901 (1987). https://doi.org/10.1021/cr00081a003
39. J. Lohrke, T. Frenzel, J. Endrikat et al., 25 years of contrast-enhanced MRI: Developments, current challenges and future perspectives. Adv. Ther. **33**, 1 (2016). https://doi.org/10.1007/s12325-015-0275-4
40. L. Vander Elst, F. Chapelle, S. Laurent, et al., Stereospecific binding of MRI contrast agents to human serum albumin: The case of Gd-(S)-EOB-DTPA (Eovist) and its (R) isomer, J. Biol. Inorg. Chem. **6**, 196 (2001). https://doi.org/10.1007/s007750000195
41. C.F.G.C. Geraldes, S. Laurent, Classification and basic properties of contrast agents for magnetic resonance imaging. Contrast Media Mol. Imaging **4**, 1 (2009). https://doi.org/10.1002/cmmi.265
42. S. Lee, R. George Thomas, M. Ju Moon, et al., Near-infrared heptamethine cyanine based iron oxide nanoparticles for tumor targeted multimodal imaging and photothermal therapy. Sci. Rep. **7**, 1 (2017). https://doi.org/10.1038/s41598-017-01108-5
43. C. Shen, X. Wang, Z. Zheng et al., Doxorubicin and indocyanine green loaded superparamagnetic iron oxide nanoparticles with PEGylated phospholipid coating for magnetic resonance

with fluorescence imaging and chemotherapy of glioma. Int. J. Nanomed. **14**, 101 (2019). https://doi.org/10.2147/IJN.S173954
44. W.C. Law, Z. Xu, K.T. Yong et al., Manganese-doped near-infrared emitting nanocrystals for in vivo biomedical imaging. Opt. Express **24**, 17553 (2016). https://doi.org/10.1364/oe.24.017553
45. X. Wang, H. Hu, H. Zhang et al., Single ultrasmall Mn2+ -doped NaNdF4 nanocrystals as multimodal nanoprobes for magnetic resonance and second near-infrared fluorescence imaging. Nano Res. **11**, 1069 (2018). https://doi.org/10.1007/s12274-017-1727-8
46. A. Abdukayum, C.X. Yang, Q. Zhao et al., Gadolinium complexes functionalized persistent luminescent nanoparticles as a multimodal probe for near-infrared luminescence and magnetic resonance imaging in vivo. Anal. Chem. **86**, 4096 (2014). https://doi.org/10.1021/ac500644x
47. R. Bardhan, W. Chen, M. Bartels et al., Tracking of multimodal therapeutic nanocomplexes targeting breast cancer in vivo. Nano Lett. **10**, 4920 (2010). https://doi.org/10.1021/nl102889y
48. V.S.R. Harrison, C.E. Carney, K.W. MacRenaris et al., Multimeric near IR-MR contrast agent for multimodal in vivo imaging. J. Am. Chem. Soc. **137**, 9108 (2015). https://doi.org/10.1021/jacs.5b04509
49. C.T. Yang, K.K. Ghosh, P. Padmanabhan et al., PET-MR and SPECT-MR multimodality probes: Development and challenges. Theranostics **8**, 6210 (2018). https://doi.org/10.7150/thno.26610
50. A. Zlitni, S.S. Gambhir, Molecular imaging agents for ultrasound. Curr. Opin. Chem. Biol. (2018). https://doi.org/10.1016/j.cbpa.2018.03.017
51. K.E. Wilson, T.Y. Wang, J.K. Willmann, Acoustic and photoacoustic molecular imaging of cancer. J. Nucl. Med. **54**, 1851 (2013). https://doi.org/10.2967/jnumed.112.115568
52. T.M. Robinson, Basic principles of ultrasound, in *Physics for Medical Imaging Applications*, ed. by Y. Lemoigne, A. Caner, G. Rahal (NATO Science, 240) (Springer, Dordrecht, 2007). https://doi.org/10.1007/978-1-4020-5653-6_9
53. R.W. Cootney, Ultrasound imaging: Principles and applications in rodent research. ILAR J. **42**, 233 (2001). https://doi.org/10.1093/ilar.42.3.233
54. L. Abou-Elkacem, S.V. Bachawal, J.K. Willmann, Ultrasound molecular imaging: Moving toward clinical translation. Eur. J. Radiol. **84**, 1685 (2015). https://doi.org/10.1016/j.ejrad.2015.03.016
55. R. Cavalli, M. Argenziano, E. Vigna et al., Preparation and in vitro characterization of chitosan nanobubbles as theranostic agents. Colloids Surf. B Biointerfaces **129**, 39 (2015). https://doi.org/10.1016/j.colsurfb.2015.03.023
56. H.Y. Huang, H.L. Liu, P.H. Hsu et al., A multitheragnostic nanobubble system to induce blood-brain barrier disruption with magnetically guided focused ultrasound. Adv. Mater. **27**, 655 (2015). https://doi.org/10.1002/adma.201403889
57. C. Sciallero, L. Balbi, G. Paradossi et al., Magnetic resonance and ultrasound contrast imaging of polymer-shelled microbubbles loaded with iron oxide nanoparticles. R. Soc. Open Sci. **3**, 160063 (2016). https://doi.org/10.1098/rsos.160063
58. X. Zhang, R. Liu, Z. Dai, Multicolor nanobubbles for FRET/ultrasound dual-modal contrast imaging. Nanoscale **10**, 20347 (2018). https://doi.org/10.1039/c8nr05488f
59. E. Huynh, M.A. Rajora, G. Zheng, Multimodal micro, nano, and size conversion ultrasound agents for imaging and therapy. Wiley Interdiscip Rev Nanomed Nanobiotechnol **8**, 796 (2016). https://doi.org/10.1002/wnan.1398
60. J. Yao, When pressure meets light: Detecting the photoacoustic effect at the origin, Light Sci. Appl. **6** (2017). https://doi.org/10.1038/lsa.2017.62
61. A.G. Bell, Upon the production and reproduction of sound by light. J. Soc. Telegr. Eng. **9**, 404 (1880). https://doi.org/10.1049/jste-1.1880.0046
62. G.D. Ludwig, The velocity of sound through tissues and the acoustic impedance of tissues. J. Acoust. Soc. Am. **22**, 862 (1950). https://doi.org/10.1121/1.1906706
63. J. Märk, H. Dortay, A. Wagener et al., Dual-wavelength 3D photoacoustic imaging of mammalian cells using a photoswitchable phytochrome reporter protein. Commun. Phys. **1**, 3 (2018). https://doi.org/10.1038/s42005-017-0003-2

64. P.K. Upputuri, M. Pramanik, Photoacoustic imaging in the second near-infrared window: a review. J. Biomed. Opt. **24**, 1 (2019). https://doi.org/10.1117/1.jbo.24.4.040901
65. M. Li, Y. Tang, J. Yao, Photoacoustic tomography of blood oxygenation: A mini review. Photoacoustics **10**, 65 (2018). https://doi.org/10.1016/j.pacs.2018.05.001
66. A.B.E. Attia, G. Balasundaram, M. Moothanchery et al., A review of clinical photoacoustic imaging: Current and future trends. Photoacoustics **16**, 1 (2019). https://doi.org/10.1016/j.pacs.2019.100144
67. I. Stoffels, S. Morscher, I. Helfrich et al., Metastatic status of sentinel lymph nodes in melanoma determined noninvasively with multispectral optoacoustic imaging. Sci. Transl. Med. **7**, 319 (2015). https://doi.org/10.1126/scitranslmed.aad1278
68. J. Weber, P.C. Beard, S.E. Bohndiek, Contrast agents for molecular photoacoustic imaging. Nat. Methods **13**, 639 (2016). https://doi.org/10.1038/nmeth.3929
69. L.V. Wang, Photoacoustic tomography: Deep tissue imaging by ultrasonically beating optical diffusion, in *Make Life Visible*, ed. by Y. Toyama, A. Miyawaki, M. Nakamura, M. Jinzaki (Springer, Singapore, 2020)
70. K. Huang, Y. Zhang, J. Lin et al., Nanomaterials for photoacoustic imaging in the second near-infrared window. Biomater. Sci. **7**, 472 (2019). https://doi.org/10.1039/c8bm00642c
71. K. Cheng, H. Chen, C.H. Jenkins et al., Synthesis, characterization, and biomedical applications of a targeted dual-modal near-infrared-II fluorescence and photoacoustic imaging nanoprobe. ACS Nano **11**, 12276 (2017). https://doi.org/10.1021/acsnano.7b05966
72. L.W. Goldman, Principles of CT and CT technology. J. Nucl. Med. Technol. **35**, 115 (2007). https://doi.org/10.2967/jnmt.107.042978
73. T. Kim, N. Lee, D.R. Arifin et al., In vivo micro-CT imaging of human mesenchymal stem cells labeled with gold-poly-l-lysine nanocomplexes. Adv. Funct. Mater. **27**, 1604213 (2017). https://doi.org/10.1002/adfm.201604213
74. J.F. Hainfeld, D.N. Slatkin, T.M. Focella et al., Gold nanoparticles: A new X-ray contrast agent. Br. J. Radiol. **79**, 248 (2006). https://doi.org/10.1259/bjr/13169882
75. O. Rabin, J.M. Perez, J. Grimm et al., An X-ray computed tomography imaging agent based on long-circulating bismuth sulphide nanoparticles. Nat. Mater. **5**, 118 (2006). https://doi.org/10.1038/nmat1571
76. S.R. Cherry, Multimodality imaging: Beyond PET/CT and SPECT/CT. Semin. Nucl. Med. **39**, 348 (2009). https://doi.org/10.1053/j.semnuclmed.2009.03.001
77. D. Vo, P.K. Nguyen, Multimodality molecular imaging in cardiac regenerative therapy. J. Nucl. Cardiol. **24**, 1803 (2017). https://doi.org/10.1007/s12350-017-0785-3
78. T. Shuai, Y. Zhou, G. Shao et al., Bimodal molecule as NIR-CT contrast agent for hepatoma specific imaging. Anal. Chem. **92**, 1138 (2019). https://doi.org/10.1021/acs.analchem.9b04212
79. H. Xu, T.Y. Ohulchanskyy, A. Yakovliev et al., Nanoliposomes co-encapsulating CT imaging contrast agent and photosensitizer for enhanced, imaging guided photodynamic therapy of cancer. Theranostics **9**, 1323 (2019). https://doi.org/10.7150/thno.31079
80. Y. Dai, D. Yang, D. Yu et al., Mussel-inspired polydopamine-coated lanthanide nanoparticles for NIR-II/CT dual imaging and photothermal therapy. ACS Appl. Mater. Interfaces **9**, 26674 (2017). https://doi.org/10.1021/acsami.7b06109
81. A. Rahmim, M.A. Lodge, N.A. Karakatsanis et al., Dynamic whole-body PET imaging: Principles, potentials and applications. Eur. J. Nucl. Med. Mol. Imaging **46**, 501 (2019). https://doi.org/10.1007/s00259-018-4153-6
82. M. Conti, L. Eriksson, Physics of pure and non-pure positron emitters for PET: A review and a discussion. EJNMMI Phys. **3**, 8 (2016). https://doi.org/10.1186/s40658-016-0144-5
83. J. Marik, M.S. Tartis, H. Zhang et al., Long-circulating liposomes radiolabeled with [18F]fluorodipalmitin ([18F]FDP). Nucl. Med. Biol. **34**, 165 (2007). https://doi.org/10.1016/j.nucmedbio.2006.12.004
84. E. Andreozzi, J.W. Seo, K. Ferrara et al., Novel method to label solid lipid nanoparticles with 64Cu for positron emission tomography imaging. Bioconjug. Chem. **22**, 808 (2011). https://doi.org/10.1021/bc100478k

85. J. Xie, K. Chen, J. Huang et al., PET/NIRF/MRI triple functional iron oxide nanoparticles. Biomaterials **31**, 3016 (2010). https://doi.org/10.1016/j.biomaterials.2010.01.010
86. Y. Sun, X. Zeng, Y. Xiao et al., Novel dual-function near-infrared II fluorescence and PET probe for tumor delineation and image-guided surgery. Chem. Sci. **9**, 2092 (2018). https://doi.org/10.1039/c7sc04774f
87. A.K. Buck, S. Nekolla, S. Ziegler et al., SPECT/CT. J. Nucl. Med. **49**, 1305 (2008). https://doi.org/10.2967/jnumed.107.050195
88. B.F. Hutton, The origins of SPECT and SPECT/CT. Eur. J. Nucl. Med. Mol. Imaging **41**, 3 (2014). https://doi.org/10.1007/s00259-013-2606-5
89. S.S. Kelkar, T.M. Reineke, Theranostics: Combining imaging and therapy. Bioconjug. Chem. **22**, 1879 (2011). https://doi.org/10.1021/bc200151q
90. J. Lamb, J.P. Holland, Advanced methods for radiolabeling multimodality nanomedicines for SPECT/MRI and PET/MRI. J. Nucl. Med. **59**, 382 (2018). https://doi.org/10.2967/jnumed.116.187419
91. B.P. Burke, C. Cawthorne, S.J. Archibald, Multimodal nanoparticle imaging agents: Design and applications, Philos. Trans. R. Soc. A Math. Phys. Eng. Sci. **375** (2017). https://doi.org/10.1098/rsta.2017.0261
92. M. Kamimura, R. Saito, H. Hyodo et al., Over-1000 nm near-infrared fluorescence and spect/ct dual-modal in vivo imaging based on rare-earth doped ceramic nanophosphors. J. Photopolym. Sci. Technol. **29**, 525 (2016). https://doi.org/10.2494/photopolymer.29.525
93. T.H. Nguyen Pham, N.A. Lengkeek, I. Greguric, et al., Tunable and noncytotoxic PET/SPECT-MRI multimodality imaging probes using colloidally stable ligand-free superparamagnetic iron oxide nanoparticles, Int. J. Nanomed. **12**, 899 (2017). https://doi.org/10.2147/ijn.s127171
94. D. Kryza, J. Taleb, M. Janier et al., Biodistribution study of nanometric hybrid gadolinium oxide particles as a multimodal SPECT/MR/optical imaging and theragnostic agent. Bioconjug. Chem. **22**, 1145 (2011). https://doi.org/10.1021/bc1005976
95. L. Frullano, C. Catana, T. Benner et al., Bimodal MR-PET agent for quantitative pH imaging. Angew. Chemie Int. Ed. **49**, 2382 (2010). https://doi.org/10.1002/anie.201000075
96. L. Farzin, S. Sheibani, M.E. Moassesi et al., An overview of nanoscale radionuclides and radiolabeled nanomaterials commonly used for nuclear molecular imaging and therapeutic functions. J. Biomed. Mater. Res. A. **107**, 251 (2019). https://doi.org/10.1002/jbm.a.36550
97. I.E.G. Morrison, A. Samilian, P. Coppo et al., Multicolour correlative imaging using phosphor probes. J. Chem. Biol. **8**, 169 (2015). https://doi.org/10.1007/s12154-015-0141-5
98. D. Glenn, H. Zhang, N. Kasthuri et al., Correlative light and electron microscopy using cathodoluminescence from nanoparticles with distinguishable colours. Sci. Rep. **2**, 865 (2012). https://doi.org/10.1038/srep00865
99. T. Furukawa, H. Niioka, M. Ichimiya et al., High-resolution microscopy for biological specimens via cathodoluminescence of Eu- and Zn-doped Y_2O_3nanophosphors. Opt. Express **21**, 25655 (2013). https://doi.org/10.1364/oe.21.025655
100. K. Miki, K. Kojima, K. Oride et al., pH-Responsive near-infrared fluorescent cyanine dyes for molecular imaging based on pH sensing. Chem. Commun. **53**, 7792 (2017). https://doi.org/10.1039/c7cc03035e
101. D. Aigner, B. Ungerböck, T. Mayr et al., Fluorescent materials for pH sensing and imaging based on novel 1,4-diketopyrrolo-[3,4-c]pyrrole dyes. J. Mater. Chem. C **1**, 5685 (2013). https://doi.org/10.1039/c3tc31130a
102. G. Despras, A.I. Zamaleeva, L. Dardevet et al., H-Rubies, a new family of red emitting fluorescent pH sensors for living cells. Chem. Sci. **6**, 5928 (2015). https://doi.org/10.1039/c5sc01113b
103. L. Salmon, G. Molnár, D. Zitouni et al., A novel approach for fluorescent thermometry and thermal imaging purposes using spin crossover nanoparticles. J. Mater. Chem. **20**, 5499 (2010). https://doi.org/10.1039/c0jm00631a
104. Z. Yang, J. Cao, Y. He et al., Macro-/micro-environment-sensitive chemosensing and biological imaging. Chem. Soc. Rev. **43**, 4563 (2014). https://doi.org/10.1039/c4cs00051j

105. S. Wang, Y. Fan, D. Li et al., Anti-quenching NIR-II molecular fluorophores for in vivo high-contrast imaging and pH sensing. Nat. Commun. **10**, 1058 (2019). https://doi.org/10.1038/s41467-019-09043-x
106. C. Chen, R. Tian, Y. Zeng et al., Activatable fluorescence probes for "turn-On" and ratiometric biosensing and bioimaging: From NIR-I to NIR-II. Bioconjug. Chem. **31**, 276 (2020). https://doi.org/10.1021/acs.bioconjchem.9b00734
107. M.Y. Berezin, K. Guo, W. Akers et al., Near-infrared fluorescence lifetime pH-sensitive probes. Biophys. J. **100**, 2063 (2011). https://doi.org/10.1016/j.bpj.2011.02.050
108. J. Zhao, Y. Li, M. Yu et al., Time-resolved activation of pH sensing and imaging in vivo by a remotely controllable DNA nanomachine. Nano Lett. **20**, 874 (2020). https://doi.org/10.1021/acs.nanolett.9b03471
109. D. Li, S. Wang, Z. Lei et al., Peroxynitrite activatable NIR-II fluorescent molecular probe for drug-induced hepatotoxicity monitoring. Anal. Chem. **91**, 4771 (2019). https://doi.org/10.1021/acs.analchem.9b00317
110. J. Zhang, Z. Li, X. Tian et al., A novel hydrosoluble near-infrared fluorescent probe for specifically monitoring tyrosinase and application in a mouse model. Chem. Commun. **55**, 9463 (2019). https://doi.org/10.1039/c9cc04714j
111. R. Guo, Y. Ma, Y. Tang et al., A novel mitochondria-targeted near-infrared (NIR) probe for detection of viscosity changes in living cell, zebra fishes and living mice. Talanta **204**, 868 (2019). https://doi.org/10.1016/j.talanta.2019.06.050
112. T. Chen, Z. Chen, R. Liu et al., A NIR fluorescent probe for detection of viscosity and lysosome imaging in live cells. Org. Biomol. Chem. **17**, 6398 (2019). https://doi.org/10.1039/c9ob01222b
113. K. Dou, W. Huang, Y. Xiang et al., Design of activatable NIR-II molecular probe for in vivo elucidation of disease-related viscosity variations. Anal. Chem. **92**, 4177 (2020). https://doi.org/10.1021/acs.analchem.0c00634
114. T.K.D. Doan, M. Umezawa, K. Nigoghossian et al., Development of Molecular Imaging Probe for Dual NIR/MR Imaging. J. Photopolym. Sci. Technol. **33**, 117 (2020). https://doi.org/10.2494/photopolymer.33.117
115. P. Caravan, A.V. Astashkin, A.M. Raitsimring, The gadolinium(III)-water hydrogen distance in MRI contrast agents. Inorg. Chem. **42**, 3972 (2003). https://doi.org/10.1021/ic034414f
116. W. Xu, K. Kattel, J.Y. Park et al., Paramagnetic nanoparticle T_1 and T_2 MRI contrast agents. Phys. Chem. Chem. Phys. **14**, 12687 (2012). https://doi.org/10.1039/c2cp41357d
117. T.K.D. Doan, S. Fukushima, T. Furukawa et al., Multispectral emissions of lanthanide-doped gadolinium oxide nanophosphors for cathodoluminescence and near-infrared upconversion/downconversion imaging. Nanomaterials **6**, 163 (2016). https://doi.org/10.3390/nano6090163
118. G. Chen, T.Y. Ohulchanskyy, S. Liu et al., Core/shell NaGdF 4: Nd 3+/NaGdF 4 nanocrystals with efficient near-infrared to near-infrared downconversion photoluminescence for bioimaging applications. ACS Nano **6**, 2969 (2012). https://doi.org/10.1021/nn2042362
119. N.J.J. Johnson, W. Oakden, G.J. Stanisz et al., Size-tunable, ultrasmall NaGdF4 nanoparticles: Insights into their T_1 MRI contrast enhancement. Chem. Mater. **23**, 3714 (2011). https://doi.org/10.1021/cm201297x
120. J. Jung, M.A. Kim, J.H. Cho et al., Europium-doped gadolinium sulfide nanoparticles as a dual-mode imaging agent for T 1-weighted MR and photoluminescence imaging. Biomaterials **33**, 5865 (2012). https://doi.org/10.1016/j.biomaterials.2012.04.059
121. G. Zhu, L. Chen, F. Zeng et al., GdVO4:Eu3+, Bi3+ nanoparticles as a contrast agent for MRI and luminescence bioimaging. ACS Omega **4**, 15806 (2019). https://doi.org/10.1021/acsomega.9b00444
122. P. Padhye, A. Alam, S. Ghorai et al., Doxorubicin-conjugated β-NaYF4:Gd^{3+}/Tb^{3+} multifunctional, phosphor nanorods: A multi-modal, luminescent, magnetic probe for simultaneous optical and magnetic resonance imaging and an excellent pH-triggered anti-cancer drug delivery nanovehicle. Nanoscale **7**, 19501 (2015). https://doi.org/10.1039/c5nr04473a

123. J. Zhou, Y. Sun, X. Du et al., Dual-modality in vivo imaging using rare-earth nanocrystals with near-infrared to near-infrared (NIR-to-NIR) upconversion luminescence and magnetic resonance properties. Biomaterials **31**, 3287 (2010). https://doi.org/10.1016/j.biomaterials.2010.01.040
124. N. Liu, R. Marin, Y. Mazouzi et al., Cubic: Versus hexagonal-effect of host crystallinity on the T 1 shortening behaviour of NaGdF 4 nanoparticles. Nanoscale **11**, 6794 (2019). https://doi.org/10.1039/c9nr00241c
125. N.D. Kim, H.J. Yun, I. Kyu Song, et al., Preparation and characterization of nanostructured Mn oxide by an ethanol-based precipitation method for pseudocapacitor applications, Scr. Mater. **65**, 448 (2011). https://doi.org/10.1016/j.scriptamat.2011.05.034
126. R. An, P. Lei, P. Zhang, et al., Near-infrared optical and X-ray computed tomography dual-modal imaging probe based on novel lanthanide-doped $K_{0.3}Bi_{0.7}F_{2.4}$ upconversion nanoparticles. Nanoscale **10**, 1394 (2018). https://doi.org/10.1039/c7nr06758e
127. F. Ren, L. Ding, H. Liu et al., Ultra-small nanocluster mediated synthesis of Nd3+ -doped core-shell nanocrystals with emission in the second near-infrared window for multimodal imaging of tumor vasculature. Biomaterials **175**, 30 (2018). https://doi.org/10.1016/j.biomaterials.2018.05.021
128. Y. Sheng, L. De Liao, N. Thakor et al., Rare-earth doped particles as dual-modality contrast agent for minimally-invasive luminescence and dual-wavelength photoacoustic imaging. Sci. Rep. **4**, 6562 (2014). https://doi.org/10.1038/srep06562
129. Y. Sun, X. Zhu, J. Peng et al., Core-shell lanthanide upconversion nanophosphors as four-modal probes for tumor angiogenesis imaging. ACS Nano **7**, 11290 (2013). https://doi.org/10.1021/nn405082y
130. X. Hu, Y. Tang, Y. Hu et al., Gadolinium-chelated conjugated polymer-based nanotheranostics for photoacoustic/magnetic resonance/NIR-II fluorescence imaging-guided cancer photothermal therapy. Theranostics **9**, 4168 (2019). https://doi.org/10.7150/thno.34390
131. J.J. Ma, M.X. Yu, Z. Zhang et al., Gd-DTPA-coupled Ag2Se quantum dots for dual-modality magnetic resonance imaging and fluorescence imaging in the second near-infrared window. Nanoscale **10**, 10699 (2018). https://doi.org/10.1039/c8nr02017e
132. L. Liu, J. Wang, Q. You et al., NIRF/PA/CT multi-modality imaging guided combined photothermal and photodynamic therapy based on tumor microenvironment-responsive nanocomposites. J. Mater. Chem. B **6**, 4239 (2018). https://doi.org/10.1039/c8tb00859k
133. H. Huang, M. Dunne, J. Lo et al., Comparison of computed tomography-and optical image-based assessment of liposome distribution. Mol. Imaging **12**, 148 (2013). https://doi.org/10.2310/7290.2012.00028
134. Y. Liu, G.F. Tian, X.W. He et al., Microwave-assisted one-step rapid synthesis of near-infrared gold nanoclusters for NIRF/CT dual-modal bioimaging. J. Mater. Chem. B **4**, 1276 (2016). https://doi.org/10.1039/c5tb02322j
135. J. Chen, X.Q. Yang, Y.Z. Meng et al., Reverse microemulsion-mediated synthesis of Bi_2S_3–QD@SiO_2–PEG for dual modal CT–fluorescence imaging in vitro and in vivo. Chem. Commun. **49**, 11800 (2013). https://doi.org/10.1039/c3cc47710j
136. M.Y. Qin, X.Q. Yang, K. Wang et al., In vivo cancer targeting and fluorescence-CT dual-mode imaging with nanoprobes based on silver sulfide quantum dots and iodinated oil. Nanoscale **7**, 19484 (2015). https://doi.org/10.1039/c5nr05620a
137. S.K. Mishra, S. Kannan, A bimetallic silver-neodymium theranostic nanoparticle with multimodal NIR/MRI/CT imaging and combined chemo-photothermal therapy. Inorg. Chem. **56**, 12054 (2017). https://doi.org/10.1021/acs.inorgchem.7b02103
138. M.L. Schipper, G. Iyer, A.L. Koh et al., Particle size, surface coating, and PEGylation influence the biodistribution of quantum dots in living mice. Small **5**, 126 (2009). https://doi.org/10.1002/smll.200800003
139. W. Cai, K. Chen, Z.B. Li et al., Dual-function probe for PET and near-infrared fluorescence imaging of tumor vasculature. J. Nucl. Med. **48**, 1862 (2007). https://doi.org/10.2967/jnumed.107.043216

140. K. Chen, Z.B. Li, H. Wang et al., Dual-modality optical and positron emission tomography imaging of vascular endothelial growth factor receptor on tumor vasculature using quantum dots. Eur. J. Nucl. Med. Mol. Imaging **35**, 2235 (2008). https://doi.org/10.1007/s00259-008-0860-8
141. A. Miyata, T. Ishizawa, M. Kamiya et al., Photoacoustic tomography of human hepatic malignancies using Intraoperative indocyanine green fluorescence imaging. PLoS ONE **9**, 112667 (2014). https://doi.org/10.1371/journal.pone.0112667
142. T. Guan, W. Shang, H. Li et al., From detection to resection: Photoacoustic tomography and surgery guidance with indocyanine green loaded gold nanorod@liposome core-shell nanoparticles in liver cancer. Bioconjug. Chem. **28**, 1221 (2017). https://doi.org/10.1021/acs.bioconjchem.7b00065
143. R. Vakili-Ghartavol, A.A. Momtazi-Borojeni, Z. Vakili-Ghartavol et al., Toxicity assessment of superparamagnetic iron oxide nanoparticles in different tissues. Artif. Cells, Nanomed. Biotechnol. **48**, 443 (2020). https://doi.org/10.1080/21691401.2019.1709855
144. N. Singh, G.J.S. Jenkins, R. Asadi et al., Potential toxicity of superparamagnetic iron oxide nanoparticles (SPION). Nano Rev. **1**, 5358 (2010). https://doi.org/10.3402/nano.v1i0.5358
145. P. Møller, N.R. Jacobsen, J.K. Folkmann et al., Role of oxidative damage in toxicity of particulate. Free Radic. Res. **44**, 1 (2010). https://doi.org/10.3109/10715760903300691
146. P. Caravan, Protein-targeted gadolinium-based magnetic resonance imaging (MRI) contrast agents: Design and mechanism of action. Acc. Chem. Res. **42**, 851 (2009). https://doi.org/10.1021/ar800220p
147. T. Jin, Critical review—Recent progress in NIR fluorophores emitting over 1000 nm for bioimaging. ECS J. Solid State Sci. Technol. **8**, R9 (2019). https://doi.org/10.1149/2.0111901jss
148. M. Casalboni, F. De Matteis, P. Prosposito et al., Fluorescence efficiency of four infrared polymethine dyes. Chem. Phys. Lett. **373**, 372 (2003). https://doi.org/10.1016/S0009-2614(03)00608-0
149. Z. Tao, G. Hong, C. Shinji et al., Biological imaging using nanoparticles of small organic molecules with fluorescence emission at wavelengths longer than 1000 nm. Angew. Chemie Int. Ed. **52**, 13002 (2013). https://doi.org/10.1002/anie.201307346
150. M. Kamimura, S. Takahiro, M. Yoshida et al., Over-1000 nm near-infrared fluorescent biodegradable polymer nanoparticles for deep tissue in vivo imaging in the second biological window. Polym. J. **49**, 799 (2017). https://doi.org/10.1038/pj.2017.59
151. H. Zhou, H. Yang, L. Tang et al., Mn-loaded apolactoferrin dots for: In vivo MRI and NIR-II cancer imaging. J. Mater. Chem. C **7**, 9448 (2019). https://doi.org/10.1039/c9tc01929d
152. B. Wu, B. Wan, S.T. Lu et al., Near-infrared light-triggered theranostics for tumor-specific enhanced multimodal imaging and photothermal therapy. Int. J. Nanomed. **12**, 4467 (2017). https://doi.org/10.2147/IJN.S137835
153. N. Beziere, N. Lozano, A. Nunes et al., Dynamic imaging of PEGylated indocyanine green (ICG) liposomes within the tumor microenvironment using multi-spectral optoacoustic tomography (MSOT). Biomaterials **37**, 415 (2015). https://doi.org/10.1016/j.biomaterials.2014.10.014
154. M.A. Bruckman, K. Jiang, E.J. Simpson et al., Dual-modal magnetic resonance and fluorescence imaging of atherosclerotic plaques in vivo using VCAM-1 targeted tobacco mosaic virus. Nano Lett. **14**, 1551 (2014). https://doi.org/10.1021/nl404816m
155. R. O'Halloran, B.H. Kopell, E. Sprooten et al., Multimodal neuroimaging-informed clinical applications in neuropsychiatric disorders. Front. Psychiatry **7**, 63 (2016). https://doi.org/10.3389/fpsyt.2016.00063
156. H.U. Voss, L.A. Heier, N.D. Schiff, Multimodal imaging of recovery of functional networks associated with reversal of paradoxical herniation after cranioplasty. Clin. Imaging **35**, 253 (2011). https://doi.org/10.1016/j.clinimag.2010.07.008
157. G.M. van Dam, G. Themelis, L.M. Crane et al., Intraoperative tumor-specific fluorescence imaging in ovarian cancer by folate receptor-α targeting: First in-human results. Nat. Med. **17**, 1315 (2011). https://doi.org/10.1038/nm.2472

158. I.S. Alam, I. Steinberg, O. Vermesh et al., Emerging intraoperative imaging modalities to improve surgical precision. Mol. Imag. Biol. **20**, 705 (2018). https://doi.org/10.1007/s11307-018-1227-6
159. C. Chi, Y. Du, J. Ye et al., Intraoperative imaging-guided cancer surgery: From current fluorescence molecular imaging methods to future multi-modality imaging technology. Theranostics **4**, 1072 (2014). https://doi.org/10.7150/thno.9899
160. J. Zheng, N. Muhanna, R. De Souza et al., A multimodal nano agent for image-guided cancer surgery. Biomaterials **67**, 160 (2015). https://doi.org/10.1016/j.biomaterials.2015.07.010
161. C. Ewelt, F.W. Floeth, J. Felsberg et al., Finding the anaplastic focus in diffuse gliomas: The value of Gd-DTPA enhanced MRI, FET-PET, and intraoperative, ALA-derived tissue fluorescence. Clin. Neurol. Neurosurg. **113**, 541 (2011). https://doi.org/10.1016/j.clineuro.2011.03.008
162. M.K. Ravoori, S. Singh, R. Bhavane et al., Multimodal magnetic resonance and near-infrared-fluorescent imaging of intraperitoneal ovarian cancer using a dual-mode-dual-gadolinium liposomal contrast agent. Sci. Rep. **6**, 1 (2016). https://doi.org/10.1038/srep38991
163. C. Zeng, W. Shang, K. Wang et al., Intraoperative identification of liver cancer microfoci using a targeted near-infrared fluorescent probe for imaging-guided surgery. Sci. Rep. **6**, 1 (2016). https://doi.org/10.1038/srep21959
164. Y. Nakamura, A. Mochida, P.L. Choyke et al., Nanodrug delivery: Is the enhanced permeability and retention effect sufficient for curing cancer? Bioconjug. Chem. **27**, 2225 (2016). https://doi.org/10.1021/acs.bioconjchem.6b00437
165. H. Lee, K. Y. Mi, S. Park, et al., Thermally cross-linked superparamagnetic iron oxide nanoparticles: Synthesis and application as a dual imaging probe for cancer in vivo. J. Am. Chem. Soc. **129**, 12739 (2007). https://doi.org/10.1021/ja072210i
166. L. Xiang, D. Xing, H. Gu et al., Real-time optoacoustic monitoring of vascular damage during photodynamic therapy treatment of tumor. J. Biomed. Opt. **12**, 014001 (2007). https://doi.org/10.1117/1.2437752
167. Y. Zheng, J. Huang, T. Zhu et al., Stem cell tracking technologies for neurological regenerative medicine purposes. Stem Cells Int. **2017**, 2934149 (2017). https://doi.org/10.1155/2017/2934149
168. E. Napoli, C.V. Borlongan, recent advances in stem cell-based therapeutics for stroke. Transl. Stroke Res. **7**, 452 (2016). https://doi.org/10.1007/s12975-016-0490-6
169. F. Chao, Y. Shen, H. Zhang et al., Multimodality molecular imaging of stem cells therapy for stroke. Biomed. Res. Int. **2013**, 849819 (2013). https://doi.org/10.1155/2013/849819
170. N. Kubis, Y. Tomita, A. Tran-Dinh et al., Vascular fate of adipose tissue-derived adult stromal cells in the ischemic murine brain: A combined imaging-histological study. Neuroimage **34**, 1 (2007). https://doi.org/10.1016/j.neuroimage.2006.09.014
171. A. Momeni, S. Neelamegham, N. Parashurama, Current challenges for the targeted delivery and molecular imaging of stem cells in animal models. Bioengineered **8**, 316 (2017). https://doi.org/10.1080/21655979.2016.1233090
172. J. Poulos, The limited application of stem cells in medicine: A review. Stem Cell Res. Ther. **9**, 1 (2018). https://doi.org/10.1186/s13287-017-0735-7
173. Z. Pei, X. Lan, Z. Cheng, et al., Multimodality molecular imaging to monitor transplanted stem cells for the treatment of ischemic heart disease. PLoS One **9** (2014). https://doi.org/10.1371/journal.pone.0090543
174. M. Liu, Y. Wang, M. Li, et al., Using tyrosinase as a tri-modality reporter gene to monitor transplanted stem cells in acute myocardial infarction. Exp. Mol. Med. **50** (2018). https://doi.org/10.1038/s12276-018-0080-7
175. M. Roca, E.F.J. de Vries, F. Jamar, et al., Guidelines for the labelling of leucocytes with (111)In-oxine. Inflammation/infection taskgroup of the European Association of Nuclear Medicine. Eur. J. Nucl. Med. Mol. Imaging **37**, 835 (2010). https://doi.org/10.1007/s00259-010-1393-5
176. E. Wolfs, T. Struys, T. Notelaers et al., ^{18}F-FDG labeling of mesenchymal stem cells and multipotent adult progenitor cells for PET imaging: Effects on ultrastructure and differentiation capacity. J. Nucl. Med. **54**, 447 (2013). https://doi.org/10.2967/jnumed.112.108316

177. M. Barczewska, J. Wojtkiewicz, A. Habich et al., MR monitoring of minimally invasive delivery of mesenchymal stem cells into the porcine intervertebral disc. PLoS ONE **8**, 74658 (2013). https://doi.org/10.1371/journal.pone.0074658

The manufacturer's authorised representative in the EU is Springer Nature Customer Service Centre GmbH, Europaplatz 3, 69115 Heidelberg, Germany. If you have any concerns regarding our products, please contact ProductSafety@springernature.com

Printed and bound by CPI Group (UK) Ltd, Croydon, CR0 4YY
06/10/2026
02225252-0006